Routledge Handbook of Public Communication of Science and Technology

Second Edition

Communicating science and technology is a high priority of many research and policy institutions, a concern of many other private and public bodies, and an established subject of training and education. Over the past few decades, the field has developed and expanded significantly, both in terms of professional practice and in terms of research and reflection.

The *Routledge Handbook of Public Communication of Science and Technology*, second edition, provides a state-of-the-art review of this fast-growing and increasingly important area through an examination of the research on the main actors, issues, and arenas involved.

In this new and fully revised edition, the book brings the reviews up to date and deepens the analysis. As well as a substantial reworking of many chapters, it gives more attention to digital media and the global aspects of science communication, with the inclusion of four new chapters. Several new contributors are added to leading mass communication scholars, sociologists, public relations practitioners, science writers, and others featured herein.

With key questions for further discussion highlighted in each chapter, the handbook is a student-friendly resource and its scope and expert contributors mean it is also ideal for practitioners and professionals working in the field. Combining the perspectives of different disciplines and of different geographical and cultural contexts, this original text provides an interdisciplinary and global approach to the public communication of science and technology. It is a valuable resource for students, researchers, educators, and professionals in media and journalism, sociology, the history of science, and science and technology.

Massimiano Bucchi is Professor of Science and Technology in Society and of Science Communication at the University of Trento, Italy, and he has been a Visiting Professor at several academic and research institutions in Asia, Europe and North America. His publications include *Science in Society* (Routledge, 2004), *Beyond Technocracy: Science, Politics and Citizens* (Springer, 2009) and essays in journals such as *Nature*, *Science* and *Public Understanding of Science*.

Brian Trench is a researcher and trainer in science communication and former Head of the School of Communications, Dublin City University, Ireland. He has published widely on topics in science communication and science in society, and he has given talks and courses in many countries. He has served on advisory committees to government, state agencies, higher education and cultural institutions regarding science communication, research ethics and evaluation.

Massimiano Bucchi and Brian Trench are co-editors of *Critical Concepts in Sociology: Science Communication* (Routledge, 2015, Major Works series) and are members of the scientific committee of the international Public Communication of Science and Technology (PCST) network; Brian Trench is president of the network from 2014.

"Massimiano Bucchi, Brian Trench, and the world-leading group of authors they have assembled give us a one-stop text vital to anyone who is seriously interested in the field of science communication. The chapters are highly informative and thought-provoking, providing a basis for well-informed discussion and debate around issues that are central to understanding how non-expert audiences can find out about and engage with science, technology, engineering, and medicine. The international scope of the book gives it an appeal to the science-communication community across the world."

Professor Steven Miller, *University College London, UK*

"This work provides a useful introduction to the study of research trends in the public communication of science and technology. It is particularly strong in showing the changes in this field [...] With editors and contributors from various parts of the world, the book is particularly sensitive to international issues... Highly recommended."

Review of the first edition in *Choice* (American Library Association)

"This informative as well as formative book will foster the knowledge of those entering the science-communication field and those already well-established in it, and may even influence their actions in such an important field."

Review of the first edition in *Public Understanding of Science*

Routledge Handbook of Public Communication of Science and Technology

Second Edition

Edited by Massimiano Bucchi and Brian Trench

LONDON AND NEW YORK

First published 2008
by Routledge
2 Park Square, Milton Park, Abingdon, Oxon, OX14 4RN

and by Routledge
711 Third Avenue, New York, NY 10017

Second Edition 2014

Routledge is an imprint of the Taylor & Francis Group, an informa business

© 2014 selection and editorial material Massimiano Bucchi and Brian Trench; individual chapters, the contributors

The right of Brian Trench and Massimiano Bucchi to be identified as authors of the editorial material, and of the individual contributors as authors of their chapters has been asserted by them in accordance with sections 77 and 78 of the Copyright, Designs and Patents Act 1988.

All rights reserved. No part of this book may be reprinted or reproduced or utilised in any form or by any electronic, mechanical, or other means, now known or hereafter invented, including photocopying and recording, or in any information storage or retrieval system, without permission in writing from the publishers.

Trademark notice: Product or corporate names may be trademarks or registered trademarks, and are used only for identification and explanation without intent to infringe.

British Library Cataloguing-in-Publication Data
A catalogue record for this book is available from the British Library

Library of Congress Cataloging in Publication Data
Routledge handbook of public communication of science and technology / edited by Massimiano Bucchi and Brian Trench.—2 Edition.
   pages cm
1. Communication in science—Handbooks, manuals, etc. 2. Technical writing—Handbooks, manuals, etc. I. Bucchi, Massimiano, 1970– II. Trench, Brian. III. Title: Handbook of public communication of science and technology.
Q223.H344 2014
501'.4—dc23   2013049610

ISBN13: 978-0-415-83461-2 (hbk)
ISBN13: 978-0-203-48379-4 (ebk)

Typeset in Bembo
by Book Now Ltd, London

Printed and bound in the United States of America by Publishers Graphics, LLC on sustainably sourced paper.

Contents

List of illustrations	vii
Notes on contributors	ix
Foreword to the second edition	xiv

1	Science communication research: themes and challenges *Massimiano Bucchi and Brian Trench*	1
2	Popular science books: from public education to science bestsellers *Alice Bell and Jon Turney*	15
3	Science journalism: prospects in the digital age *Sharon Dunwoody*	27
4	Science museums and centres: evolution and contemporary trends *Bernard Schiele*	40
5	Public relations in science: managing the trust portfolio *Rick E. Borchelt and Kristian H. Nielsen*	58
6	Scientists as public experts: expectations and responsibilities *Hans Peter Peters*	70
7	Scientists in popular culture: the making of celebrities *Declan Fahy and Bruce V. Lewenstein*	83
8	Science and technology in film: themes and representations *David A. Kirby*	97
9	Environmentalists as communicators of science: advocates and critics *Steven Yearley*	113
10	Publics and their participation in science and technology: changing roles, blurring boundaries *Edna F. Einsiedel*	125

Contents

11 Public understanding of science: survey research around the world   140
    Martin W. Bauer and Bankole A. Falade

12 Risk, science and public communication: third-order thinking
    about scientific culture   160
    Alan Irwin

13 Engaging in science policy controversies: insights from
    the US climate change debate   173
    Matthew C. Nisbet

14 Communicating the social sciences: a specific challenge?   186
    Angela Cassidy

15 Health campaign research: enduring challenges and new
    developments   198
    Robert A. Logan

16 Global spread of science communication: institutions and
    practices across continents   214
    *Brian Trench and Massimiano Bucchi, with Latifah Amin,
    Gultekin Cakmakci, Bankole A. Falade, Arko Olesk, Carmelo Polino*

17 Assessing the impact of science communication: approaches
    to evaluation   231
    Federico Neresini and Giuseppe Pellegrini

Index   246

Illustrations

Figure

4.1	Evolution of SMCs	46

Tables

4.1	Presentation of ready-made science versus science-in-the-making	54
8.1	Dominant stereotypes, scientific themes, and representative films across time	102
10.1	Categorising public participation	126
11.1	Nationally representative surveys on general attitudes to science	141
11.2	Periods, problems and proposals	145
11.3	Examples of knowledge and attitude items in literacy research	146
12.1	Characteristics of first-, second- and third-order thinking about scientific and risk communication	167
17.1	Evaluation and experimental design	239

Contributors

Martin W. Bauer is Professor of Social Psychology and Methodology at the London School of Economics and Political Science, UK, where he researches the relations between science and modern common sense in a comparative perspective and the role of public resistance in techno-scientific developments. His recent book publications include *Journalism, Science and Society* (Routledge, 2007, co-edited with M. Bucchi), *The Culture of Science: How the Public Relates to Science across the Globe* (Routledge, 2012, co-edited with R. Shukla and N. Allum) and *Atoms, Bytes and Genes: Public Resistance and Techno-Scientific Responses* (Routledge, 2014). He has been Editor of *Public Understanding of Science* since 2009.

Alice Bell is a freelance writer and researcher specialising in the politics of science and was previously Lecturer in Science Communication at Imperial College London, UK. She edits *The Guardian* science policy blog, has a monthly column in *Popular Science* and has been published by *The Observer*, *Times Higher Education* and *Research Fortnight*. She has degrees in history of science and sociology of education and did her PhD on the rhetorics of children's science books; popular science publishing is a long-standing interest of hers.

Rick E. Borchelt is Director for Communications and Public Affairs at the US Department of Energy's Office of Science and was previously Special Assistant to the Director at the US National Cancer Institute. He also practiced science and technology public affairs at the US Department of Agriculture, the National Aeronautics and Space Administration, the University of Maryland, the Genetics and Public Policy Center of The Johns Hopkins University and the National Academy of Sciences, and he was a White House Special Assistant for Science and Technology Public Affairs in the Clinton Administration. He has taught science communication and science policy at Vanderbilt University and The Johns Hopkins University.

Massimiano Bucchi is Professor of Science and Technology in Society at the University of Trento, Italy, and has been a Visiting Professor in several academic and research institutions in Asia, Europe and North America. He is author of several books published in English, Italian, Chinese, Korean and Portuguese, including *Science and the Media* (Routledge, 1988), *Science in Society* (Routledge, 2004), *Beyond Technocracy: Science, Politics and Citizens* (Springer, 2009), *Il Pollo di Newton. La Scienza in Cucina* (Guanda, 2013), and has published essays and papers in journals such as *Nature*, *Science* and *Public Understanding of Science*. More information can be found at www.massimianobucchi.it.

Angela Cassidy is a research academic working in the sociology and history of the life and human sciences. She has a particular interest in how scientific knowledge is constructed, communicated and acted upon in public scientific controversies, and she has studied this process in the context of debates over popular evolutionary psychology, food risk, animal health and wildlife management. She is currently a Wellcome Trust Research Fellow in the Department of History, King's College London, UK, where she is investigating the late-twentieth-century history of the bovine TB problem in Britain.

Sharon Dunwoody is Evjue-Bascom Professor of Journalism and Mass Communication, Emerita, at the University of Wisconsin-Madison, USA. She studies components of the mediated science-communication process, from the attitudes and behaviours of journalists and scientists who generate messages to the efforts of audiences to process them. Her co-edited books (both with S. Friedman and C. Rogers) include *Scientists and Journalists* (Free Press, 1986) and *Communicating Uncertainty* (Lawrence Erlbaum Associates, 1999). She has been a communication adviser to oversight committees of The National Academies and the American Association for the Advancement of Science.

Edna F. Einsiedel is University Professor and Professor of Communication Studies in the Department of Communication and Culture at the University of Calgary, Canada. Her research interests are in the social issues around emerging technologies, focusing on public engagement and participation and their institutional arrangements. She has led Canadian participation in international public participation initiatives on climate change and biodiversity. Her publications have appeared in journals such as *Science, Nature Biotechnology, Public Understanding of Science, Science Communication*, and *Science and Engineering Ethics*. She served as Editor of *Public Understanding of Science,* 2004–2009.

Declan Fahy is Assistant Professor in Health, Science and Environmental Journalism at the School of Communication, American University, Washington, DC. His research examines scientists as celebrities and public intellectuals, and new methods and models of science journalism. His work has been published in *Journalism, Journalism Studies, Science Communication, Nature Chemistry, BMC Medical Ethics, Health Promotion Practice* and *Irish Communications Review*. He is a former newspaper reporter and his recent journalism has appeared in *The Scientist* and the *Columbia Journalism Review*. He is currently completing a book on scientists as celebrities.

Bankole A. Falade is a researcher at the Department of Social Psychology of the London School of Economics and Political Science, UK. His research interests include science communication, science and the media, and the role of religion, politics and ethics in the transformation of common sense. His current research looks at vaccination resistance, religion and attitudes to science in Nigeria. In his previous career he was a science journalist, a university lecturer and Editor of both the *Sunday Punch* and *Sunday Independent* newspapers in Lagos, Nigeria.

Alan Irwin is professor in the Department of Organisation at Copenhagen Business School, Denmark. His research deals with scientific governance and science–public relations. His books include *Risk and the Control of Technology* (Manchester University Press, 1985), *Citizen Science* (Routledge, 1995), *Misunderstanding Science?* (Cambridge University Press, 1996, co-edited with Brian Wynne), *Sociology and the Environment* (Polity Press, 2001) and *Science, Social Theory and Public Knowledge* (Open University Press, 2003, co-authored with Mike Michael). He is a member of the Strategy Advisory Board for the UK Global Food Security programme.

David A. Kirby is Senior Lecturer in Science Communication Studies at the University of Manchester, UK, and was previously an evolutionary geneticist. His experiences as a bench scientist have informed his internationally recognised studies into the interactions between science, media, and cultural meanings. His book *Lab Coats in Hollywood: Science, Scientists and Cinema* (MIT Press, 2013) examines collaborations between scientists and the entertainment industry in the production of movies. His next book, *Indecent Science: Film Censorship and Science, 1930–68*, will explore how movies served as a battleground over science's role in influencing morality.

Bruce V. Lewenstein is Professor of Science Communication in the Departments of Communication and of Science and Technology Studies at Cornell University, USA. He works primarily on the history of public communication of science, with excursions into other areas of science communication, such as informal science education. He seeks to document the ways that the public communication of science is fundamental to the process of producing reliable knowledge about the natural world. He was Editor of *Public Understanding of Science*, 1998–2003, and Co-Chair of a US National Research Council study on *Learning Science in Informal Environments* (2009).

Robert A. Logan is Professor Emeritus at the University of Missouri-Columbia, USA, School of Journalism, where he was Associate Dean for Undergraduate Studies and Director of the Science Journalism Center. He is on the senior staff of the US National Library of Medicine (NLM) where he assists in the evaluation and production of NLM's comprehensive health informatics services to the public. He has published extensively on issues in science journalism and is a member of the editorial boards of the *Journal of Mass Media Ethics* and *Science Communication*.

Federico Neresini teaches science, technology and society, and the sociology of innovation at the University of Padua, Italy. His main research interests are in the sociology of science, particularly the construction of scientific knowledge, the public communication of science and the social representations of science. His research activities have been focused on biotechnology and nanotechnology. He is Coordinator of the PaSTIS (Padova Science, Technology, Innovation Studies) research unit and a member of the Observa – Science in Society research centre. He has published articles in journals such as *Nature, Science, Public Understanding of Science, Science Communication* and *New Genetics and Society*.

Kristian H. Nielsen is Associate Professor in the History of Science and Science Communication at the Centre for Science Studies at Aarhus University, Denmark. He has degrees in physics, philosophy and history of science and technology and has studied in Japan, France, Britain and the US, as well as in Denmark. Among his research interests are the history of popular science and contemporary science communication. He has published in *Annals of Science, British Journal for the History of Science, Environmental Communication, International Journal of Science Communication, Public Understanding of Science*, and *Science Communication*.

Matthew C. Nisbet is Associate Professor of Communication at American University, Washington, DC, where he studies the role of communication and advocacy in debates over science, technology and the environment. He has published more than 70 peer-reviewed studies, scholarly book chapters and reports and he has served as a Shorenstein Fellow on Press, Politics and Public Policy at Harvard University, as a Google Science Communication Fellow, as an Osher Fellow at the Exploratorium science centre and as a Health Policy Investigator at the Robert Wood Johnson Foundation. More information about his research, teaching and writing can be found at www.climateshiftproject.org.

Contributors

Giuseppe Pellegrini teaches social research methodology at the University of Padua, Italy. His research focuses on social policy, citizenship rights and public participation with specific regard to food issues. He is Coordinator of the Science and Citizens research area at the Observa – Science in Society research centre. His most recent publications include *Women and Society: Italy and the International Context* (Observa, 2013, co-edited with Barbara Saracino) and *Os Jovens e a Ciencia* (CRV, 2013, co-edited with Nelio Bizzo) and articles in journals such as *Food Policy* and *Comparative Sociology*.

Hans Peter Peters is a senior researcher at the Institute of Neuroscience and Medicine, Forschungszentrum Jülich, Germany, and Adjunct Professor for Science Journalism at the Free University of Berlin. His research focuses on public sense-making of science, technology and the environment under the conditions of a *media society*, and on the interactions of science and the media. He has directed several projects on public communication of climate change, genetic engineering, biomedicine and neuroscience, analyzing the participation of scientists in public communication as well as the implications for science governance.

Bernard Schiele is Professor in the Faculty of Communication, University of Québec in Montréal, Canada, and a researcher at the Interuniversity Research Centre on Science and Technology, CIRST/IRCST. He is co-editor of the books *Science Communication Today* (CNRS Editions, 2013, with P. Baranger), *Science Communication in the World* (Springer, 2012, with M. Claessens and S. Shi) and *Communicating Science in Social Contexts* (Springer, 2008, with D. Cheng, M. Claessens, T. Gascoigne, J. Metcalfe and S. Shi), author of *Le Musée de Sciences* (L'Harmattan, 2001) and co-author of *Science Centers for this Century* (Editions Multimondes, 2000, with E. Koster). He is the former President of the International Advisory Scientific Committee for the Science Museum in Beijing and he has been a member of the Public Communication of Science and Technology Network scientific committee since 1989.

Brian Trench is a researcher, trainer and evaluator in science communication, formerly Senior Lecturer and Head of School in the School of Communications, Dublin City University, Ireland. He has given talks or courses in 20 countries on topics in science communication and science in society, and he has published widely in these areas. He served on the Irish Council for Science Technology and Innovation, 1997–2003, the advisory committee for Euroscience Open Forum/Dublin City of Science 2012 and the Public Communication of Science and Technology Network scientific committee since 2000. He is co-editor with M. Bucchi of *Critical Concepts in Sociology: Science Communication* (Routledge, forthcoming in 2015 in the Major Works series).

Jon Turney is a writer and editor, and a frequent reviewer of books about science, and has taught science writing and science communication at University College, Birkbeck College, Imperial College and City University, all in London, UK. He is Editorial Manager of the European Union-funded project Nanopinion, 2012–2014, and he was earlier a journalist with the *Times Higher Education Supplement*. His books include *Frankenstein's Footsteps: Science, Genetics and Popular Culture* (Yale University Press, 1998), *Rough Guide to Genes and Cloning* (Rough Guides, 2007, co-authored with Jess Buxton) and *Rough Guide to the Future* (Rough Guides, 2010). He co-edited *A Quark for Mister Mark: 101 Poems about Science* (Faber, 2000, with M. Riordan).

Steven Yearley is Professor of Sociology of Scientific Knowledge at the University of Edinburgh, UK. He has worked extensively on the environmentalists' approach to scientific issues and on the analysis of controversies with a scientific component, from climate change to GMOs, from synthetic biology to the management of wild deer. He is currently working on a project on the monitoring of climate change in Britain and with the University of Oslo on science advice for policies relating to climate change. His books include *Making Sense of Science* (Sage, 2005), *Cultures of Environmentalism: Empirical Studies in Environmental Sociology* (Palgrave Macmillan, 2005, 2009) and *Sage Dictionary of Sociology* (2005, with Steve Bruce).

Foreword to the second edition

By presenting the first edition of this Handbook in 2008, we expressed the hope that it might contribute to greater stability and clarity in thinking about and planning science communication and that it might encourage more reflective practice and more formal study. While we cannot say what impacts the Handbook may have had, we can observe that, over the intervening period, science communication, as professional practice and academic subject matter, has continued to become more articulate and self-aware. It has also become an increasingly busy and diverse area of practical activity and research.

Training, education and publication in science communication have continued to increase and have spread across more countries and more media. The number of university-level courses has grown. The number of academic journals has grown, as has the number of papers published by each of them. There has been a spate of books presenting collections of essays on topics in science communication. The international Public Communication of Science and Technology conferences organised by the PCST network and held in India (2010), Italy (2012) and Brazil (2014) have attracted increasing numbers of participants and of research contributions and reflections on practice.

All of this activity and the positive response to the Handbook's first edition have made a new edition necessary. The first edition sought, in the words of our Introduction, to 'offer a state-of-the-art map of a field that has developed substantially, not only at the level of practice, but also in terms of research and reflection, as well as in terms of the diversity and richness of points of view'. In all of these dimensions, there have been further developments since 2008. But, along with taking account of new publications, the revised chapters also incorporate the insights of further reflection on the topics and the benefits of cooperation with new co-authors. They all also reflect two trends much more strongly in evidence now than six years ago: the global spread and the digitalisation of science communication. In some cases, the chapters have been significantly rewritten as the authors have reworked their argumentation.

In this new edition, specific chapters on developing countries and on the Internet have given way to a broader treatment of globalisation and the consideration in almost all chapters of applications and implications of online media. We have also commissioned new chapters on emerging themes such as celebrity scientists and science controversies, as well as offering our own highly synthesised view of the conceptual stability of the field and some of the theoretical challenges it faces.

The Handbook is addressed to practitioners, educators, researchers and students in science communication. It is particularly with students in mind that we include three or four 'key questions' with each chapter, hoping that these might be considered suitable or adaptable for seminars and assignments, and as prompts to further reading.

Foreword to the second edition

We thank the established contributors and we particularly welcome the new contributors, who have all committed so willingly to this project. Massimiano Bucchi would like to thank the Carlos III University of Madrid for the support received in the context of a 'chair of excellence' research fellowship in collaboration with Banco Santander.

Massimiano Bucchi, Trento
Brian Trench, Dublin
March, 2014

1
Science communication research
Themes and challenges

Massimiano Bucchi and Brian Trench

Science and society: a theatrical dialogue[1] and an introduction

A lane in a European city, one afternoon in early spring.

SOCIETY: *(Walks along, talking on his mobile.)*
SCIENCE: *(Approaches, out of breath, with an armful of papers and a laptop.)*
SOCIETY: *(Puts away his mobile.)* Hey, Science, is it really you? I didn't recognise you without your white coat ... Where are you off to?
SCIENCE: Oh, hello ... excuse me, I'm in a bit of a hurry. I'm on my way to an international symposium in Sweden. We are going to discuss the construction of a new particle accelerator.
SOCIETY: Another one? Didn't you set one up a little while ago in Geneva? The famous experiment in which the Earth was to disappear down a black hole? But there wasn't really a black hole, was there?
SCIENCE: Nothing escapes you, does it! But you know, the black hole was the usual press hype ... And remember, traffic to the physics institutes' websites broke all records at the time ...
SOCIETY: Well, they will have been interested too because it brought a lot of publicity ... I remember reading an interview with that famous physicist ... whatshisname ... you know who I mean ... But tell me about this new accelerator: what's it for exactly?
SCIENCE: What's it for, what's it for ... excuse me, but what's your auntie for? It's used for experiments with neutrons that help us to understand the nature of certain materials better, with extremely important theoretical implications and applications ... but why waste time explaining things to you? You don't listen to me, you don't understand me, you have never understood me! We've known each other for 400 years, and our relationship just gets worse and worse! At least I have the courage to admit it: you are not really interested in me! And you are a little bit frightened of me too, aren't you?
SOCIETY: Frightened, me? I listen to your talks at the Science Festival every year! And on television, I never miss a science programme, not one ...
SCIENCE: All right ... but you are more resistant than staphylococcus when it comes to getting your wallet out ...

SOCIETY: Listen ... well, what about you, eh? I already have to perform miracles just to balance the books. I should like to see you paying for health, education, social security ... And it's not true that I don't spend money on research ... do you know how much the nanotechnologies cost, in Europe as a whole? Well, do you? Give me a figure ...
SCIENCE: I know, I know ... but you must understand that the money you spend on me is money well spent ... I bring innovation, development, jobs, technology ...
SOCIETY: I am sure you do ... By the way, I have read that research in genetics may extend life, people may live to the age of 150 ... is it true?
SCIENCE: Well, yes, we are working on it but it will take years of work, investment, resources ...

The relationship between science and society is often represented in terms of misunderstandings, gaps to be filled and bridges to be built. This traditional stereotype posits science as distinct and separate from society in terms of content, organisational practices, institutional aims and communication processes. In this light, some *translation* is required to establish connection between science and society at large, making elements of the science domain approachable, understandable and eventually appealing.

This traditional vision has some historical grounds and still bears the influence of socio-historical processes which, between the seventeenth and nineteenth centuries, defined science as a distinctive social institution with increasing political, economic and cultural relevance (Ben-David 1971; Merton 1973; Ezrahi 1990).

However, during the last few decades scholars and commentators have pointed to relevant transformations in research practice and organisation as well as in its dynamic interaction with society (Ziman 2000; Metlay 2006; Bucchi 2014). In particular, recent research and reflection have drawn attention to the increasing intersection and permeability of boundaries between science and society. For example, in areas like biomedicine or information technology, heterogeneous networks connecting scientific experts with non-experts and quasi-experts (patient organisations, citizen groups, users) are increasingly replacing traditional expert communities (Callon 1999; Callon *et al.* 2001; Bucchi 2009; see also Einsiedel in this volume).

These transformations encompass and reflect the very dynamics of science communication. Later, we outline some of the challenges these transformations present to researchers, but first, we offer a highly synopsised review of the current state of the art through an examination of the usage of terms that recur frequently as demanding the attention of practitioners and researchers.

A conceptual review in ten keywords

This conceptual review of theoretical reflections and research in science communication is presented through an exploration of ten frequently used terms: popularisation; model; deficit; dialogue; engagement; participation; publics; expertise; visible scientists; scientific culture. We outline how these terms have acquired a range of meanings, including distinctly different meanings, some of which coexist in current usage. With a firm fix on these terms, and an appreciation of how they may be deployed normatively, descriptively or analytically, the reader should be able to navigate much of the field of science communication research. It will be seen that all of these terms recur at different places through the other chapters of this Handbook. In many cases, the nuances of these concepts have been explored further. We have included some references to the literature but it would have made for very difficult reading if we had attached references at all possible points to our observations on the spread and trend of discussion. Readers may find this section useful as a resource or dictionary to consult for general conceptual clarification when reading the following chapters.[2]

Science communication research: themes and challenges

Popularisation is the term with the longest tradition among those used to describe a wide range of practices in making scientific information accessible to general, non-expert audiences. Early examples of popularisation include Fontenelle's *Entretiens sur la pluralité des mondes* (1686), a series of conversations between a philosopher and a marquise. During the eighteenth century, science popularisation gradually defined itself as a distinctive narrative genre, often targeting, in particular, female readers as supposedly ignorant and curious – 'symbols of ignorance, goodwill and curiosity' (Raichvarg and Jacques 1991: 39) – as in the classic Algarotti's *Newtonianism for Ladies* (1739) or de Lalande's *L'Astronomie des Dames* (1785). Other relevant channels of popularisation emerged later with scientific discoveries frequently featured in the daily press, science museums, public lectures and the great exhibitions and fairs that showed visitors the latest marvels of science and technology. Particularly during the second half of the nineteenth century, popularisation and populariserss profited from changes in the publishing business and the increasing reading audience to become influential voices, but their success also testified to the increasing relevance of science as a cultural force. The sales figures of books like Brewer's *Guide to the Scientific Knowledge of Things Familiar* (195,000 copies up to 1892) are impressive even by contemporary standards. Through their books and public lectures, populariserss (*showmen of science*) like J. H. Pepper and J. G. Wood in England or Paolo Mantegazza in Italy became public celebrities of their time (Lightman 2007). In the following century and particularly after World War II, the new global and policy landscape redefined popularisation in conceptual and even ideological terms, particularly in the US and western Europe. With science's social and political role significantly captured by Vannevar Bush's metaphor of the *goose laying golden eggs* (e.g. delivering economic wealth, social progress and military power if appropriately fed), popularisation was expected to 'sell science' to the broader public to strengthen social support and legitimation (Lewenstein 2008).[3] This fuelled the development of new popularisation strategies and channels, including interactive science centres and partnerships between science institutions and Hollywood studios. When a new phase of critical reflection on the role of science in development and (more broadly) in society opened, the concept of popularisation also came under criticism as embodying a paternalistic, diffusionist conception of science communication (Hilgartner 1990).[4] More recent conceptualisations have reappraised the term, considering it suitable to describe specific types and contexts of communicative interactions among science and the public.[5] In China, for example, *popularisation* is the preferred term to refer to a wide range of science-in-society activities.

Model of communication is one of the key theoretical concepts in science communication. Despite that, very few explicit models of science communication have been designed and proposed. Over 20 years ago, sociologists and communication scholars identified problems of theory and conceptualisation in the dominant science popularisation practices of the time (e.g. Dornan 1991; Wynne 1991). They referred, in this context, to the model of communication underlying such practices, meaning the mental construction of relations between the actors in the communication process. They identified the dominant model in terms such as *top-down* and *hierarchical* and pointed to the assumption that the target public was defined by a deficit (see *Deficit* below) of some kind.

Over the past two decades, science communication communities in research and practice have sustained a discussion about the limits of inherited models and the characteristics of models that are more appropriate for the present day. Part of that discussion and research has been explicitly prescriptive and binary: it labels some models of communication as old and discredited and others as new and appropriate. In this context, the shift in preferences from one model to another is represented as evolutionary and irreversible.

But another part of that discussion and research is more descriptive and analytical: it has been aimed at understanding better the range of possible models, how different models are applied, how the language used to describe a practice may disguise the model that effectively shapes the practice (Wynne 2006), how different models can coexist (Miller 2001; Sturgis and Allum 2005) and what governs the choices made among them. Some attempts have been made to set out a wide spectrum of models, incorporating more narrowly defined options that might apply in specific and changing circumstances (Trench 2008b).

Deficit is a central concept in identifying the intellectual (or ideological) foundations of certain science-in-society ideas and practices and enabling their critique. Two assumptions often underlie this concept: public opinion and political decision-makers are misinformed about science and the issues raised by its development; this misinformation is fuelled by inadequate and sensationalist media coverage of technoscientific topics. This situation is seen as being exacerbated by poor training in basic science and a general disinterest among the institutions and the cultural intelligentsia in scientific research. Consequently, citizens and political decision-makers are seen to fall prey to *irrational* fears which fuel their hostility and suspicion towards entire sectors of research and technological innovation (e.g. nuclear energy, GM foods, stem cells).

From this perception arises the need to propose initiatives covering the gap between experts and the general public, reversing public attitudes towards science and technology or at least attenuating their hostility. Such emphasis on the public's inability to understand the achievements of science – according to a model of linear, pedagogical and paternalistic communication – has warranted the label of 'deficit model' to this view of the public understanding of science (e.g. Wynne 1991; Ziman 1991).

From the early 1990s, several scholars have criticised the deficit approach by highlighting the weak empirical foundations of its assumptions and the limited results achieved by the communicative actions it has inspired. Critics of the deficit-based approach do not deny that relevant awareness problems may exist across publics (see *Publics* below) on issues related to science but suggest that this is not the best starting point: focus instead, they say, on what the audiences do know and on their questions and concerns.

Discussion has continued over many years on what kinds of knowledge about science the public generally lacks and needs to have: knowledge of scientific fact, of scientific theory, of scientific methods or of the organisation and governance of science (see Durant 1993). A more traditional notion of missing knowledge of facts (scientific literacy/illiteracy) remains widely assumed in contemporary science-in-society practice, notably in contexts where there are perceived problems of anti-science, pseudoscience and superstition, as, for example, in Indian programmes of science awareness (Raza and Singh 2013).

Dialogue came to be presented as the acceptable alternative to the deficit model from the late 1990s as public concern over certain science and technology issues became evident – despite significant communication efforts – and demand for involvement in such issues increased. Multiplying examples of non-experts or alternative experts actively contributing to shape the agenda of research in fields like biomedicine have led to rethinking the very meaning of science communication in several arenas. For example, a frequently cited report of the House of Lords (2000) in Britain acknowledged the limits of science communication based on a paternalistic, top-down science-public relationship and detected a 'new mood for dialogue'. In many countries and at the European level, funding schemes and policy documents shifted their keywords from *public awareness of science* to *citizen engagement*, from *communication* to *dialogue*, from *science and society* to *science in society*.

The claimed shift from deficit to dialogue remains a powerful narrative in public communication of science. The two approaches are widely seen as distinct and one as inherently superior to the other. The shift is often stated as an irrefutable fact: commentaries speak of the 'dialogical turn' as a historical change that has taken effect across Europe, and more widely (e.g. Phillips *et al*. 2012). Dialogue and related approaches are now much more frequently proposed and enacted than those that might be defined as deficit based, at least in Europe, Australasia and North America. However, closer examination reveals a complex picture; for example, the striking case of Denmark – long associated very strongly with pioneering dialogical initiatives – where there is an apparent reversal of the trend (Horst 2012).

The study of this case links to a thread running through the research and reflection of the last decade of scepticism about the scale, or even the reality, of the claimed shift to dialogue. It has been suggested, for example, that dialogical approaches may be used in order more effectively to remedy public deficits. It has been argued that some dialogue methods are not genuinely two-way, in that the original sponsors of the communication (generally scientific or policy institutions) stay in control and the citizens taking part have no significant influence on the final outcomes (Davies *et al.* 2008; Bucchi 2009). There is yet another strand to the discussion of the communication and cultural practices that draws attention to the possibilities and pleasures of dialogical events which are not oriented to specific political or informational end goals but rather to the process of taking part (e.g. Davies *et al.* 2008). In science cafés, a spreading form of science communication (see Einsiedel and Trench *et al.* in this volume), the satisfaction for those involved may reside in the exchange itself rather than anything beyond it.

Engagement has become, in many countries, a prevalent term to describe a wide range of science-in-society practices in policy, education, information or entertainment contexts. *Engagement* can refer to the actions and attitudes both of knowledge producers and of various sectors of the public. When researchers, for example, go to the streets to talk about their work, this may be called *public engagement*. Equally, the attention and interest shown by their audiences may also be called *public engagement*. In some cultural contexts, especially in Britain, public engagement is as comprehensive a term as public communication; the acronym PEST (public engagement with science and technology) is used as the catch-all term in preference to PCST (public communication of science and technology) or PUS (public understanding of science). The change of vocabulary carries with it, at least implicitly, a shift to an understanding of relations between the partners in the process as more equal and more active.

Different levels and modes of engagement are envisaged, for example, by reference to downstream and upstream engagement (Wilsdon and Willis 2004). The latter has been proposed for priority attention on the basis that early involvement of the public in discussion and eventually negotiation of new developments in science and technology will likely lead to more satisfactory outcomes for all involved, and specifically to knowledge that has earned public trust. The case of genetically modified foods and crops is cited as an example of late, or downstream, public engagement; citizens in many countries across the world were presented with products ready for use and, in many cases, they reacted in a hostile manner. In Europe, in particular, governments, researchers and businesses applied what they saw as the lessons of that experience when they sought to ensure earlier (upstream) engagement with nanotechnology.

Public engagement activities are nowadays regarded, in several countries, as a relevant dimension of the mandate – as well as a responsibility – of research institutions in the context of the so-called *third mission* of universities. On this basis, scholars and policy-makers are discussing the most appropriate indicators to identify and analyse the range and impact of such activities (Bauer and Jensen 2011; Bucchi and Neresini 2011).

Participation has come to represent a stronger form of engagement by the public both with scientific ideas and with the governance of science. The term has acquired specific meaning in science in society through association with ideas of participatory democracy and participatory communication. In these contexts, participation implies strongly active citizens who can engage at many levels, including in deliberation on the very topics for negotiation and communication. Thus, participation tends to be used in science-in-society to refer to a third option that goes beyond the deficit-dialogue binary split and overcomes the need to refer, for example, to 'real dialogue' in order to insist on the authenticity of the process (e.g. Riise 2008). If deficit and related modes of communication can be considered one-way, and dialogue two-way, then participation can be represented as three-way because it implies publics or citizens talking with each other as well as talking back to science and its institutions. In the European Commission's framework programme of research, Horizon 2020,[6] support will be given to exploration of participatory mechanisms for deliberating on science, including on agendas for science, where the main agents of public participation are civil society organisations. Increasing interest is also being given, in some contemporary science centres that are based on articulation of relations between arts and sciences, to cultural representations of science as open-ended and also available for interpretation and critique (see Schiele in this volume). In this context, public participation in science is equivalent to that of critical audiences at the theatre or in the concert hall.

Yet other forms of public participation in science are represented by 'citizen science' and 'open science' (Bonney *et al.* 2009; Delfanti 2013). In the first, citizens may contribute to scientific research as collectors or contributors of data – for example, adding observations of certain animal species to an online database to be later analysed by researchers; in the second, researchers make all protocols, data, analyses and publications available online for public scrutiny, allowing the interested public to access not only *ready-made science* (as was typically the case in popularisation) but also *science-in-the-making*. In some cases, this accessibility paves the way for an actual contribution in terms of content: for example, in the Fold-It project the form of certain proteins was identified through open collaborative modes between experts and non-experts (see Einsiedel in this volume).

Publics has become a common term in discussion and study of science in society, indicating in shorthand that *the public* is diverse, even fragmented. Because it is not a common, much less everyday, word, *publics* often has to carry the quote marks around it that draw attention to its deliberate use. Adopting the plural form was an important part of recognising that generalisations about the public – specifically in terms of its deficits – are very rarely valid, and often seriously misleading (Einsiedel 2000). Referring to publics has been associated with the proposal of a contextual model of communication, according to which the communicators inform themselves about, and are attentive to, the various understandings, beliefs and attitudes within the public.

Beyond the obvious differentiation of publics as young or old, male or female and scientifically educated or not, the plural-publics approach has been supported by the accumulation of data on the widely varying interest, attention and disposition towards scientific matters in the populations of individual countries and, comparatively, across countries and continents. From surveys of public knowledge of scientific facts initiated over 50 years ago, these studies have become increasingly sophisticated and nuanced. They measure fine distinctions within and between national populations on, for example, levels of trust towards scientists and scientific institutions and attitudes to emerging technologies. They allow such attitudes to be correlated with educational experiences and world views. On the basis of cross-country analysis of survey findings, the patterns of national cultures of science (see *Scientific Culture* below) can be sketched (e.g. Allum *et al.* 2008; Bauer *et al.* 2012). A strong focus on publics is almost standard now in the training of scientists for public communication; short courses offered to researchers by research councils, universities, professional organisations and others very often start by asking: who are the publics you want to communicate with, and why? (Miller *et al.* 2009).

Expertise is one of the most common forms through which scientific knowledge and actors enter the public domain, i.e. when scientists take on public roles validating, interpreting and commenting on developments in science, and advising governments and other social institutions on their implications. As producers of knowledge, scientists tend to operate in tightly circumscribed and increasingly specialised spaces. When scientists are called on to be experts in public arenas, they are expected to take a broader view and answer media questions or offer policy advice on themes in which they may not be strictly competent (see Peters in this volume).

Studies of science in society have often focused on how scientific expertise is expressed and accorded authority in public. Increasingly, expertise of several kinds is involved when complex scientific issues are played out in public arenas. Contemporary developments in science typically happen at the interfaces of several scientific and technological specialist practices. Sometimes they also have political, economic or ethical implications which invoke contributions from experts in those fields. Scientists active in public communication are increasingly required to relate their own expertise to that of scholars and practitioners in topics that were previously considered remote, sometimes even antagonistic. When complex environmental and medical matters are negotiated through the legal or parliamentary systems, perhaps with a view to establishing constitutional ground rules or setting down regulations, scientific expertise may be scrutinised in contexts and by criteria very different from those of the scientific communities.

Scientific expertise has come to be further problematised by reference to the tacit, less formal knowledge that various social groups possess through their experience or culture. In case studies in health and agriculture in the 1980s and 1990s, the term *lay expertise* (or *lay knowledge*) was coined to refer to the knowledge that patients and farmers brought to a particular issue and that qualified the definition of that topic given by scientific experts (Wynne 1992; Epstein 1995). On the other hand, there have been ripostes to that approach, insisting on the attribution of expertise only to those with formal qualifications (e.g. Durodié 2003).

Scientific expertise in contemporary societies is facing relevant challenges in connection with factors like expanded accessibility of specialist information to non-experts, increasing questioning of the choice and competence of experts, and public exposure to controversial specialist debates and competing expertise.

Visible scientists or public scientists have been present in every generation since modern science emerged in the seventeenth century. Some of the founders of modern science were visible public figures, and some of the earliest institutions of modern science such as professional societies and academies dedicated themselves, at least in part, to making the achievements of science visible and public. However, those who did science were not defined as *scientists* until the nineteenth century, and, up to then, the potential public for science was restricted to a shallow layer of the highly educated. With the professionalisation of science, rapid growth in the number of scientists and the development of a mass public, a particular concern grew about the relative invisibility of science: the vast majority of science and scientists were invisible to the vast majority of society. A classic American study (Goodell 1977) drew attention to named scientists in psychology, anthropology, molecular biology and other fields who had achieved public visibility as informers and explainers of contemporary science. But it also highlighted institutional constraints which meant that scientists might as often be punished as rewarded for seeking such visibility.

In this period, developments in society required scientific expertise to be more accessible. The space race, engaging the two major geopolitical blocs, drove efforts to increase public investment and interest in the new scientific and technological discoveries and conquests. Rapid developments in medical science and in information technologies needed explainers. The most successful popularisers exploited the opportunities of the rapidly spreading medium of television to become household names. In astronomy, new technologies and natural history, in particular, photogenic or otherwise charismatic scientists developed highly visible careers as TV presenters. Some others, called on to be expert sources for the political and media systems, became public scientists in myriad ways, as newspaper contributors, TV show panellists, advisory committee or expert group members and as politicians.

From the 1970s government around the world created ministries of science, technology or research, and individual scientists were drawn into the political systems as policymakers or advisers. The strength of presence of such public scientists – whether in media, politics or public affairs more generally – and the features of their visibility may be taken as a relevant dimension to analyse a country's scientific culture (see *Scientific Culture* below). Fuelled by further developments of mass media, the celebrity culture that grew up around entertainment and sport has affected many other sectors, and there are celebrity scientists in many societies, just as there are celebrity actors, authors and economists (see Fahy and Lewenstein in this volume). Their views are sought and broadcast on topics well beyond their areas of recognised expertise, and their private lives become public affairs: it is also through such dynamics that the deepening interpenetration of science and society takes place, which characterises contemporary scenarios.

Scientific culture or *Culture of Science* is used in several variations to refer to the standing of science in the general culture of a country or other cultural context. Two interconnected uses of the term have largely dominated debate in the past few decades. One use, significantly influenced by Snow's concept of 'two cultures', contrasts scientific culture with that of the humanities and the arts, and it deprecates their separation and the lack of public attention for scientific culture (Snow 1959). The second use has been almost interchangeable with *public understanding of science* in its more traditional and limited meaning. This equates scientific culture with public attention to and interest in scientific topics and levels of scientific literacy and thus, through a deficit and diffusionist perspective, to public acceptance and support of different science and technology developments. Such usage has been extended to encompass technology explicitly, as in the French term *culture scientifique, technique et industrielle*, generally shortened to CSTI, or the European Commission's chosen term for a short period, *RTD culture*, referring to research and technological development (Miller *et al.* 2002).

The narrow, diffusionist interpretation of scientific culture has been widely criticised as limited and unfounded on several grounds (see *Deficit, Models* above). Empirical studies have shown that concern for and scepticism about certain scientific developments may actually be associated with higher levels of literacy and information (thus, in one usage, stronger scientific culture) and vice versa, that blind trust – and in some cases even expectations of *miracles* – with regard to science can be largely disconnected from actual knowledge and understanding (e.g. Bucchi 2009; Bauer and Falade in this volume). The narrow vision also takes for granted, in a similar vein to Snow, that scientific culture can be defined as a distinct, coherent and monolithic object that can be infused or injected into general culture and society through appropriate communications.

A broader view underscores increasing diversity and fragmentation within science practice; significant permeability of the boundaries between contemporary science and society; cross-fertilisation between images and narratives in general culture and scientific concepts and ideas; significant visibility and presence of scientific figures and concepts in the public sphere as well as in contemporary arts. This culture of science in society encompasses not just understanding of specific scientific content but also an awareness and social intelligence of science as part of society and culture, thereby able to discuss and evaluate science's role, priorities and implications in an open, balanced and critical fashion. Also more recently, but more technically oriented, a discussion has started on defining indicators to *measure* scientific culture as a combination of traditional indicators (e.g. R&D investments and output), indicators of science communication activities (e.g. media coverage intensity, science museum visitors) and of public attitudes to science.

Challenges for science communication research

While the use of concepts such as those examined above will continue to evolve and greater clarity about the range of their applications will contribute to the development of research in this field, there are, as mentioned earlier, some significant shifts in relations that require new approaches and possibly new terms. The concept of *mediatisation*, for instance, has been proposed to describe the proximity and sensitivity of science actors and institutions to the practices and logics of the general media (Weingart 1998; Peters *et al.* 2008; Rödder *et al.* 2012). This may be one productive way to think about the reshaping of communicative relationships and, above all, to move away from conceptualisations of science and society as separate and distinct from each other. This remains perhaps the central challenge for contemporary science communication research, but there are related challenges that arise from the co-evolution of science, society and communication media. Here, we seek to identify some of the most important of these challenges.

Plural science, plural public

Permeability and heterogeneous networking between science and society intersect with increasing fragmentation of publics, of media and of their social uses. Science institutions and actors are diversifying their attitudes and practices, also in the communication domain, which makes it problematic to continue using traditional expressions like *scientific community*, implying internal homogeneity and a shared commitment to specific norms and values (Bucchi 2009, 2014). But it is no less important to reflect on and investigate the diversity and articulation of *publics* of science communication. The traditional usage of *public* evoked a notion of passive and target-like readers and spectators, often addressed and defined paternalistically. Although one should not neglect that significant portions of the public may remain potentially disenfranchised from interactions and participatory processes with regard to science, it is nevertheless clear that social transformations – as represented in characterisations of contemporary society as pervaded by a sense of uncertainty, risk or distrust – along with changes in media technology and use are playing relevant roles in redefining and multiplying public spaces for science communication. These changes require science communication research to develop more complex maps of the relations between sciences and publics.

New mediations

Digital media allow, among other things, research institutions and actors to supply to end users an unprecedented amount and variety of materials, e.g. videos, interviews with scientists, selected news items. In the broader context of ever stronger public relations efforts by research institutions, this contributes to processes that could be summarised as *the crisis of mediators*. This crisis is not specific to science communication, or even to science journalism where it has been actively discussed, but is particularly relevant to this field. Traditional mediators of science communication like newspapers, magazines, television and radio programmes, and science museums and centres are losing their traditional centrality as filters and guarantees of the quality of information. Some of the expectations of the processes of science opening up to public view through digital media (Trench 2008a) may have been misplaced or premature, but the increasingly pervasive use of digital media requires researchers to think of media as significantly more than channels of scientific information.

Quality and evaluation

The above considerations pave the way for a thorough reflection on the theme of quality in science communication. Professional mediators used to guarantee quality through *brands* and the reputation of their medium. By and large, readers, viewers and visitors could confidently assume that content printed in the science sections of the *New York Times* or *Il Corriere della Sera*, or broadcast by the BBC or displayed in a major science exhibition would be a high-quality extract of findings and ideas filtering from the scientific community. But contemporary information overload requires the user to become more competent, and it demands new definitions of quality. Public communication of science should now be mature enough to pass from a *heroic phase*, in which *everything goes* for the sake of communicating science to a phase in which quality criteria are central for all parties involved. This implies developing indicators and standards of performance, particularly for institutions, and assigns added importance to the issue of evaluation (see Neresini and Pellegrini in this volume). As social networks of evaluation develop in other domains, notably in travel, new relations of trust may develop with regard to assessment of scientific information. This signals that reliance on peer-reviewed science as *the* guarantee of authenticity and validity is unlikely to be effective. It also represents a major challenge for the public relations practice of scientific institutions and thus for the analysis of their place in society (see Borchelt and Nielsen in this volume).

Collapsing communication contexts

Above all, it is the traditional sequence of the communicative process (specialist discussion/didactic exposition/public communication or *popularisation*) that has been disrupted. The didactic and public exposition of science is no longer, as in Kuhn's theory, a mere static and petrified page written by the winners in the struggle to establish a new scientific paradigm (Kuhn 1962). Even science museums, the places par excellence of *fossilised* science, increasingly hold exhibitions on current and controversial science issues[7]. Users of scientific information increasingly have access to science in its making and highly controversial debates among specialists. Some of the implications of this new scenario have been spectacularly highlighted by recent cases like Climategate in 2009, when email exchanges among climate change researchers became available on the web, exposing internal communication dynamics that traditionally were confined to the *backstage* of knowledge production processes; or in the discussion about the discovery of the so-called *fast neutrinos* in 2012 – a controversy among specialists unfolding in real time and open to public view[8]. Analysis of public communication is required, more and more, to consider how and by whom the substance and the mode of such communication are determined in exchanges within and between sciences.

Science in society and science in culture

Understanding these situations may benefit from reappraising the object of science communication research as *how society talks about science*. This implies researching the cultural contexts – scientific, artistic, everyday and other – of such talk. The increasingly blurred boundaries of communication contexts should also encourage researchers to explore with more courage conceptual affinities and potential inspiration in the humanities, arts and culture, largely neglected by science communication scholars despite the growing science/art practice. For example, concepts such as *style* may be relevant to understanding variety in science communication as well as addressing the challenge of quality (Bucchi 2013). This resonates with long-standing invitations to 'put science into culture' (e.g. Lévy-Leblond 1996), emphasising its connections with other domains rather than its separation from society and culture as expressed in models and visions of knowledge translation and transfer. It also invites us to recognise the importance of a broader culture of science in society that goes beyond familiarity with technical contents to include an awareness of its role, implications, aims, potentialities and limits. It eventually demands that not only society, the public and culture are problematised in their relationship with science but that science problematises its own cultural premises. In this way, science communication – both practice and research – can contribute to increased reflexivity within society and within science.

Global trends and challenges

Public communication of science has become a global enterprise with common denominators as well as distinctive regional characterisations (see Trench *et al.* in this volume). This certainly expands opportunities for experimenting with communication formats and for comparative analysis of, for example, the application of similar approaches in different contexts. It also makes increasingly visible the strong contextual interaction of science communication patterns with broader cultural, policy and sociopolitical landscapes. Finally, it further highlights how difficult and even misleading it would be to expect a single, straightforward response to contemporary challenges of science communication, such as those outlined above, or to fulfil the expectation of eventually finding the *best* and most appropriate, one-size-fits-all model of science/public interaction. Taking the global view reduces the temptation to see different analytical models of interactions among experts and the public as a chronological sequence of stages in which the emerging forms obscure the previous ones. Arrangements traditionally invoked in the past for science communication were largely seeking uniformity and standardisation of practices, mostly by anchoring and flattening quality to a single or principal requisite or criterion, such as accuracy in transporting the message, adherence to scientific sources or independence of mediators. Focusing on science in culture, and in cultures, helps us to account for the continuing coexistence of different patterns of science communication that may coalesce or diverge depending on specific conditions and on the issues at stake. This should lead us to reappraise, for example, national differences in terms other than being more or less distant from an abstractly defined gold standard.

Key questions

- What trends can you discern in the changing usage of historically central terms such as *popularisation* and *dialogue*?
- In what ways might scientists' and scientific institutions' uses of media be said to contribute to a crisis of mediators?
- What specific approaches to research might be used to analyse the *culture of science in society* and to enhance understanding of science in culture?

Notes

1. An extended version of this dialogue is in M. Bucchi (2010); an adapted animated version is available at http://www.youtube.com/watch?v=X__D1eWBkXo
2. Some limitation of such a lexical exercise is inevitable when based on terms that are prevalent in the English-language literature, although these are also widely used in the international discussion. A version of this review of key terms appeared in Italian in *Annuario Scienza Tecnologia e Societa* (2014), published by il Mulino and Observa – Science in Society.

3   Bush was scientific advisor to the US government during World War II, and delivered an influential report, *Science: The Endless Frontier*, in 1945.
4   This type of critique is often associated with the observation that in some languages, like French or Italian, the equivalent of popularisation is *vulgarisation, divulgazione* – which may sound less neutral than popularisation, already incorporating an implicit value judgement of its modest relevance in comparison with more elevated scientific communication and practice.
5   For example, situations characterised by low public sensitivity or mobilisation, moderate perception of controversy among experts, and great visibility of science actors and institutions involved (Bucchi 2008).
6   This programme covers the period 2014-2020 and is mainly focused on developing research into technological innovations; see http://ec.europa.eu/research/horizon2020/index_en.cfm?pg=h2020
7   An early reflection on controversial exhibitions involving science issues is in Gieryn (1998); see also Schiele in this volume.
8   The concept of *backstage* was introduced by Goffman (1959); for an application with regard to science communication contexts, see Bucchi (1998), Trench (2012). For a recent overview and analysis of the Climategate case, see Grundmann (2013).

References

Allum, N., P. Sturgis, D. Tabourazi and I. Brunton-Smith (2008) 'Science, knowledge and attitudes across cultures: a meta-analysis', *Public Understanding of Science*, 17, 1: 35–54.
Bauer, M. and P. Jensen (2011) 'The mobilisation of scientists for public engagement', *Public Understanding of Science*, 20, 1: 3–11.
Bauer, M., R. Shukla and N. Allum (eds) (2012) *The Culture of Science: How the Public Relates to Science across the World*, London and New York: Routledge.
Ben-David, J. (1971) *The Scientist's Role in Society: A Comparative Study*, Englewood Cliffs: Prentice Hall.
Bonney, R., C. B. Cooper, J. Dickinson, S. Kelling, T. Phillips, K.V. Rosenberg and J. Shirk (2009) 'Citizen Science: A developing tool for expanding science knowledge and scientific literacy', *BioScience*, 59, 11: 977–984.
Bucchi, M. (1998) *Science and the Media: Alternative Routes in Scientific Communication*, London and New York: Routledge.
Bucchi, M. (2008) 'Of deficits, deviations and dialogues: theories of public communication of science', in M. Bucchi and B. Trench (eds) *Handbook of Public Communication of Science and Technology*, London and New York: Routledge, 57–76.
Bucchi, M. (2009) *Beyond Technocracy: Science, Politics and Citizens*, New York: Springer.
Bucchi, M. (2010) *Scientisti e antiscientisti. Perché scienza e società non si capiscono*, Bologna: il Mulino.
Bucchi, M. (2013) 'Style in science communication', *Public Understanding of Science*, 22, 8: 904–915.
Bucchi, M. (2014) 'Norms, competition and visibility in contemporary science: the legacy of Robert K. Merton', *Journal of Classical Sociology* (in press).
Bucchi, M. and F. Neresini (2011) 'Which indicators for the new public engagement activities? An exploratory study of European research institution', *Public Understanding of Science*, 20, 1: 64–79.
Callon, M. (1999) 'The role of lay people in the production and dissemination of scientific knowledge', *Science, Technology and Society*, 4, 1: 81–94.
Callon, M., P. Lascoumes and Y. Barthe (2001) *Agir Dans un Monde Incertain. Essai sur la Démocratie Technique*, Paris: Editions du Seuil.
Davies, S., E. McCallie, E. Simonsson, L. Lehr and S. Duensing (2008) 'Discussing dialogue: perspectives on the value of science dialogue events that do not inform policy', *Public Understanding of Science*, 18, 3: 338–353.
Delfanti, A. (2013) *Biohackers – The Politics of Open Science*, London: Pluto Books.
Dornan, C. (1990) 'Some problems in conceptualising the issue of "science in the media"', *Critical Studies in Media Communication*, 7, 1: 48–71.
Durant, J. (1993) 'What is scientific literacy?' in J. Durant and J. Gregory (eds) *Science and Culture in Europe*, London: Science Museum, 29–138.
Durodié, B. (2003) 'Limitations of public dialogue in science and the rise of the new "experts"', *Critical Review of International Social and Political Philosophy*, 6, 4: 82–92.
Einsiedel, E. (2000) 'Understanding "publics" in public understanding of science', in M. Dierckes and C. von Grote (eds) *Between Understanding and Trust: The Public, Science and Technology*, London and New York: Routledge, 205–215.

Epstein, S. (1995) 'The construction of lay expertise: AIDS, activism and the forging of credibility in the reform of clinical trials', *Science, Technology and Human Values*, 20, 4: 408–437.
Ezrahi, Y. (1990) *The Descent of Icarus*, Cambridge: Harvard University Press.
Gieryn, T. (1998) 'Balancing acts: science, Enola Gay and history wars at the Smithsonian', in S. Macdonald (ed.) *The Politics of Display: Museums, Science, Culture*, London: Routledge, 197–227.
Goffman, E. (1959) *The Presentation of Self in Everyday Life*, Garden City: Doubleday.
Goodell, R. (1977) *The Visible Scientists*, Boston: Little, Brown.
Grundmann, R. (2013) '"Climategate" and the scientific ethos', *Science, Technology and Human Values*, 38, 1: 67–93.
Hilgartner, S. (1990) 'The dominant view of popularisation: conceptual problems, political uses', *Social Studies of Science*, 20, 3: 519–539.
Horst, M. (2012) 'Deliberation, dialogue or dissemination: changing objectives in the communication of science and technology in Denmark', in B. Schiele, M. Claessens and S. Shi (eds) *Science Communication in the World: Practices, Theories and Trends*, Dordrecht: Springer, 95–108.
House of Lords Select Committee on Science and Technology (2000) *Science and Technology: Third Report*, London: Stationery Office, at http://www.parliament.the-stationeryoffice.co.uk/pa/ld199900/ldselect/ldsctech/38/3801.htm
Kuhn, T. S. (1962), *The Structure of Scientific Revolutions*, Chicago: The University of Chicago Press.
Lévy-Leblond, J-M. (1996) *La Pierre de Touche – la science a l'épreuve*, Paris: Gallimard.
Lewenstein, B. (2008) 'Popularisation', in J. L. Heilbron (ed.) *The Oxford Companion to the Hiistory of Modern Science*, Oxford: Oxford University Press: 667–668.
Lightman, B. (2007) *Victorian Popularizers of Science. Designing Nature for New Audiences*, Chicago: The University of Chicago Press.
Merton, R. K. (1973), *The Sociology of Science*, Chicago: The University of Chicago Press.
Metlay, G. (2006) 'Reconsidering renormalisation: stability and change in 20th-century views on university patents', *Social Studies of Science*, 36, 4: 565–597.
Miller, S. (2001) 'Public understanding of science at the crossroads', *Public Understanding of Science*, 10, 1: 115–120.
Miller, S., Caro, P., Koulaidis, V., de Semir, V., Staveloz, W. and Vargas, R. (2002) *Report from the Expert Group, Benchmarking the Promotion of RTD Culture and Public Understanding of Science*, Brussels: European Commission.
Miller, S., D. Fahy and ESConet Team (2009) 'Can science communication workshops train scientists for reflexive public engagement?', *Science Communication*, 31, 1: 116–126.
Peters, H. P., Heinrichs, H., Jung, A., Kallfass, M. and Petersen, I. (2008) 'Medialisation of science as a prerequisite of its legitimisation and political relevance', in D. Cheng, M. Claessens, T. Gascoigne, J. Metcalfe, B. Schiele and S. Shi (eds) *Science Communication in Social Contexts: New Models, New Practices*, Dordrecht: Springer, 71–92.
Phillips, L., A. Carvalho and J. Doyle (eds) (2012) *Citizen Voices: Performing Public Participation in Science and Environment Communication*, Bristol and Chicago: Intellect.
Raichvarg, D. and J. Jacques (1991) *Savants et Ignorants – une histoire de la vulgarisation de la science*, Paris: Seuil.
Raza, G. and S. Singh (2103) 'Science communication in India at a crossroads, yet again', in P. Baranger and B. Schiele (eds), *Science Communication Today: International Perspectives, Issues and Strategies*, Paris: CNRS Éditions, 243–262.
Riise, J. (2008) 'Bringing science to the public', in D. Cheng, M. Claessens, T. Gascoigne, J. Metcalfe, B. Schiele and S. Shi (eds) *Communicating Science in Social Contexts: New Models, New Practices*, Dordrecht: Springer, 301–310.
Rödder, S., M. Franzen and P. Weingart (eds) (2012) *The Sciences' Media Connection: Public Communication and its Repercussions, Sociology of the Sciences Yearbook,* Springer: Dordrecht.
Snow, C. P. (1959) *The Two Cultures*, Cambridge: Cambridge University Press.
Sturgis, P. and N. Allum (2005) 'Science in society: re-evaluating the deficit model of public attitudes', *Public Understanding of Science*, 13, 1: 55–74.
Trench, B. (2008a) '*Internet: turning science communication inside-out?'*, in M. Bucchi and B. Trench (eds) *Handbook of Public Communication of Science and Technology*, London and New York: Routledge, 185–194.
Trench, B. (2008b) '*Towards an analytical framework of science communication models'*, in D. Cheng, M. Claessens, T. Gascoigne, J. Metcalfe, B. Schiele and S. Shi (eds) *Communicating Science in Social Contexts: New Models, New Practices*, Dordrecht: Springer, 119–138.

Trench, B. (2012) 'Scientists' blogs: glimpses behind the scenes', in S. Rödder, M. Franzen and P. Weingart (eds) *The Sciences' Media Connection: Public Communication and Its Repercussions, Sociology of the Sciences Yearbook,* Dordrecht: Springer, 273–290.

Weingart, P. (1998) 'Science and the media', *Research Policy,* 27, 8: 869–879.

Wilsdon, J. and R. Willis (2004) *See-Through Science: Why Public Engagement Needs to move Upstream,* London: Demos.

Wynne B. (1991) 'Knowledges in context', *Science, Technology and Human Values,* 16, 1: 111–121.

Wynne, B. (1992) 'Misunderstood misunderstanding: social identities and public uptake of science', *Public Understanding of Science,* 1, 3: 281–304.

Wynne, B. (2006) 'Public engagement as a means of restoring public trust in science: hitting the notes but missing the music?', *Community Genetics,* 9, 3: 211–220.

Ziman, J. (1991) 'Public understanding of science', *Science, Technology and Human Values,* 16, 1: 99–105.

Ziman, J. (2000) *Real Science: What It Is and What It Means,* Cambridge: Cambridge University Press.

2
Popular science books
From public education to science bestsellers

Alice Bell and Jon Turney

Introduction

The term *popular science* – in its broadest sense – is a piece of science communication which assumes interest, but no particular expertise, on the part of the reader. Today, it is often a publishing tag, printed on the backs of books or emblazoned across the tops of booksellers' shelves. But popular science has a long history with broader applications. Historians of science with an interest in popular science (e.g. Fyfe and Lightman 2007) have been keen to extend research from the study of books to include lectures, songs, museums, part-works, magazines, radio and television. In terms of changes in practice, increasingly we see blogging having an impact on books and other print forms, though there have long been books spinning-off (or working cross-platform) with television programmes, exhibitions and lecture series. There is also a growing use of comic art as a form of popular science.

Even if we limit ourselves to books, popular science is rather difficult to pin down. This is probably partly because what we call *science* and *popular* are ambiguous entities themselves. From a historical perspective, Topham (2007) suggests the term emerged out of a mix of agendas surrounding the development of cheaper educational publishing and more popular forms of journalism in the early nineteenth century. It was by no means a term of singular meaning or origins. As Myers (2003) suggests, popular science is perhaps largely defined by what it is not. It is not the scientific talk that scientists engage in with each other. Neither, many would argue, is it fiction, though many will explore the boundaries of fiction with such work. Nor does *popular science* include, for many, guides to technology, health advice, arguments over the politics of climate change, nature writing, educational manuals or works one might find in the Mind, Body and Spirit sections of bookshops (though some would argue we should extend our definition to include at least some of these categories too). As Mellor (2003) puts it, simply by virtue of their content and their labelling as popular science, books may help build a sense of which ideas and activities count as science. This is perhaps one of the reasons popular science can be such a politically contentious topic, implicated in controversies over what is or is not science; although, as we shall see, there is a politics to its existence between expertise and the public too.

This chapter[1] starts with a brief overview of models of popular science before moving on to sketch a brief history of popular science publishing. We finish by outlining some of the ways in which people have sought to understand the content of the form with some questions for further analysis.

The *popular* in popular science

It is worth questioning what is meant by the *popular* in popular science. This is partly a basic matter of considering the first principles of our study, but it also serves as a prompt to question the nature of assumed audiences in popular science. It seems that popular science books need not be popular in the sense of commanding large audiences; many items remain understandably – sometimes proudly – niche in scope. The term then perhaps refers more to intent, though we might debate the reach and validity of such intent. We can also consider how such intents vary within subgenres and types of popular science book. For example, children's science books reflect a range of ideas about the assumed relationship between young people and science (Bell 2008).

Questions of pitching to an appropriate *level* are often applied, where higher level is assumed to mean more technical, closer in some way to the actual science, and lower level, less technical and more distant from the science. Another metaphorical way of looking at it may be to consider the degree of dilution of the science. The term *dumbing down* gets used too, and this is perhaps where the value judgements involved in the process start to become more apparent. As is often the case in science communication, the implication is that popularisation is a generous act on the part of science, done for the benefit of the public; but we might equally argue popular science is more a performance to serve the status of the writer and characters in the book, not the readers who, it should also be remembered, are expected to remain quiet consumers of the text.

Many see popular science as the epitome of top-down deficit-model thinking, labelling publics as simply the recipients of knowledge. By implying the need for a separate category of popular science for the laity and working from an assumption that science flows from experts to publics, the very existence of popular science helps articulate a boundary between those who can articulate true, reliable knowledge and those who consume it. Hilgartner (1990: 534) puts it neatly when he suggests much popular science acts to endow the scientific establishment with 'the epistemic equivalent of the right to print money'. This paper on the 'dominant view' of popular science remains rightfully a classic in analysis of the form. Popular science has become unfashionable in many areas of more recent (*post-PUS*) science communication practice and analysis, though, perhaps for the same reason, it is still highly regarded by sectors of the scientific establishment.

We might thus read the traditional model of popular science as a *courteous translation*, less a matter of enabling public interaction with science and more about keeping the public at arm's length. Popular science takes the more esoteric academic science, tied up in paywalled journals and complex jargon, and offers it to the public via metaphor, analogy or similar, packaged in mass-market paperbacks which are more accessible. But, rather like decontextualised hands-on presentations in museums, it delivers a largely *read-only* experience of science; one that, on the surface, offers science to the people but also keeps them out. A good example of this is provided by physicist Russell Stannard's series of *Uncle Albert* books for children. Inspired by George Gamow's use of fiction to explain modern physics as well as a sense – drawn from studying educational psychology – that young people need to learn through some sort of hands-on experience, Stannard (1999) uses fictional approaches based on fact to build worlds of the very fast (relativity), the very big (black holes) and the very small (quantum physics). Fictional and non-fictional elements are clearly demarcated by Stannard, who aims to provide the reader with a hands-on experience by proxy through the main protagonist who is allowed to travel around such new and exciting spaces via the magic thought bubble of her 'Uncle Albert'. A more extended example of this approach can be found in the *Magic School Bus* series, and arguably metaphors and analogy do something similar in much popular science writing. They may paint a clear expository picture of what science tells us about the world, but in doing so they often lose a sense of how such ideas came about or allow readers a chance to intervene or debate them.

However, we should be careful of a too-simple critique of popular science. Explanatory text might well seem rather one-way, but it should also be seen as part of a network of many other opportunities to connect with science and, therefore, not necessarily that problematic in itself as long as a more diverse ecosystem of public engagement is maintained, as suggested by Lewenstein's (1995) web model of communication. Moreover, most texts will do many things at once and make compromises along the way. In his children's books, Stannard is very careful to base his stories around a child asking questions as well as highlighting some of the debates within the history of modern physics. Texts are also open to be ignored, remixed and critiqued in society at large, even if such responses do not make their way to their original author (see Jenkins 2006 for discussion on web culture – but more broadly applicable). Richard Dawkins' word 'meme' is an especially interesting case study in this (Brown 2013; Salon 2013).

Recent work from historians of science has further critiqued the idea that the popularisation of science is simply a form of cultural hegemony done to passive publics. Fyfe and Lightman's (2007) collection of essays on nineteenth-century popular science, *Science in the Marketplace*, characterises the audiences of popular science as consumers, happily buying popular science products be they books, magazines, exhibitions, shows, songs or toys. It is important to note here that they are applying a sense of consumer identity as a relatively powerful one. To Fyfe and Lightman, the consumers of scientific culture in the nineteenth century were increasingly aware of the range of forms of expertise and the different, competing, ideas on offer. As well as choosing which products to partake of, such consumers had the ability to choose which ones to trust and how far. There is much power in Fyfe and Lightman's analysis, although we should be careful of their arguably slightly romantic view of consumer power. True, customers are not simply passive dupes, but neither are they all-powerful; there is space for more nuanced views of power applied to popular science, as well as for more empirical audience studies.

An outline history of popular science

The simplest sketch of the history of popular science is one that could, at first, look rather like one led by the works of great men.[2] As the cult of Newton grew in the century after his death, many popular lectures and books, including a celebrated account of his ideas by Voltaire, set out versions of the Newtonian system which were more accessible. Our view of this history is easily distorted by the prominence of particular books or authors. Much discussion centres on key works like *The Origin of Species* in 1859, which were expounding path-breaking science but were still accessible to a generally educated readership. Or it tends to focus on periods when well-known scientists wrote for popular audiences, as in the 1930s when Arthur Eddington and James Jeans published bestselling versions of Einsteinian physics and astrophysics. There were even children's versions, such as the pseudonymous effort by Tom Telescope of 1861, attributed to the pioneering children's author John Newbery but possibly written by Oliver Goldsmith or Christopher Smart (Secord 1985).

Such works grew in number and variety as disciplines developed and divided, and more science became susceptible to an apparent need for popularisation by intellectual gaps (such as mathematical formalisation) or social/cultural ones (through professionalisation of science). Closer historical study reveals that there were many writers, whose books are now largely forgotten but were important at the time, active as expositors of science throughout the last two centuries. Over this time, values changed around popular science; at one point it was common for texts, especially those for children, to synthesise scientific and religious education through the lens of nature study.

By Thomas Henry Huxley's time in Victorian England it was becoming possible to support oneself (precariously) as a writer specialising in natural history and science. But, as Fyfe (2005) puts it, 'by the 1850s an increasingly clear division had developed between writings that contributed to a scientific reputation (and usually paid little) and those that paid the bills (but did nothing for one's reputation)'. Fyfe's study – set against the backdrop of the shift from the 'public sphere' of the eighteenth century to the mass audience of the nineteenth – shows how a market emerged for a wide range of science books, aimed at instruction, moral improvement or even entertainment, many written by women.

Increasingly, authors specialised in science or its exposition, and many of these new authors were not active as researchers but rather specialised in writing for popular audiences. Among historians, as Bowler (2006) argues, there arose a widespread assumption that the vast majority of scientists turned their back on writing for lay readers in the early twentieth century as professionalisation took hold. But he suggests this was not really true; some scientists were happy to write for the wider public, and publishers valued them because the authors' credentials allowed them to sell the books as *educational*. In this mixed economy, now little-known authors, usually without professional scientific training, were writing more *entertaining* works. It is perhaps true, though, that few prominent scientists wrote for the public in this period. But by the 1930s, a cadre of 'visible scientists' (Goodell 1977) was again active. Retrospective reviews generally lump together the conservative British physicists Eddington and Jeans on the one hand and the more liberal or radical biologists, such as Huxley and Haldane, on the other. But there were also notable popular works by Einstein, for example, whose global fame boosted the appeal of a number of expositions of special and general relativity which remain in print today.

After World War II, there was not only a large literate public but a larger cohort who had accessed higher education. Again, accounts of the general development of popular science books differ according to the focus of the author. Lewenstein (2005) detects a major shift in the prominence of science books in the 1970s, but his verdict is based on study of Pulitzer Prize winners and *New York Times* listed bestsellers in this period. As he puts it, 'beginning with Carl Sagan's *Dragons of Eden* in 1978, then every year or every other year the Pulitzers begin honouring a science book.... Clearly something happens in the late 1970s to make science books more central to American culture. Science becomes a part of the general public discussion'. The influence of other media was powerful, with Sagan's TV series *Cosmos* leading to a book which became a bestseller on both sides of the Atlantic. The modern popular science book boom was under way.

Again this broad-brush story is slightly too simple, and there were also more gradual changes which were visible earlier. In the US, the advent of mass higher education brought a new reading public to non-fiction. A close study of the bestsellers that were published by Knopf about human history and evolution in the 1950s and 1960s shows how they came to depend less on old models. In the early post-war period this publisher relied on a biographical style closely akin to that used by researcher-turned-writer Paul deKruif in his successful pre-war *Microbe Hunters*. But over time the publishers found that books with considerably more technical content could be successfully promoted and could sell strongly. The overall story of the subject, incorporating up-to-date results and told by an expert or a journalist in close touch with the experts, proved as appealing as the stories and struggles of the discoverers (Luey 1999).

Both models continued to exist through the popular science boom – now declared by some to be over (Tallack 2004) – and beyond. So we now have a profusion of titles in print on a wide range of subjects and offering many different treatments of similar subjects. Some books are regarded as classics and remain strong sellers over many years, although otherwise popular science can be a reasonably evanescent area of publishing. An example of the more durable classics is Dawkins' *Selfish Gene*, first published in 1976, expanded into a successful second edition in 1989, and republished in a thirtieth anniversary edition along with a companion volume of essays considering its influence in 2006. Children's books can have a particular cultural durability, perhaps because they are more likely to focus on more settled scientific principles but also because adults will share the writings of their childhood with younger generations. Some books would stay in publication partly due to having become a traditional school prize, as with John Henry Pepper's *Playbook of Science* (Secord 2003). There was something of a nostalgia market in youth popular science in the early twenty-first century, following the success of *The Dangerous Book for Boys*, although it seems to be fading. The web has also allowed for some republishing of magazines. There is a prestigious archive of *New Scientist* viewable through Google, and *Popular Science* magazine has an impressive 'word frequency visualizer' to encourage browsing through their archive; type a word in and see the distribution of its use over 140 years of publications.

Books and beyond

Books are an attractive focus for study of science popularisation partly because of their long-established significance in cultural production. The technology of the printed book has been around in the West since the fifteenth century, so its history spans that of modern science. But it's also true that books are easy to study. Historians of science have pointed out that popular science is more than just books, perhaps as a self-critique but also as a prompt to search out archives for traces of other examples of the mediation of science. Such a critique is applicable to contemporary researchers; studies based on books can be very illuminating, as long as their specific social contexts are remembered and they are not made to act as a proxy for science communication in general.

Books occupy a specific position, economically and socially, compared with other forms of science communication. There is an upfront cost in contrast to outlets that are free at the point of consumption such as much of the web, much television and many museums. Reading a book also implies some time and varying degrees of commitment. It can also be a reasonably independent involvement compared with social or family events such as a science show or festival, although there can be a sociability to reading (e.g. book clubs, online review and discussion forums). Whereas a degree course in a science subject takes significant time and money, a book – especially since the advent of paperbacks – is relatively cheap and portable, easy to keep in a pocket and dip into during the day. All of these issues, in turn, have an impact on the relationships books may assume with readers, and the relationships between science and a public they may help produce.

As already mentioned, there is a history of seeing popular science books as a way of bringing otherwise rarefied knowledge to the people. This discourse runs through science communication in general – and might be generally problematised – but we should also consider specific social-class dimensions of science books when exploring this point. Paperback publishing has a history as a political attempt to offer the advantages of education to the masses, not just those who can afford to go to universities. Many scientists have taken to books and magazines with a sense that knowledge is power and they might help redistribute such power – Haldane's writing for the *Daily Worker* being a notable example (e.g. Haldane 1940). But it is also true that books and magazines can be both marketed and consumed as luxury items, with the role of recent children's non-fiction in 'topping up the middle-class child' of particular interest (see Vincent and Ball 2007; Buckingham and Scanlon 2005). In a sociological analysis of the sites of contemporary bookselling, Wright (2005) argues that much of our literary consumption embodies a style of 'soft capitalism' which aims to quietly perform respectability. Of course, books are not only sold, but borrowed, lent, even stolen. There is also an argument to be made that science books are qualitatively different from the sorts of literary cultures on sale in most bookshops, with science playing a more instrumental role in social mobility. But with evidence that science increasingly plays a role in the construction of middle-class identities (see Savage *et al.* 2013) and the rise of 'geek chic' (Corner and Bell 2011), analysts of contemporary popular science would do well to pay attention to the role it plays in twenty-first-century identity politics, with a particular focus on issues of social status.

Just as we might ask if there really are fewer quality female writers, we should question the historiography that offers a story dominated by texts in English. English has become the language of much professional science, but there is no need to assume that English texts should set an international pattern for popularisation. Children's literature offers some interesting case studies here. As Gillieson (2008) notes in her study of the *Eyewitness* guides, these are designed around images, with spaces left for text to be inserted in a variety of languages. But the foundational nature of science for young people does not necessarily make it applicable across national borders, and image-based publications do not necessarily have multicultural appeal. The *Horrible Science* books, for example, have been translated into several languages but struggled to make an impact in the United States. As scholarship in popular science grows, there are greater opportunities to connect disparate cultural studies and consider popular science literature in comparative contexts. A special issue of the journal *Public Understanding of Science* on popular science publishing, for example, included work on China (Wu and Qiu 2013) and Spain (Hochadel 2013).

When considering the larger social context of the popular science book, it is worth noting its aim to do more than just present science. Many are also political projects, though how overt that political project is varies. Recent examples include Mark Henderson's *Geek Manifesto* and Ben Goldacre's *Bad Pharma* (which worked alongside a well-orchestrated online campaign), as well as several books on climate change, which often apply both science and historical exposition for political ends (e.g. Oreskes and Conway 2010; Hansen 2009). As Buckingham (2000) writes of the early 1990s wave of environmental media for young people, it might be suggested cynically that much climate-related literature is directed at young people as a way of putting off dealing with the issue until the next generation. As the children of the early 1990s become parents, we can add further questions on the ethics of targeting young people, the age-appropriateness of some visions of the future and, more generally, whether science writers should be doing political activism under the cover of exposition. Children are subject to lobbying from a range of perspectives in and around science and technology, from the Primary Science Teaching Trust (formerly AstraZeneca Science Teaching Trust) to Terry, the Friendly Fracosaurus, a cartoon character produced by an energy company active in the controversial gas extraction technology of fracking (Hickman 2011). Projects that not only promote science and engineering to young people and the public at large but also promote particular forms of it can be very powerful, and the various semiotics of science education may offer a rather attractive public relations opportunity. That said, it is noticeable that the Royal Society's science book prize has found it hard to retain a sponsor in recent years as industrial sponsorship – along with much public funding – is either simply cut or shifts to the more potentially interactive and locally based network of science festivals.

The biggest change to popular science writing in recent years has, arguably, come through the web. One might worry about the quantity of free content, meaning people no longer pay for popular science publishing, or decry the shortening of attention spans, but both concerns may also be overstated. Moreover, there are opportunities for popular science here. With agents increasingly looking to blogs, rather than senior common rooms, to find new writers, there is the possibility of greater diversity of writers, although such a model does also assume writers are able to work for free as they build a reputation.

E-book *shorts* provide authors with the option of working through an idea that might, after all, be best suited to a form 8,000 words long; or allowing a trial version of an idea which, after feedback and the chance to articulate an audience, may be commissioned and developed into a full book which is much more successful and richer due to prior publicity. There are also new opportunities with the niche audiences along the so-called *long tail* of online markets, although there are concerns this may also provide silos of content for already highly interested readers (as discussed in Fahy and Nisbet 2011). There are new opportunities for funding too, although the tendency towards making material free at the point of access brings many challenges. The variety of publishing possibilities is great: for every user-funded or micro-funded site aiming to publish one long-form piece of science writing a month there is another which may rely on a rich benefactor to churn out an essay a day. As with science journalism, science education and more, who pays for popular science in the globalised, liberalised age of the web (and therefore, whom it might serve) remains an open, and important, question.

Ways to read a book

In terms of historical studies, one does not have to be limited to biographical sketches of authors or details of what books existed when (though these may also be rich pieces of research). The most difficult to realise is to recover a sense of how a book was read, as Jim Secord (2000) does for Chambers' early epic of science, *Vestiges of the Natural History of Creation*. The core of his study is an analysis of readers' responses to *Vestiges*' natural-historical account of the evolution of the universe culled from diaries, letters, reviews, and newspaper and journal commentaries. Priscilla Murphy's (2005) book-length study of Rachel Carson's *Silent Spring* offers an interesting comparison in the modern era. This is concerned with the way a book can influence a political debate; although, as this and other considerations of *Silent Spring* remind us, the fact that Carson was already a nationally celebrated author, *that Silent Spring* was serialised in the *New Yorker* before publication, and that it inspired a CBS *60 Minutes* TV programme were all key factors in increasing the book's impact (see also Kroll 2001). There is scope for more studies which integrate work on books with examination of other forms of science communication, such as the web of text which Fred Hoyle wove over many years across different media (radio, books) and genres (popularisation, science fiction) to promote his cosmological theories and his ideas about the origins of life (Gregory 2005).

More sociologically, Jurdant (1993) offers the useful metaphor of popular science as the autobiography of science, a way in which the scientific community writes a collective story of its life, with all the specialist insights and reasons to be sceptical such narratives imply. Books also contain implied epistemologies (Turney 2001a: 49–55) or can incrementally build boundaries of what is or what is not science (Mellor 2003; see also Gieryn 1999 on 'boundary work'). Gregory and Miller (1998) also offer an idea of popularised science as science in public, which not only focuses on the form as a way of presenting science but can help us consider the occasions where the popularisation process is part of the making of science and/or of science policy.

There may be normative views about science in society embedded in popular science. Mary Midgley (1992) offers perhaps the classic example of such critique, and, in a more detailed study, Hedgecoe (2000) suggests that the popularisation of genetics contributed to the overreaching claims for genetic explanation which have been labelled *geneticisation*. Nisbet and Fahy's (2013) study of bioethics in Skloot's *The Immortal Life of Henrietta Lacks* provides a fresher approach, including a methodology that incorporates related publications such as book reviews which thus considers more than just the text itself. Another route into the analysis of popular science is to reflect upon it in a more literary manner. This allows a focus on semiotics and ways in which realism is constructed, or looking for precise spaces where language offers opportunities to open or close areas for debate. We might explore the ways metaphor are used, where boundaries between fictionalisation and non-fiction exposition are drawn – even assonance, rhythm and other wordplay, depending on the case study. Elizabeth Leane (2007) offers an extended example of such an approach applied to physics and is especially interesting in her application of work previously applied to fiction, such as Haynes's (1994) taxonomy for considering the characterisation of scientists in literature. This shows how even if popular science is rooted in our ideas of reality, it is still a form of storytelling. Other studies have folded in analysis of visual culture (e.g. Eisner 1985, 1996; McCloud 1993; Kress and van Leeuwen 2006; Barker 1989). Just as techniques in literary analysis may be useful in understanding images, the opposite is true: visual analysis techniques can be useful for exploring text as well as the many images popular science publishing offers us.

Popular science somehow has to emulate the persuasive power of empiricism with words on a page. The writer has to describe observations, experiments, reports, even demonstrations, in such a way that the readers believe in them (Turney 1999). It is an act of persuasion that is different from any act of discovery; although the two are often linked, they should not necessarily be confused. Humans live in a medium-sized world, only directly experiencing the range between a few millimetres and a couple of hundred meters. Scientific research transcends these limitations and considers speeds and sizes only brought within our reach via specialised equipment or mathematically based models. What the more literary aspects of science writing can do is take the reader into 'realms beyond the normal human senses' (Turney 2001a: 55). Perhaps most useful here is Shapin's (1984) classic study of the 'literary technology' of Robert Boyle, which explores how the sense of real in the passing-on of science is constructed within certain contexts, including philosophical ideas of science and broader political agendas.

Another approach to the more story-like aspects of popular science is to focus on narrative structure. White (1981, 1992) argues in the context of writing history that the tendency to relay events as narrative arises out of a desire for coherence and closure; he also argues that these are both illusionary. Moreover, the process of organising events coherently and imposing a neat ending imprints a moral view on the events described in a text; the moral or political standpoint of a writer acts as the organising principle for the narration. Applying this idea to scientific contexts, Mellor (2007: 501) argues that the 'inexorable movement of a narrative towards a predetermined end' allows assumptions made by scientific writing to be glossed over and to go unchallenged (see also Brown 2006).

Curtis (1994) applies White's approach to science writing and argues that by structuring texts as a form of detective story, where revealing the truth provides narrative closure, writers employ a powerful rhetorical tool which may help imbue accounts of scientific work with an appearance of certainty. Curtis advocates what he sees as a more *Lakatosian* view of science, as an *ebb and flow* of questions and continual research. He puts it memorably when he suggests that with narrative accounts of science 'we begin with unanswered questions. We end with unquestioned answers' (Curtis 1994: 431).

Perhaps in response to this sense of science as continual questioning, children's popular science has often been structured around presenting questions and answers. An interesting example of these dialogic structures is Murphy's (2007) *Why is Snot Green?*, based on questions child visitors asked the writer when he worked in the interactive galleries of London's Science Museum.

What makes this book especially interesting in terms of its narrativity (or lack thereof) is the way Murphy includes cross-referencing footnotes and citations to further reading to allude to continual learning and/or discovery outside the perimeters of a book plotted to a conclusion. Stannard also builds his stories around the questioning of young children (Bell 2007), conveying a sense of science's *ebb and flow* to conclude his tightly plotted narratives on a message of uncertainty. Artful science writers are not about to be limited by something as dull as a narrative theory, and neither are their audiences.

We can think of the story of science as a grand story of the universe, and such a view offers us a chance to consider something of the cultural role popular science plays. The historical sciences may be particularly suited to provide accounts of change over time (Turney 2001b; see also O'Hara 1992). Eger suggests that the emerging canon of popular science makes up a grand narrative in the form of the 'new epic' of science, but it does so collectively:

> From Darwin's original theory, the lines of extension radiate downward to prebiotic (chemical) evolution as expounded by Prigogine and Eigen; to cosmic evolution as described by Weinberg, Paul Davies, and the astrophysicists; to human culture as Wilson explains in his theories of sociobiology; and finally, through the work of brain physiologists and AI researchers, to consciousness itself.
>
> (Eger 1993: 197)

A narrative of nature is woven into a narrative of science: the natural world is ordered through a reductive framework which imagines one set of scientific detail as inherently behind another (e.g. biology reduces to chemistry, and then to physics). Eger may suggest a multi-volume epic, but it is worth noting that single popular science books have tried to at least allude to such a scientific epic, selling themselves on science's ability to present a coherent, reduced narrative. The titles of some of the most popular books are indicative: from the seminal *Brief History of Time* (Hawking 1987) to Bryson's *A Short History of Nearly Everything* (2000). These stories of the past and future development of the world, be they cosmological or genetic, are narratives of nature and of science which perhaps offer secular alternatives to religious texts (see also Beer 2000; Midgley 2002). This perhaps offers another example of popular science books being superseded by live events, such as the Rationalist Association's *Nine Lessons and Carols for Godless People* or the Sceptics in the Pub movement; though both may also work to promote or inspire books, just as Sagan's television shows did before them. The place for popular science books in larger ecosystems of science in popular culture and public policy remains to be worked out.

Concluding remarks

This chapter has considered the long-standing question of whom popular science is supposed to serve. It also reflected upon the make-up of popular science in terms of focusing on books or not (and if so, how books may sit in larger media contexts), but also in terms of the different types of books and ways in which books might connect as well as compete with other media. We raised issues of class, gender and culture, all topics which are relatively under-studied, and offered these as analytical lenses along with other sociological questions, including those based on the ideas of philosophy of science implied in popular science books, images of scientists presented, and the idea of *boundary work* and use in influencing public policy. We might imagine that popular science tells a sort of multi-authored *autobiography of science*.

We also offered the texts as literary objects to be considered, and suggested some sources for the analysis of visuals as well as language. With reference to the disruptive effect of the web, the future of popular science as a publishing category is uncertain, but it has shown itself to be malleable to a range of cultural, political and scientific agendas for centuries. Future studies should track such changes in the form, but also explore more fully the role of popular science books as political and cultural objects, the part they play influencing policymaking and the ways in which readers articulate a sense of self through their consumption of such books.

Key questions

- Consider a set of books tagged 'science' by an online retailer. Which amongst them would you call 'popular science'? What concepts of *science* and *popular* are you using?
- Is popular science an inherently top-down approach to science communication?
- What are the philosophies of science implicit in a chosen example of popular science? What views on the ethics or application of science are most overt here?
- Select an example of non-fiction popular science and examine what characterisation of the scientist is being applied in it. What rhetorical purposes does this characterisation serve?

Notes

1 This new version of the chapter, as compared with the version in the first edition of this book in 2008, includes broader reflections on the impact of the Internet on popular science books publishing and reading. It also looks at developments in historical scholarship reflecting ideas about the meanings of popular science. Sections are added on books for young people, the analysis of images and some reference to global trends in publications, issues of class and the role of women authors.
2 Marchant (2011) rightly points out in a contemporary context that the privileging of male authors is worth questioning.

References

Barker, M. (1989) *Comics, Ideology, Power and the Critics*, Manchester: Manchester University Press.
Beer, G. (2000) *Darwin's Plots: Evolutionary Narrative in Darwin, George Eliot, and Nineteenth-Century Fiction*, second edition, Cambridge: Cambridge University Press.
Bell, A. (2007) 'What Albert did next: the Kuhnian child in science writing for young people', in P. Pinsent (ed.) *Time Everlasting: Representations of Past, Present and Future in Children's Literature*, Lichfield: Pied Piper Publishing, 250–267.
Bell, A. R. (2008) 'The childish nature of science: exploring the child/science relationship in popular non-fiction', in A. R. Bell, S. R. Davies and F. Mellor (eds) *Science and Its Publics*, Newcastle: Cambridge Scholars Publishing, 79–98.
Bowler, P. J. (2006) 'Experts and publishers: writing popular science in early twentieth-century Britain, writing popular history now', *British Journal for the History of Science*, 39, 2: 159–187.
Brown, A. (2013) 'Richard Dawkins and the meaningless meme', *Guardian Comment is Free* (24 June); online at http://www.theguardian.com/commentisfree/2013/jun/24/richard-dawkins-meaningless-meme-viral
Brown, N. (2006) 'Shifting tenses – from "regimes of truth" to "regimes of hope"', *SATSU Working Paper No 30*; online at www.york.ac.uk/org/satsu/OnLinePapers/OnlinePapers.htm
Buckingham, D. (2000) *The Making of Citizens: Young People, News and Politics*, London and New York: Routledge.
Buckingham, D. and Scanlon, M. (2005) 'Selling learning: towards a political economy of edutainment media', *Media, Culture and Society*, 27, 1: 41–58.
Corner, A. and Bell, A. (2011) 'Specsaviours', *Times Higher Education Magazine*, (25 August); online at www.timeshighereducation.co.uk/417188.article
Curtis, R. (1994) 'Narrative form and normative force: Baconian story-telling in popular science', *Social Studies of Science*, 24, 3: 419–461.
Eger, M. (1993) 'Hermeneutics and the new epic of science', in M. W. McRae (ed.) *Literature of Science: Perspectives on Popular Science Writing*, Athens, GA: University of Georgia Press, 186–209.

Eisner, W. (1985) *Comics and Sequential Art,* Tamarac, FL: Poorhouse Press.
Eisner, W. (1996) *Graphic Storytelling and Visual Narrative,* Tamarac, FL: Poorhouse Press.
Fahy, D. and Nisbet, M. C. (2011) 'The science journalist online: shifting roles and emerging practices', *Journalism,* 12, 7: 778–779.
Fyfe, A. (2005) 'Conscientious workmen or booksellers' hacks? The professional identities of science writers in the mid-nineteenth century', *Isis,* 96, 2: 192–223.
Fyfe, A. and Lightman, B. (eds) (2007) *Science in the Marketplace: Nineteenth-Century Sites and Experiences,* Chicago and London: University of Chicago Press.
Gieryn, T. F. (1999) *Cultural Boundaries of Science: Credibility on the Line,* Chicago and London: University of Chicago Press.
Gillieson, K. (2008) *A Framework for Graphic Description in Book Design,* PhD thesis, University of Reading.
Goodell, R. (1977) *The Visible Scientists,* Boston: Little, Brown.
Gregory, J. (2005) *Fred Hoyle's Universe,* Oxford: Oxford University Press.
Gregory, J. and Miller, S. (1998) *Science in Public: Communication, Culture and Credibility,* Cambridge, MA: Basic Books.
Haldane, J. B. S. (1940) *Science in Peace and War,* London: Lawrence and Wishart.
Hansen, J. (2009) *Storms of My Grandchildren: The Truth About the Coming Climate Catastrophe and Our Last Chance to Save Humanity,* London: Bloomsbury.
Haynes, R. (1994) *From Faust to Strangelove: Representations of the Scientist in Western Literature,* Baltimore: Johns Hopkins University Press.
Hedgecoe, A. (2000) 'The popularisation of genetics as geneticization', *Public Understanding of Science,* 9, 2: 183–189.
Hickman, L. (2011) '"Fracking" company targets US children with colouring book', *Guardian* Environment blog (14 July); online at www.theguardian.com/environment/blog/2011/jul/14/gas-fracking-children-colouring-book
Hilgartner, S. (1990) 'The dominant view of popularization: conceptual problems, political uses', *Social Studies of Science,* 20, 3: 519–539.
Hochadel, O. (2013) 'A boom of bones and books: the "popularization industry" of Atapuerca and human-origins research in contemporary Spain', *Public Understanding of Science,* 22, 5: 530–537.
Jenkins, H. (2006) *Convergence Culture: Where Old and New Media Collide,* New York: New York University Press.
Jurdant, B. (1993) 'The popularization of science as the autobiography of science', *Public Understanding of Science,* 2, 4: 365–373.
Kress, G. and van Leeuwen, T. (2006) *Reading Images: The Grammar of Visual Design,* second edition, London and New York: Routledge.
Kroll, G. (2001) 'The "silent springs" of Rachel Carson: mass media and the origins of modern environmentalism', *Public Understanding of Science,* 10, 4: 403–420.
Leane, E. (2007) *Reading Popular Physics: Disciplinary Skirmishes and Textual Strategies,* Farnham, Surrey: Ashgate.
Lewenstein, B. (1995) 'From fax to facts: communication in the cold fusion saga', *Social Studies of Science,* 25, 3: 403–436.
Lewenstein, B. (2005) *Science Books Since World War II: History of the Book in America, Vol 5;* online at www.people.cornell.edu/pages/bvl1/books2004.pdf
Luey, B. (1999) '"Leading the public gently": popular science books in the 1950s', *Book History,* 2, 1: 218–253.
McCloud, S. (1993) *Understanding Comics: The Invisible Art,* Kitchen Sink Press.
Marchant, J. (2011) 'Why are so few popular science books written by women?', *Guardian* Notes and Theories (4 October); online at www.theguardian.com/science/blog/2011/oct/04/popular-science-books-women
Mellor, F. (2003) 'Between fact and fiction: demarcating science from non-science in popular physics books', *Social Studies of Science,* 33, 4: 509–538.
Mellor, F. (2007) 'Colliding worlds: asteroid research and the legitimatization of war in space', *Social Studies of Science,* 37, 4: 499–531.
Midgley, M. (1992) *Science as Salvation: A Modern Myth and its Meaning,* London and New York: Routledge.
Midgley, M. (2002) *Evolution as a Religion: Strange Hopes and Stranger Fears,* London and New York: Routledge.
Murphy, G. (2007) *Why Is Snot Green?* London: Macmillan Children's Books.

Murphy, P. (2005) *What a Book Can Do: The Publication and Reception of Silent Spring*, Amherst, MA: University of Massachusetts Press.

Myers, G. (2003) 'Discourse studies of popular science: questioning the boundaries', *Discourse Studies*, 5, 2: 265–279.

Nisbet, M. and Fahy, D. (2013) 'Bioethics in popular science: evaluating the media impact of The Immortal Life of Henrietta Lacks on the biobank debate', *BMC Medical Ethics*, 14: 10.

O'Hara, R. J. (1992) 'Telling the tree: narrative representation and the study of evolutionary history', *Biology and Philosophy*, 7, 2: 135–160.

Oreskes, N. and Conway, E. (2010) *Merchants of Doubt: How a Handful of Scientists Obscure the Truth on Issues from Tobacco Smoke to Global Warming*, London: Bloomsbury.

Salon, O. (2013) 'Richard Dawkins on the internet's hijacking of the word "meme"', Wired.co.uk (20 June); online at www.wired.co.uk/news/archive/2013-06/20/richard-dawkins-memes

Savage, M., Devine, F., Cunningham, N., Taylor, M., Li, Y., Hjellbrekke, J., Le Roux, B., Friedman, S. and Miles, A. (2013) 'A new model of social class? Findings from the BBC's Great British Class Survey', *Sociology*, 47, 2, 219–250.

Secord, J. (1985) 'Newton in the nursery: Tom Telescope and the philosophy of tops and balls, 1761–1838', *History of Science*, 23: 127–151.

Secord, J. (2000) *Victorian Sensation: The Extraordinary Publication, Reception and Secret Authorship of Vestiges of the Natural History of Creation*, Chicago and London: University of Chicago Press.

Secord, J. (2003) 'Introduction to The Boy's Playbook of Science', in A. Fyfe (ed.) *Science For Children*, Vol. 6, Bristol: Thoemmes Press, v–ix.

Shapin, S. (1984) 'Pump and circumstance: Robert Boyle's literary technology', *Social Studies of Science*, 14, 1: 481–520.

Stannard, R. (1999) 'Einstein for young people', in E. Scanlon, E. Whitelegg and S. Yates (eds) *Communicating Science: Contexts and Channels*, London: Routledge, 134–145.

Tallack, P. (2004) 'Echo of the big bang: an end to the boom in popular science books may actually raise standards', *Nature*, 432: 803–804.

Topham, J. R. (2007) 'Publishing "popular science" in early nineteenth-century Britain', in A. Fyfe and B. Lightman (eds) *Science in the Marketplace: Nineteenth-Century Sites and Experiences*, Chicago and London: University of Chicago Press, 135–168.

Turney, J. (1999) 'The word and the world', in E. Scanlon, E. Whitelegg and S. Yates (eds) *Communicating Science: Contexts and Channels*, London and New York: Routledge, 120–133.

Turney, J. (2001a) 'More than story-telling: reflecting on popular science' in S. M. Stocklmayer, M. M. Gore and C. Bryant (eds) *Science Communication in Theory and Practice*, Dordrecht: Kluwer Academic, 47–62.

Turney, J. (2001b) 'Telling the facts of life: cosmology and the epic of evolution', *Science as Culture*, 10, 2: 225–247.

Vincent, C. and S. J Ball (2007) 'Making up the middle class child: families, activities and class dispositions', *Sociology*, 41, 6: 1061–1078.

White, H. (1981) 'The value of narrativity in the representation of reality', in W. J. T. Mitchell (ed.) *On Narrative*, Chicago, IL: University of Chicago Press, 1–23.

White, H. (1992) 'Historical employment and the problem of truth', in S. Friedlander (ed.) *Probing the Limits of Representation: Nazism and the 'Final Solution'*, Cambridge, MA: Harvard University Press, 37–53.

Wright, D. (2005) 'Commodifying respectability: distinctions at work in the bookshop', *Journal of Consumer Culture*, 5, 3: 295–314.

Wu, G. and H. Qiu (2013) 'Popular science publishing in contemporary China', *Public Understanding of Science*, 22, 5: 521–529.

3
Science journalism
Prospects in the digital age

Sharon Dunwoody

Introduction

Science journalism is an increasingly imperilled occupation that, perversely, is needed now more than ever. In a world where both citizens and advertisers increasingly control their own delivery of information via online channels, the kind of legacy mass media that have long served as the principal employers of science journalists – newspapers and magazines – are faltering in many countries. Journalists cut loose from these media organisations are scrambling to find their footing elsewhere. It will be years before successful models for delivery of substantive science journalism emerge from the bevy of experiments now under way.

And yet, science journalism has never been more important. Citizens of the globe are buffeted by one issue after another – the potential impacts of GM crops; the mysterious die-off of bees; individualised medical treatment via genomics; climate disruption; the prospect of bringing extinct species back to life – and have few places to turn for independent, evidence-based information. Historically, most people have depended on mediated channels, those ubiquitous packagers of information intended for large numbers of readers/listeners/viewers, where they typically encounter science information almost inadvertently as they watch TV news, read their morning newspaper or page through a magazine from the corner news stand. While that is still the case in many countries, today's citizens rely increasingly on volitional searching of the Internet for their information. The science journalists are there, blogging and placing stories in a variety of web-only outlets. But finding that good information requires effort on the part of the individual searcher, effort that the typical individual rarely expends.

This chapter[1] discusses these conundrums and what they portend for the future of science journalism. It first tracks the historical evolution of the field, then moves to characteristics of modern science journalists and their media outlets. It ends by returning to the challenges that lie ahead.

A brief history

Science stories have appeared in the mass media for as long as these channels have existed. Who wrote those stories, on the other hand, has varied over time and across cultures. Scholars in a number of countries have sought to track the evolution of *popular science* in their respective cultures (see, for example, Bauer and Bucchi 2007; Broks 2006; Burnham 1987; Golinski 1992). What they find is a process initially characterised by scientists' efforts to share knowledge as widely as possible, followed by a retrenchment that moved scientists away from direct contact with publics and, in Broks's words, transformed the public 'from participants to consumers' (Broks 2006: 33). In this characterisation of the process in Britain, scientists in the late eighteenth century sought to diffuse scientific understanding throughout the culture, assuming substantial benefits would accrue in the course of the integration of science with the workaday world of ordinary folks. By the nineteenth century, however, the relentless advance of specialised knowledge began to create a chasm between scientists and society. Broks describes this as evolution from 'the Enlightenment ideal of "experience"' to 'the early nineteenth-century construction of "expertise"' with scientists morphing even further by the end of the nineteenth century into an even less accessible category of beings called the 'professional expert' (Broks 2006: 28). As scientists withdrew from the world of popularisation, the construction of popular science narratives was turned over increasingly to journalists.

Burnham (1987) captures the same trend in the United States. By the late nineteenth century, several popular science magazines were already established – pre-eminent among them *Scientific American* and *Popular Science Monthly* – and newspaper editors were happy to reprint texts of science lectures and to publish scientists' reflections on natural phenomena such as meteor showers. The scientists themselves were equally willing to invest time and energy in public communication endeavours. Scientists in the latter part of the nineteenth century tended to view popularisation as part of their job.

In the early twentieth century, however, increasing specialisation and professionalisation pushed scientists to see themselves as apart from everyday people. As scientists developed their own languages, their own training regimens and their own reward systems, communication with others outside the occupation became less of a priority. To make matters worse, major scientific societies began to punish scientists for daring to popularise by ostracising offending individuals and even denying them access to rewards, such as membership in honorific societies. Goodell's classic book *The Visible Scientists* (1977) is replete with examples of how even senior, accomplished scientists were subjected to sustained repercussions as a result of their popularising efforts. Although, as I argue later in this chapter, popularisation has again become *au courant* for many scientists, residual hostility within the scientific culture makes it a risky behaviour even today. But, back in the early twentieth century, too much of an investment in popularisation could ruin a scientist's career, so many scientists left the world of popularisation to journalists and the mass media.

The mass media's interest in science has remained steady throughout the centuries. The technology of warfare, discoveries of planets and entire galaxies (not to mention Martian canals!) and advances in medical care were easy for journalists to *sell* to their editors. These editors did not care that a topic was scientific, only that it was novel and likely to grab the attention of their readers. Canvas the issues of any newspaper of the eighteenth, nineteenth and early twentieth centuries and you are likely to find stories that we would today classify as 'science' in the broadest sense.

Still, few journalists by the mid-twentieth century would have defined themselves as science writers. Specialist reporters are expensive and, consequently, rare in most media organisations. Editors believed strongly in the ability of a good generalist to cover anything and worried more about the by-products of cozy relationships between journalists and their sources than about the need to apply specialised knowledge to complex topics. Through much of the twentieth century, in fact, a common practice in American news media was to rotate reporters across beats every few years to prevent the pitfalls of reporter/source intimacy.

A few specialised science reporters did gain a foothold in newspapers and wire services early in the twentieth century in Britain and the United States. But it took the technological innovations catalysed by World War II, post-war decisions by governments in several countries to invest in scientific research, the space race of the 1960s and the growing environmental concerns of the 1970s and 1980s to galvanise media organisations into finding science and environmental reporters to cover what loomed as some of the major stories of the century. Gregory and Miller (1998) characterise this post-war period as the time when science journalism became an organised, visible and increasingly powerful presence in journalism.

The numbers of science reporters burgeoned in many countries over the course of the twentieth century (see, for example, Metcalfe and Gascoigne 1995 on Australia). In addition to the establishment of country-specific organisations of science writers, global associations such as the World Federation of Science Journalists arose, and formal science journalism training was provided at universities around the world. With increased numbers of journalists, coverage also increased as a number of longitudinal studies demonstrated for the latter part of the twentieth century (Metcalfe and Gascoigne 1995; Bucchi and Mazzolini 2003).

Despite the flowering of science journalism during this time, it is important to remember that science reporters – like most classes of specialist reporters – have always constituted a small subset of all journalists in media organisations around the world. Thus, science stories remained relatively minor components of media coverage. An analysis of science coverage by four Greek newspapers, for example, found that the proportion of the news hole given over to science ranged from 1.5 to 2.5 per cent (Dimopoulos and Koulaidis 2002); similar to what Pellechia (1997) found in the United States and to what Metcalfe and Gascoigne (1995) found in Australia. In Greek newspapers, political coverage accounted for some 25 per cent while sports made up 15 per cent of stories (Dimopoulos and Koulaidis 2002).

By the end of the twentieth century, a sea change was under way. New communication channels were cropping up that permitted readers/viewers to implement their own information-seeking practices. While the legacy media – newspapers, television, radio – continue to play important roles in the science information diet of many consumers around the world, today's lay person relies increasingly on the Internet. A Wellcome Trust survey of British adults and young people found the Internet to be the channel of choice for information about medical research for 23 per cent of adults and 35 per cent of young people; adults were more likely to indicate a preference for television (29 per cent) while young people were less likely to prefer these channels (27 per cent and 13 per cent respectively).[2] American data from 2010 showed that, while television has long been the preferred channel for science information, for the first time the Internet was running neck and neck with television (National Science Board 2012).

The increasing popularity of the Internet as a channel for information meant something had to give. And that *something*, in many countries, has been citizens' reliance on newspapers. The decline in newspaper advertising and the slower but steady decline in buyers over the years have led newspapers to shed staff and, in many places in the United States, even to reduce frequency of publication. Figures from the US Department of Labor's Bureau of Labor Statistics indicate that the newspaper industry as a whole in the United States declined by 40 per cent over the course of a decade (Zara 2013). There has been a correspondingly large drop in dedicated science sections. In 1989, weekly science sections in US newspapers numbered 95; by early 2013, only 19 survived. Since the primary employer of science journalists in the United States has long been newspapers, this change has forced many journalists to redefine what it means to be a journalist (Zara 2013).

Science journalism today

So where does all this leave the science journalist? In some countries, these journalists feel embattled. But in other cultures, they continue to thrive and the occupation, by all accounts, continues to grow. Systematic data are hard to find, but anecdotal accounts suggest that science journalists in the United States increasingly find themselves on their own, in the ranks of freelancers as their former media organisations downsize (Brumfiel 2009). Conditions in Canada and Britain, while not yet critical, show similar patterns. Faced with the need to become entrepreneurs, science journalists in these countries have embraced new media as a cheap and sometimes effective way of reaching publics. Additionally, the crisis in the United States has led journalists to experiment with new information delivery structures; I address this topic later in the chapter.

Elsewhere in the world, science journalists seem to be holding their own, according to an analysis of data from hundreds of science reporters from around the world. Taking advantage of data from four surveys of science journalists archived at SciDev.Net, Bauer and colleagues sought to construct a picture of 'global science journalism' in the twenty-first century (Bauer *et al.* 2013). The researchers used data from a survey of 179 participants in the 2009 World Conference of Science Journalists, held in London; a survey of 320 journalists from Latin America conducted in 2010 and 2011; a subset of data from a larger survey project from six regions, primarily developing countries; and original survey data from 93 additional journalists primarily from Africa and Asia, gathered in 2012. While the researchers caution that the complex nature of this aggregated analysis makes it hard to argue for the generalisability of the sample, global comparative data are so rare that this study deserves some attention.

Bauer *et al.* (2013) found that, while men continued to hold the majority of science journalism positions in Europe, Africa and Asia, women accounted for fully 45 per cent of the sample and actually trumped men in Latin America (55 per cent women versus 45 per cent men). University degrees were common attributes, as was journalism training; 26 per cent indicated receiving science writing training specifically while 19 per cent had general journalism training. One in ten held a doctorate. More than half had worked as science journalists for ten or fewer years, and half reported being in full-time positions. While these journalists reported they were writing more stories for the Web, they also noted that their work for more traditional print sites had increased. Among these working journalists, job satisfaction remained high. That is, respondents were reasonably satisfied with their autonomy, with access to scientists and with their ability to serve their audiences responsibly. What that last factor meant, according to these journalists, was the opportunity to inform and explain.

So science journalism sounds like great work if you can get it. But like all occupations, it is beset by its own sets of issues, some of them grounded in journalism generally and others driven by the idiosyncracies of science. Below, I discuss a few of them before returning to further consideration of the shift to the Internet and its implications for the roles of science journalists.

Science news is overwhelmingly about medicine and health

For media outlets in many countries, the bulk of what passes for science writing is all about medicine and health. Bauer tracked what he called 'The medicalisation of science news' (1998) in the British press in the latter half of the twentieth century, and Pellechia (1997) found that a set of elite US newspapers focused on medicine and health in more than 70 per cent of their stories during the same period. Einsiedel (1992) encountered a similar dominance of health topics in an analysis of science stories in seven Canadian newspapers. Television is a more eclectic medium in most countries, with an often strong focus on natural history and environmental issues. But here, too, medicine and health often dominate (Gregory and Miller 1998; León 2008; Lehmkuhl *et al.* 2012).

In a study of science coverage in a leading Italian newspaper over the course of 50 years, Bucchi and Mazzolini (2003) also found that biology and medicine accounted for more than half of the stories. But they noted that the medicalisation of science was particularly pronounced in stories written for the newspaper's special supplements and sections, while science news featured on the front page was dominated by physics and engineering stories. This suggests that science journalists may be making a conceptual distinction between news and *news you can use*, with the latter focusing more heavily on health and medicine topics.

Science news on television remains scarce

Analyses of science news on television in Europe find not much of it (de Cheveigné 2006; León 2008). Television news typically attends only sporadically to science topics, and broadcast stories emphasise the entertainment aspects of scientific discoveries and processes at the cost of in-depth, explanatory and critical treatment (Metcalfe and Gascoigne 1995; LaFollette 2002; León 2008). A recent analysis of science in BBC news programming offered a slightly more positive picture of the situation in the UK. Analysis of news coverage over the course of three months in both 2009 and 2010 found that one in four news programmes included at least one science news item, as well as that fully half of the main television news bulletins contained science news reporting (Mellor *et al.* 2011).

What about television programmes dedicated to science? In an analysis of television science programmes in 11 European countries, Lehmkuhl *et al.* (2012) found great variation in the number and nature of such programmes and concluded that market structure was a major predictor of that diversity. For example, with the exception of Britain, most science programming occurred on public service channels. The more such channels available in a country, the study found, the more science programming. However, few science programmes in any of these countries were dedicated to science *news*. The most common types of programming were either longer-form, magazine-style coverage of science issues (such as Britain's *Horizon*, Germany's *Terra X* or Austria's *Newton*) or what the team called 'advice' programmes, often health-related with question and answer structures.

An early analysis of a set of British documentary science programmes noted the heavy overlay of certainty that accompanied the programmes: 'Television presents science as producing unambiguous and intractable knowledge' (Collins 1987: 709). Recent studies also found that TV coverage of science – like much of science journalism – neglects uncertainty. A content analysis of BBC science coverage, for example, noted that only one in five sources in broadcast news stories urged caution in evaluating scientific claims (Mellor *et al.* 2011).

Drama plays a major role in much of science television programming and, according to scholars, can often trump public understanding of science goals. Silverstone (1985) studied the making of a science documentary for the BBC *Horizon* series and tracked the gradual takeover of the storyline by producers. Scientists eventually lost control of the narrative, he concluded, via a production process that privileged the kind of dramatic tension that only skilled film-makers could provide. In a similar vein, Hornig concluded of *Nova* documentaries that the programmes maintained the 'sacredness' (1990: 17) of science by portraying scientists as special and distinct from others.

Coverage of science follows journalistic norms

Media coverage of science looks a lot like coverage of other arenas, principally because the primary drivers of coverage patterns are not the content areas on which stories are focused but, instead, the production infrastructure through which that content must pass.

For example, science stories – like all journalistic accounts – tend to be *episodic* in nature. That is, journalists are more likely to produce shorter stories about concrete happenings than longer, thematic stories about issues. Underlying this pattern is the rapid pace of most media production processes. Daily or, in the case of Internet news sites, hourly production cycles cannot wait for months-long scientific processes to spool out. Journalists produce stories about pieces of processes and hope that faithful readers will be able to knit together a larger picture from these bits of narrative fabric.

Episodic coverage does not lend itself well to discussions of process. So, not surprisingly, analyses of science stories find few descriptions of the research methods employed. Dimopoulos and Koulaidis (2002) found that nearly 75 per cent of the science stories they analyzed from four Greek newspapers contained no reflection on the methodological *how* of the scientific process, and discussion of that dimension in the remaining stories was brief and superficial. Einsiedel (1992), too, noted that most of the Canadian stories her team analyzed virtually ignored process details. And a study of the coverage of scientific research in Dutch newspapers (Hijmans *et al.* 2003) found, similarly, that most stories eschewed complex process information.

Science journalism, again in ways typical of other types of journalism, seeks to hang stories on *traditional news pegs*, characteristics of real-world processes that are proven audience attention-getters. Those pegs include timeliness, conflict and novelty. Thus, for example, rather than dip into a scientific research process at some haphazard stage, the science journalist waits until the completed work is on the cusp of publication in a scientific journal. That moment of publication offers a prized timely angle, an opportunity to grab the attention of a reader/viewer with the words: 'In today's issue of *Nature* ...'.

These moments also tend to coincide with points in a process recognised – or designated – as salient by the scientific culture. Journalists typically *buy* into the legitimising structures of sources (Fishman 1980), uncritically accepting sources' designation of what is important and worthy of notice. Scientists, thus, can easily sell the argument that journalists must respect scientific process and, for example, wait for peer review to take place before embarking on a wider dissemination of research results. Scientists often complain that journalists pay undue attention to mavericks and outliers, but studies of media coverage of contested science suggest that those stories overwhelmingly reflect the views of the scientific mainstream (Goodell 1986; Nelkin 1995).

This reliance on news pegs also means that coverage of a long-running issue waxes and wanes with the presence/absence of pegs. Scientists and policymakers will struggle for decades, for example, to understand the mechanisms of cloning and to explore means for society to adapt to the technique's many tantalising and alarming possibilities. But coverage of the issue will erupt only when *something happens* in a journalistic sense – when a prime minister formally announces a new initiative, when a team of scientists unveils the first cloned cat, when a religious group lodges a complaint. While the disjunction between coverage and process can be disconcerting to some scientists, others have learned to take advantage of reporter dependence on news pegs and have become facile at guiding coverage. For example, if an important paper is about to be published in a journal, scientists may hire consultants to help them *market* their discovery to the press by appealing to the demand for news pegs. The resulting press conferences and reporter *exclusives* may be more influential in generating coverage than the original papers themselves.

The most important *audiences* for journalists have long been their editors and their sources. While their *real* audiences – members of the public – have historically had only sporadic access to the newsroom, a science writer is in daily touch with her sources and her bosses. Thus, coverage is more likely to be responsive to the priorities of these individuals. This may seem unpersuasive to scientists, who feel that journalists often run roughshod over them and treat their information in cavalier manner. But studies of media coverage of science have demonstrated repeatedly that the scientific culture is a powerful driver of what becomes news about science. Dorothy Nelkin, in her seminal book *Selling Science* (1995), reflected that media science stories overwhelmingly represent scientists as successful problem-solvers. Such coverage is not accidental, she notes; the scientific culture actively cultivates its image as society's major tool for reducing uncertainty. The invisibility of audiences is changing with the increased access that the Internet affords readers/viewers to journalistic work; I return to this topic at the end of this chapter.

Two long-standing journalistic norms – objectivity and balance – have come under intense scrutiny in the twenty-first century. Both arose as surrogates for validity, that is, as ways of compensating for journalists' inability to determine whether their sources' assertions are true or not. They are particularly salient in science journalism as much of science is contested terrain. What is a journalist to do when credible scientists make contradictory claims about a particular issue? The occupation responds: default to objectivity and balance (Dunwoody 1999; Dunwoody and Konieczna 2013).

In a world where the science journalist cannot declare what is most likely to be true, *objectivity* demands that the reporter go into *neutral transmitter* mode and focus not on validity but on accuracy. That is, rather than judging the veracity of a truth claim, the journalist concentrates instead on representing the claim accurately in her story. The issue is no longer whether the claim is supported by evidence but, rather, the goodness of fit between what a source says and what a journalist presents.

Similarly, when a science reporter cannot determine who is telling the truth, the norm of *balance* suggests that he represent as many truth claims as possible in the story. When validity is impossible, in other words, a good fall-back position is comprehensiveness. The journalist is, in effect, telling the reader: 'The truth is in here somewhere'.

But Boykoff and Boykoff (2004), for example, argue that balance too often means giving truth claims equal space even when they are not, in fact, equally valid. They use the example of global warming coverage in US newspapers to demonstrate that, even in the face of burgeoning consensus among scientists that humans are making a substantive contribution to warming, many media accounts still give significant play to global warming outliers who dispute the trend. Mooney and Nisbet (2005) find similar patterns in coverage of the debate in the United States over teaching evolution in the biology classroom; attempts to *balance* the arguments of biologists and creationists, they claim, confers legitimacy on both sides in the minds of readers.

At least one American study indicates that journalists are keenly aware of the problems created by objective, balanced accounts but feel that journalistic norms prevent them from abandoning these behaviours. Dearing (1995) found the expected balancing of extreme points of view in coverage of several scientific issues where healthy majority viewpoints were being contested by outliers. In interviews, the journalists readily acknowledged the likely bogus nature of the mavericks' positions but then indicated that both editors and audiences expected their stories to treat those positions with respect.

Training remains contentious, under-studied

Should science writers be formally trained in science, or should they more properly come up through the ranks of journalism? If one looks across countries, the former seems to trump the latter. In some countries, doctoral degrees are highly sought by newsrooms; in others, science writing training programmes increasingly privilege those applicants who have science credentials. The argument embedded in these preferences is not that journalistic training is irrelevant but that a marriage of scientific and journalistic skills will yield better results than will journalistic skills alone.

The value of formal science training seems obvious and, not surprisingly, is strongly endorsed by the scientific culture, which feels that such grounding will produce more accurate and responsible stories. Many science graduate students in search of alternative careers find that science writing has intuitive appeal. Given the robustness of these training beliefs, it is interesting to note that there is little empirical evidence to support them. Only a few studies have been conducted in the United States to explore differences in journalistic quality that can be pinned to differences in training, and none of those studies have found formal science training to be strongly predictive of that quality. For example, Wilson (2000) gave American environmental journalists a global warming knowledge test and then compared the answers of those journalists with formal science training to the answers of those without such training. While formal science education made a modest difference in reporter knowledge, it was trumped by another variable: number of years on the job. Years on the job has proved the best quality predictor in a variety of studies of journalistic work in the United States (Dunwoody 2004). As is the case for most skilled occupations, experiential learning is probably the most critical predictor of job performance.

The great and general shift to the Internet

The availability of the Internet as an information channel has profoundly affected audiences' patterns of information-seeking. In many countries, traditional, mediated channels are either in holding patterns (television) or are in decline (newspapers) as the public adjusts to the enormous amount of information available to it electronically.

A dominant Internet environment does not, however, necessarily mean an *anything goes* pattern of information-seeking. The worldwide popularity of sites such as Yahoo! News, CNN, MSNBC, Google News and the New York Times[3] suggests an enduring need for a credible, initial filter on information. We are hungry to keep up with current events, but we continue to depend on journalism to make reasoned choices and to craft readable narratives.

Science journalists have also embraced the Internet as, among other uses, a primary site for story searches. Respondents in one recent study reported spending more than three hours a day, on average, on the Internet. This survey of science writers in 14 European countries found that the journalists relied on a relatively small group of sites – among them EurekAlert!, Nature, BBC News, and New Scientist – for story ideas, and they overwhelmingly agreed that 'the Internet has made my job easier' (Granada 2011: 802). Many of them also admitted, however, that such reliance increases their focus on breaking news, a trend that may exacerbate the dominance of episodic narratives over more thematic ones.

Scientists' use of the Internet, on the other hand, remains more muted. Although many scientists have embraced Internet communication and its promise to link them directly to audiences (see below), others continue to rely on more mediated paths. A recent survey of neuroscientists in Germany and the United States, for example, found that although the respondents believed that *new media* such as blogs and online social networks do influence public opinion and policy decisions, they reported that they themselves use more traditional outlets – newspapers, television, magazines (both legacy and online) – to keep up with scientific developments (Allgaier *et al.* 2013).

Finally, the Internet has opened direct communication lines between audience members and both scientists and journalists. Several scholars have begun to study the online interactions between journalists and audiences, particularly through the lens of comments by audience members in response to online science journalism stories. Both Secko *et al.* (2011) and Laslo *et al.* (2011) characterise this process as the evolution of *unfinished stories*. The initial science story, rather than deemed a final product, serves instead as a catalyst for an ongoing narrative construction process in which both journalists and readers participate. Riesch (2011), among others, documents the dynamic nature of such narratives in a couple of case studies in which stories whose narratives become contested disappear from the online sites of major media organisations.

Are scientists losing or gaining control?

For much of the twentieth century, scientists avoided public contact and, as a result, knew much less about public communication processes than did the journalists who contacted them. That gave journalists an edge in their relationships with their sources, but that has begun to lessen as scientists have come to realise the value of public visibility and take active steps to structure their own public images. Twenty-first-century scientists increasingly come equipped with media training, and have begun to communicate directly with publics on their own through popular science books, blogs and websites.

Such visibility can be harmful, as many *burned* scientists still ruefully report, but the social and scientific legitimacy that can attend such visibility is luring many scientists into acquiring greater communicative expertise. Several studies demonstrate that media coverage makes a scientist's work look more important not only to members of the public (including funders) but also to other scientists; for example, media visibility of peer-reviewed, published work increases the number of citations of that research in the scientific literature (Phillips *et al.* 1991; Kiernan 2003). As a result, scientists in all disciplines are acquiring communication skills and are learning to take advantage of communication professionals employed by their organisations. These scientists report not only regular interactions with journalists but also beliefs that those interactions are good for their careers (Peters *et al.* 2008).

On the other hand, the onslaught of new information channels and the increasing ubiquity of user control means that all information creators are increasingly finding themselves buffeted by audience reactions. Scientists have always chafed at their perceived lack of control over public representations of their work. In 2002, scientists in the UK established the Science Media Centre, initially through the Royal Institution of Great Britain. The centre defines its mission as one of helping scientists to become better communicators, but it also seeks to intervene early in the course of media coverage of science issues by staging briefings and otherwise providing expert reactions to breaking news, compiling fact sheets about specific science topics that are becoming newsworthy and even by engaging in independent analyses of scientific papers for journalists. These efforts have been welcomed by many but have led some journalists to suggest that the centre functions as a large-scale science PR agency that tries to control the science agenda, as seen in a series of articles in *Columbia Journalism Review*'s online version.[4] Despite this controversy, similar centres now exist in other countries and are proposed for yet more.

Whither the science journalist?

Have we entered an era in which science journalists gradually lose their media platforms and find themselves increasingly eclipsed by savvy scientists keen to promote their research *brands*? Not yet. But as legacy media platforms struggle to maintain audience share, science journalists are being forced to become more entrepreneurial and to look for new ways to explain to their audiences the profound scientific developments under way. These journalists have embraced social media channels – Facebook, Twitter – not only to maintain contact with sources and peers but also to build their own personal brands. Today's successful science writer may work from home, where she maintains a highly visible blog (ideally hosted by a legacy media website), tweets regularly about topics that fit within her declared area of expertise (specialisation is the name of the game), freelances articles (again focused on her niche) to magazines and online sites and hopes that the synergistic effect of these activities will give her visibility, credibility and a book contract.

In some countries, journalists are banding together in non-profit organisations in order to maintain traditions of investigative and explanatory journalism. The groups rely on a wide variety of funding mechanisms, principally foundation and private donations, and often give their work free of charge to media organisations willing to publish the stories. One of those non-profits, InsideClimate News – which specialises in covering energy issues and environmental science – was awarded a Pulitzer Prize in 2013 for its reporting on an oil spill in the Midwestern United States that journalists broadened into an analysis of national pipeline safety issues (InsideClimate News Staff 2013).

In a special issue of the journal *Journalism*, devoted to 'Science Journalism in a Digital Age', contributors pondered the impact of the channel revolution on science journalism. Issue editor Stuart Allan (2011) notes that the those impacts may be both salutary and daunting. The Internet *wild west* offers science journalists the opportunity to engage directly and transparently with a variety of audiences, from everyday people to scientists; the interactive nature of social channels makes it possible for users to understand science in more profound ways; science journalists who build storytelling skills across platforms have the potential to communicate science in ways far more powerful than before. The *New York Times*' multi-platform story, 'Snowfall: The Avalanche at Tunnel Creek', chronicling the death of a group of world-class skiers caught in an avalanche in the mountains of the north-west United States, offers an example of that potential.[5]

But Allan also asks us to be aware of the possible downsides of this brave new world. The Web, like a black hole, demands constant feeding. Journalism becomes a 24/7 occupation in which stories become rapid-fire processes with no obvious end points. Building science news stories for Internet consumption presents many challenges, among them the need for constant updating, managing the speed with which information must be turned into narrative and maximising the brevity of those narratives, so critical to audiences with only seconds to spare.

Fahy and Nisbet (2011) suggest that journalistic roles will expand to accommodate these twenty-first-century changes. While some journalists will continue to embrace such long-standing roles as the need to analyse and then explain, the need to illuminate wrongdoing, the need to monitor the landscape in order to alert audiences to important changes, new communication modes will draw science journalists into new roles. Among them, note Fahy and Nisbet, are the role of curator, who aggregates and makes sense of existing news and commentary; the role of civic educator, who uses the news of science as a means of informing audiences 'about the methods, aims, limits and risks of scientific work' (Fahy and Nisbet 2011: 780); and a role that they label the 'public intellectual', journalists who not only synthesise but also interpret via a point of view.

We are too early in the process of change to determine what occupational modifications will strengthen and which ones will fade. Scholars are just beginning to explore the impacts of these changes, making it difficult to assess the societal risks and benefits that accompany them.

Other difficulties also will attend future efforts to study the behaviours and products of science journalists. Prominent among them is the question of just who constitutes a science journalist. Studies in the past have relied on organisational affiliation as an important component of that definition. But in a world filled with freelancers, many of whom work for a magazine on one day and for a government research laboratory on another, separating the science journalist from the non-journalist will prove daunting.

Similarly, what is a science news story? Does a tweet count? A blog post? And even when a story looks like the traditional stereotype of a news narrative, when is it a *finished* narrative? In an electronic publishing environment where reporters and editors can make corrections, fiddle at will with content and even remove a story altogether (Riesch 2011), how do scholars determine the point at which story evaluation should take place?

What has not changed, however, is the commitment and passion of science journalists. In my home town of Madison, Wisconsin, the local newspaper's long-time science writer Ron Seely retired, in part, because he was disheartened by the decline of journalism in these smaller urban areas. While the newspaper, regretably, has no plans to replace him, Seely quickly found a new home with the non-profit Wisconsin Center for Investigative Journalism, where he will continue a 30-plus-year career of covering complicated science and environmental issues (Fuhrmann 2013). His excitement about this next stage of his career is infectious and serves as a reminder that science journalism done well can have tremendous societal value. Societies now need to figure out how to maintain this capacity.

Key questions

- How can science journalism survive the decline of legacy media such as newspapers and magazines?
- In what ways do Internet publshing conditions change what constitutes a science news story?
- How would you define a science journalist in the twenty-first century? How is this definition different from that of 50 years ago?

Notes

1 This chapter has been substantially revised with particular attention to better capturing the work of science journalists globally and updating our understanding of how Internet and market forces are affecting journalistic practice.
2 See survey reports at www.wellcome.ac.uk/About-us/Publications/Reports/Public-engagement/WTX058859.htm; accessed 31 July 2013
3 *Top 15 Most Popular News Websites*, compiled by eBiz/MBA – The eBusiness Knowledgebase; www.ebizmba.com/articles/news-websites, accessed 30 July 2013.
4 The series begins with Fox and St. Louis 17 June 2013, continues with Eliott *et al.* 19 June 2013 and culminates with Brainard and Winslow 21 June 2013; www.cjr.org/the_observatory/
5 www.nytimes.com/projects/2012/snow-fall/#/?part=tunnel-creek

References

Allan, S. (2011) 'Introduction: science journalism in a digital age', *Journalism* 12, 7: 771–777.
Allgaier, J., S. Dunwoody, D, Brossard, Y-Y. Lo and H. P. Peters (2013) 'Journalism and social media as means of observing the contexts of science', *Bioscience*, 63, 4: 284–287.
Bauer, M. (1998) 'The medicalization of science news – from the "rocket-scalpel" to the "gene-meteorite complex"', *Social Science Information*, 37, 4: 731–751.

Bauer, M. and Bucchi, M. (eds) (2007) *Journalism, Science and Society,* London and New York: Routledge.
Bauer, M. W., Howard, S., Romo, R., Yulye, J., Massarani, L. and Amorim, L. (2013) *Global science journalism report: working conditions and practices, professional ethos and future expectations*, London: Science and Development Network.
Boykoff, M. T. and Boykoff, J. M. (2004) 'Balance as bias: global warming and the US prestige press', *Global Environmental Change*, 14, 2: 125–136.
Broks, P. (2006) *Understanding Popular Science*, Maidenhead, Berkshire: Open University Press.
Brumfiel, G. (2009) 'Supplanting the old media?', *Nature* (18 March), 458: 274–277.
Bucchi, M. and Mazzolini, R. G. (2003) 'Big science, little news: science coverage in the Italian daily press, 1964–1997', *Public Understanding of Science*, 12, 1: 7–24.
Burnham, J. C. (1987) *How Superstition Won and Science Lost*, New Brunswick, NJ: Rutgers University Press.
Collins, H. M. (1987) 'Certainty and the public understanding of science: science on television', *Social Studies of Science*, 17, 4: 689–713.
Dearing, J. (1995) 'Newspaper coverage of maverick science: creating controversy through balancing', *Public Understanding of Science*, 4, 4: 341–361.
De Cheveigné, S. (2006) 'Science and technology on TV news', in J. Willems and W. Göpfert (eds) *Science and the Power of TV*, Amsterdam: VU University Press, 85–100.
Dimopoulos, K. and Koulaidis, V. (2002) 'The socio-epistemic constitution of science and technology in the Greek press: an analysis of its presentation', *Public Understanding of Science*, 11, 3: 225–242.
Dunwoody, S. (1999) 'Scientists, journalists, and the meaning of uncertainty', in S. M. Friedman, S. Dunwoody and C. L. Rogers (eds) *Communicating Uncertainty: Media Coverage of New and Controversial Science*, Mahwah, NJ: Erlbaum, 59–79.
Dunwoody, S. (2004) 'How valuable is formal science training to science journalists?', *Communicacao e Sociedade*, 6: 75–87.
Dunwoody, S. and Konieczna, M. (2013) 'The role of global media in telling the climate change story', in S. J. A. Ward (ed.) *Global Media Ethics: Problems and Perspectives*, Oxford: Blackwell Publishing, 171–190.
Einsiedel, E. F. (1992) 'Framing science and technology in the Canadian press', *Public Understanding of Science*, 1, 1: 89–101.
Fahy, D. and Nisbet, M. C. (2011) 'The science journalist online: shifting roles and emerging practices', *Journalism*, 12, 7: 778–793.
Fishman, M. (1980) *Manufacturing the News,* Austin, TX: University of Texas Press.
Fuhrmann, L. (2013) 'Investigative journalism center hires prominent journalist Ron Seely as reporter, editor, mentor', *WisconsinWatch.org*; online at www.wisconsinwatch.org/2013/06/20/investigative-journalism-center-hires-prominent-journalist-ron-seely-as-reporter-editor-mentor/; accessed 31 July 2013.
Golinski, J. (1992) *Science as Public Culture: Chemistry and Enlightenment in Britain, 1760–1820*, Cambridge: Cambridge University Press.
Goodell, R. (1977) *The Visible Scientists*, Boston, MA: Little, Brown.
Goodell, R. (1986) 'How to kill a controversy: the case of recombinant DNA', in S. M. Friedman, S. Dunwoody and C. L. Rogers (eds) *Scientists and Journalists: Reporting Science As News*, New York: Free Press, 170–181.
Granada, A. (2011) 'Slaves to journals, serfs to the web: the use of the internet in newsgathering among European science journalists', *Journalism*, 12, 7: 794–813.
Gregory, J. and Miller, S. (1998) *Science in Public: Communication, Culture, and Credibility*, Cambridge, MA: Basic Books.
Hijmans, E., Pleijter, A. and Wester, F. (2003) 'Covering scientific research in Dutch newspapers', *Science Communication*, 25, 2: 153–176.
Hornig, S. (1990) 'Television's *NOVA* and the construction of scientific truth', *Critical Studies in Mass Communication*, 7, 1: 11–23.
InsideClimate News Staff (2013) 'InsideClimate News team wins Pulitzer Prize for national reporting'; online at http://insideclimatenews.org/news/20130415/insideclimate-news-team-wins-pulitzer-prize-national-reporting; accessed 3 June 2013.
Kiernan, V. (2003) 'Diffusion of news about research', *Science Communication*, 25, 1: 3–13.
LaFollette, M. (2002) 'A survey of science content in US broadcasting, 1940s through 1950s', *Science Communication*, 24, 1: 34–71.
Laslo, E., Baram-Tsabari, A. and Lewenstein, B. V. (2011) 'A growth medium for the message: online science journalism affordances for exploring public discourse of science and ethics', *Journalism*, 12, 7: 847–870.

Lehmkuhl, M., Karamanidou, C., Mora, T., Petkova, K., Trench, B. and AVSA-Team (2012) 'Scheduling science on television: a comparative analysis of the representations of science in 11 European countries', *Public Understanding of Science,* 21, 8: 1002–1018.

León, B. (2008) 'Science related information in European television: a study of prime-time news', *Public Understanding of Science*, 17, 4: 443–460.

Mellor, F., Webster, S. and Bell, A. R. (2011) *Content Analysis of the BBC's Science Coverage.* Appendix A of BBC Trust Review of Impartiality and Accuracy of the BBC's Coverage of Science; online at www.bbc.co.uk/bbctrust/assets/files/pdf/our_work/science_impartiality/appendix_a.pdf; accessed 30 July 2013.

Metcalfe, J. and Gascoigne, T. (1995) 'Science journalism in Australia', *Public Understanding of Science*, 4, 4: 411–428.

Mooney, C. and Nisbet, M. C. (2005) 'Undoing Darwin: as the evolution debate becomes political news, science gets lost', *Columbia Journalism Review*, 2, 20 (September/October): 30–39.

National Science Board (2012) *Science and Engineering Indicators 2012,* Arlington VA: National Science Foundation.

Nelkin, D. (1995) *Selling Science*, revised edition, New York: W.H. Freeman and Company.

Pellechia, M.G. (1997) 'Trends in science coverage: A content analysis of three US newspapers', *Public Understanding of Science*, 6, 1: 49–68.

Peters, H.P., Brossard, D., de Cheveigne, S., Dunwoody, S., Kallfass, M., Miller, S. and Tsuchida, S. (2008) 'Interactions with the mass media', *Science*, 321, 5886: 204–205.

Phillips, D. P., Kanter, E. J., Bednarczyk, B. and Tastad, P. L. (1991) 'Importance of the lay press in the transmission of medical knowledge to the scientific community', *New England Journal of Medicine*, 325, 16: 1180–1183.

Riesch, H. (2011) 'Changing news: re-adjusting science studies to online newspapers', *Public Understanding of Science,* 20, 6: 771–777.

Secko, D. M., Tlalka, S., Dunlop, M., Kingdon, A. and Amend, E. (2011) 'The unfinished science story: journalist-audience interactions from the *Globe and Mail*'s online health and science sections', *Journalism*, 12, 7: 814–831.

Silverstone, R. (1985) *Framing Science: The Making of a BBC Documentary,* London: British Film Institute.

Wilson, K. (2000) 'Drought, debate, and uncertainty: measuring reporters' knowledge and ignorance about climate change', *Public Understanding of Science*, 9, 1: 1–13.

Zara, C. (2013) 'Remember newspaper science sections? They're almost all gone', *International Business Times*, 10 January; online at www.ibtimes.com/remember-newspaper-science-sections-theyre-almost-all-gone-1005680; accessed 31 May 2013.

4

Science museums and centres
Evolution and contemporary trends

Bernard Schiele

Introduction: what is a SMC?

In this chapter we focus our attention on science museums, science centres and discovery centres, which we group under the acronym SMC. Three broad questions guide our effort: What knowledge do SMCs convey to their visitors? What interactions do they emphasise with their visitors? How do they represent the science they present to their visitors?[1]

A SMC is a museum devoted to science. While such a definition readily distinguishes SMCs from art museums or history museums, how are we to consider zoos, aquariums, planetariums, observatories, conservatories, botanical gardens, arboretums, nature parks, exhibition centres, not to mention space centres, transportation museums, railway museums, and so on, all of which share the common denominator of disseminating scientific knowledge or technological applications or both? Similarly, should ethnology, anthropology or society museums be included among SMCs? A priori no, if we maintain the traditional distinction between the human and social sciences and the natural sciences. However, if we consider the recent evolution of SMCs which increasingly present science and technology in the context of their social uses, resorting to the discourse and methods of the humanities and social sciences, the distinction is less clear; all the more so since ethnology, anthropology and society museums are incorporating science and technology elements into their exhibitions.

Thus, the demarcation line between a genuine SMC and, for example, a genuine society museum is all the more difficult to establish now that exhibitions on the sciences are no longer satisfied with a discourse that dissociates them from their impact on society. One of the reasons is that the history of the complex relationships between science and society can be – perhaps simplistically – summarised as a growing integration over time, to the point where the development of present society cannot be conceived without the development of science and technology (see, e.g., Schiele 2011).

This is why museums in all areas, always sensitive to the evolving social situation, reflect in their exhibitions, programmes and activities the constant adjustments imposed on society by the development of the sciences and above all the force that they exert on its evolution. They echo the fact that a 'major source of tension in the science-society relationship arises from the increasing encroachment of science on issues related to core human values and strongly held beliefs' (Leshner 2007: 1326). In other words, if the search for knowledge can be presented as an adventure, a pleasure or a useful activity, museums can no longer dissociate it from its interactions with society; they can no longer forego a critical discourse on the role of science and technology while pursuing a mission to promote them.

Keeping these remarks in mind, we will limit ourselves to science and technology museums and begin our reflection with the definition proposed by Althins (1963) at the time when the museum community was seeking a way out of the crisis of first- and second-generation science museums. For Althins, science and technology museums '(a) are primarily concerned with the whole or part of the field of science and technology; (b) are not always sharply distinguished from natural science museums, especially so far as biology, the management of natural resources, etc., are concerned; (c) lay stress on the latest developments of the studies concerned, whose past history is nevertheless outlined in so far as is appropriate; and (d) whose subject matter is very rightly dealt with in other categories of museums, such as history, regional, and specialist museums' (Althins 1963: 132). Up to the present time, still according to Althins, their chosen mission has been to make the general public aware of the latest discoveries; to show that the development of applied sciences results from advances in the pure sciences; to pay tribute to the inventors and discoverers; to encourage young people to become inventors and discoverers; to contribute to science teaching through the subjects treated and by training of staff; to develop a critical sense and an independent mind; to show that improving living conditions always depends on progress in science and technology; and, in general, to promote adaptation to an expanding industrial society without jeopardizing the rights and cultural heritage of human beings. This vision persists, even if SMCs are concerned with science-society relationships.

Subsequent definitions did not substantially enrich this, but instead focused on describing their functions. In defining science centres, Danilov describes them as:

> basically contemporary, participatory, informal educational instruments, rather than historic, 'hands-off' repositories of artifacts. Unlike many museums that are quiet and elitist, science and technology centres are lively and populist. They seek to further public understanding of science and technology in an enlightening and entertaining manner and do not require any special interest or background to be understood or appreciated by the average person.
>
> (Danilov 1982: 2)

Geared towards service to the community, they acquired a public education objective consisting of presenting the evolution of science and technology from the beginning to the latest advances, with emphasis on the latter; communicating about science and technology to promote the acquisition of pertinent information about it, to raise the general level of knowledge, and to spark an interest in science and technology among the general public; placing the impact of science and technology in context to emphasise their role and importance in modern society; democratising knowledge, that is, offering exhibitions, programmes and activities aimed at the overall community, regardless of their prior education or their expectations and interests.

It is also significant that the number of SMCs has grown steadily to reach 3,300 at the turn of the twenty-first century (Beetlestone *et al.* 1998; Persson 2000); and this movement continues, with China, for example, deciding to acquire an extensive network of SMCs to support its development (see Trench *et al.* in this volume). It follows that a certain mastery of science and technology culture is now demanded of social actors. The SMCs present it as a legitimate, desirable and useful cultural aspiration, the means for those who possess it to secure social insertion and eventually greater social mobility.

However, museums in general, including SMCs, have no monopoly on imparting knowledge even if they are by definition spaces devoted to knowledge, be it scientific knowledge in the SMCs, historical knowledge in history museums, or anthropological or ethnological knowledge in museums focused on those subject matters. Popular science magazines, television programmes, documentaries, science fiction and futuristic films, websites, schools and universities are all places where knowledge is constantly in movement to readers, listeners, spectators, researchers, students and laypersons. For Jorge Wagensberg, the influential museum director who designed *CosmoCaixa*, the SMC established in 2004 in Barcelona:

> a museum can voluntarily teach the visitors, provide them with information, educate, develop research, preserve heritage .... But somewhere else, there is always an institution that does it better than a museum: school and universities are better at teaching, family and social milieus have a lasting impact on education, Internet is a wealth of information about anything, scientists are very good at research .... So 'What are museums good at?'
>
> [...] A museum of science is a space devoted to providing stimuli, for any citizen whatsoever, in favour of scientific knowledge, scientific method and scientific opinion, which is achieved by firstly using reality (real objects and phenomena) in conversation with itself and with the visitors.
>
> (Wagensberg 2006: 26–27)

But what kind of conversation is this? For Macdonald (2001), in the spirit of Bennett (1995) and Foucault (1970, 1977), 'museums of science can be regarded as cultural technologies which define both certain kinds of "knowledge" (and certain knowledges as "knowledge" or "science") and certain kinds of publics' (Macdonald 2001: 5). In other words, contrary to what the Althins and Danilov definitions suggest, it is not just a matter of wanting to cover a field of knowledge (the science field), nor of being confined to one form of disseminating this knowledge (informal dissemination), nor of one method of doing so (by providing stimuli, as Wagensberg maintains); but, rather, it is to understand how an institution – the SMCs – present knowledge: 'museums which deal with science are not simply putting science on display; they are also creating particular kinds of science for the public, and are lending to the science that is displayed their own legitimizing imprimatur' (Macdonald 2001: 2).

This leads me to clarify my starting questions: What were the successive devices for presenting sciences to the public? What modes of mediation characterised them? What representations of science resulted?

The contexts of SMCs

To answer these questions, and to try to anticipate emerging forms of SMCs, one must keep in mind that the various strategies previously adopted by SMCs, like the new strategies that seem obvious today, are part of a larger context than actions chosen and implemented by particular SMCs. The global environment of museums and their particular cultural environment both influence what the museums show and say, and how they do it.

The mediation strategies adopted by the SMCs take their cue from the evolution of science-society relationships. When western societies experienced profound social and cultural changes in the 1960s, they were felt in the museum field as a demand for democratisation that ultimately resulted in a radical transformation of museum-visitor relationships. Some factors that provoked these changes include higher levels of education and living standards, the urbanisation of modern society, the development of communication technologies, the rise of mass tourism (Hobsbawm [1994] 2004) and, starting in the 1990s, the impetus of the neo-liberal movement (Landry and Schiele 2013). Similarly, when the House of Lords Select Committee on Science and Technology declared in its third report, '[s]ociety's relationship with science is in a critical phase' (2000: Chapter 1), it mentioned a 'crisis of trust' in Britain that affected the entire society, a collective state of mind that the SMCs could not ignore and had to confront daily in their relationships with visitors. Museums don't exist in isolation: when society changes, they change too.

While society's constraints are generally diffuse and indirect, they are no less effective. But sometimes they are direct. The exhibition, Sexe: l'expo qui dit tout (Sex: a Tell-All Exhibition),[2] was due to show in 2012 at the Canadian Museum of Science and Technology (CMST)[3] after an appearance at the Montreal Science Centre, where no one complained. Quite the contrary, it received two awards in 2011 as best exhibition of the year from the Canadian Association of Science Centers and the Société des Musées Québécois. However, Heritage Canada Minister James Moore, who visited the exhibition prior to its opening, sparked a controversy when, sensitive to pressures from religious groups,[4] he deemed the content 'insulting to taxpayers' (Mercier 2012). In reaction, the CMST, while asserting its independence, nonetheless raised the minimum age for admission to the exhibition from 12 to 16.

Similarly, the Science in American Life exhibition presented at the National Museum of American History in 1995 drew the wrath of the scientific community: the American Physical Society demanded changes, 'a situation virtually unprecedented at the Smithsonian' (Molella 1997: 131), even though the American Chemical Society played a major role on the advisory committee throughout the four years of preparation of this show and had given its approval. For Molella, the exhibition's chief curator, the reaction of physicists to an exhibition 'presenting contemporary science in full social dress' (ibid.) is explained by its attention to social changes and values induced by the impact of science on society, whereas traditionally science exhibitions (at least in the United States) dissociated science and society and presented research as a self-focused and independent effort detached from any social involvement. Molella interprets this position by the scientific community as a reaction to scientists' loss of prestige and power in contemporary society. While his interpretation opens an interesting path of thought, one can imagine others, such as a loss of legitimacy of the museum institution in the context of the general questioning of the role of authority figures. In the end, the National Museum of American History altered the exhibition. Nonetheless, a visitor survey conducted by an independent organisation, the Office of Institutional Studies, showed that visitors maintained a positive attitude toward science, contrary to the concerns of the scientists, alarmed by a contextualising they felt could tarnish the image of science (Molella 1997).

Molella also links the pressures exerted on Science in American Life to a bid for private funding which can only lead to greater interference: 'So long as we depend on these sources we will be vulnerable to outside pressures, which inevitably worsen in stressful times' (Molella 1997: 135). Hudson (1988) also anticipates the consequences of such a situation on the museum mission, writing that 'an industrial or commercial sponsor can impose a discipline quite as strict as that which results from political dogma' (Hudson 1988: 112). Aware of the need for context and its inherent challenges for SMCs, he added:

> In today's world, a museum of science and technology which does not encourage its visitors to think of the human and social consequences of new developments is acting in a singularly irresponsible and out-of-date fashion. To worship Progress uncritically may suit the manufacturers and advertisers but it is not in the best interests of humanity.
>
> (Hudson 1988: 112)

The global environment thus makes itself felt as a set of external constraints that act either diffusely – the spirit of the age – or directly on the SMCs. Turning to their particular or relevant environment, that constituted by museum institutions, the constraints are now called *internal* since they are proper to the museum field.

While their objective is to disseminate scientific knowledge, and while they maintain links with the science field to keep abreast of advances in research, SMCs still cannot claim to be either part of the field nor even on the periphery of science. They can justifiably claim to be part of the general movement that brings science into the core of our modern world, and in this respect they contribute to the social appropriation of science and technology (Godin and Gingras 2000). So SMCs are expected to make science and technology present in the social imagination and public space. Indeed, this is what they intend to do, since they all make it the primary goal of their mission. But while their reference is the world of science, they are first and foremost connected to the museum field, in which they form a specific sub-field; and the issues of the museum field are not those of the science field. SMCs gear their strategies to social demands and the dominant practices in the museum field because, like other museums, they confront the same constraints of attendance, funding, sponsoring, renewal of exhibitions, producing programmes, and so on.

Hence the necessity to consider the structuring effect of the museum field on how the mission, objectives and practices of each museum are expressed. This environment is described as relevant because SMCs, like all museums, maintain direct relationships with each other and constantly try to adjust how they do things, based on the others. Museums can be said to be in a dynamic balance vis-à-vis each other: consider, for example, the dual role of the associations to integrate and regulate the choice of museums.[5] This field effect explains how, since the neo-liberal shift of the 1980s, museums including SMCs have borrowed their organisational mode from the business world (Landry and Schiele 2013). They have espoused the ideals and adopted the operating methods of the managerial culture of companies (Paquette 2009). Thus the SMCs launched into dissemination activities that target the largest possible public (Jacobi 1997). This movement was accompanied by an administrative rationalisation of human and financial resources so that each contributes directly to the success of the communication mission. Furthermore, this rationalisation extended to the themes covered, the objects displayed and the knowledge produced and mobilised. In other words, the imperative to communicate and maximise audience size took precedence over all other considerations, including the enhancement of science although this remains the theme behind their discourse and topics.

Phases of development of SMCs

The science centre movement took off in the late 1960s. Even though the Science Center of Pinellas County (1959) and Seattle's Pacific Science Center (1962) were the first two to declare themselves science centres, it was only in 1969 with the opening of the Exploratorium in San Francisco and the Ontario Science Centre in Toronto that a new phase in the history of science centres really began. It represented such a break with science museums that the Association of Science-Technology Centers (ASTC) was founded in 1973. (Today it includes nearly 600 SMCs.)

To understand just how decisive the change made by the arrival of science centres was, a brief historical review is necessary. Following Danilov (1982) and Hudson (1988), we divide the evolution of the science museum movement into four main phases. The first phase is characterised by the history of technology, the second and third by an emphasis on contemporary sciences and the fourth by science-society interactions. The fifth phase is now taking form. It will be dealt with separately. While each phase marked a new development, this does not imply that the earlier phases are irrevocably gone. On the contrary, each new stage must be conceived as bringing new potentialities that successively graft onto the arsenal of means available to SMCs. Thus, while the SMCs reflect the developments of contemporary science primarily in interactive environments, they do not exclude exhibitions on science history, such as the Léonard de Vinci: Projets, Dessins, Machines show (October 2012–August 2013) at the Cité des Sciences et de l'Industrie (Paris); nor do science museums with an important collection reject the use of interactive approaches, such as the Science Museum's (London) Pattern Pod gallery, designed for groups aged five to eight who are offered essentially hands-on activities. Figure 4.1 on the following page offers a synopsis of the four phases of development to the present.

Phase 1: displaying the history of technology

This phase began with the creation of the London Science Museum in 1857, whose collection consisted of the legacy from London's Great Exhibition (1851). Like the Conservatoire National des Arts et Métiers (National Conservatory of Arts and Trades) in Paris, founded in 1794 and installed in 1799 at the Prieuré de Saint-Martin-des-Champs where it is still located, the London Science Museum at the time 'was primarily an educational institution, attempting to provide teaching in basic principles to teachers and skilled workers. It was an instructional body and its collections were gathered together mainly for this purpose' (Hudson 1988: 91). It should be noted that the educational function already established in France and England was to remain a constant in the field; and even if their understanding of it today differs from what prevailed up to the 1970s, the SMCs constantly refer to it today to justify their productions.

Of prime importance at that period was the enrichment and presentation of the collection. Museums displayed their collections and the public went there to view and admire them. By presenting their collections of remarkable objects, they set the example of public education and enrichment of the mind. The National Conservatory of Arts and Trades' collection consisted originally of scientific instruments and was later enriched with technical objects such as watches and clocks, while the more diverse London Science Museum included technical, industrial and artistic objects. These museums emphasised technologies more than science, which had been the preserve of enlightened amateurs until then but was becoming professionalised, with research henceforth concentrated in universities, academies and museums (mainly natural science museums). In the eyes of today's SMCs, these museums were looking backwards to the degree that their collections reconstituted the history of technologies, including the scientific instruments collected by the National Conservatory of Arts and Trades; they evoked the development of laboratory techniques rather than the frame of mind required by science itself. It was almost 75 years later, with the opening of the Palais de la Découverte (Palace of Discovery) in Paris in 1937, that the idea of giving a stage to pure science asserted itself in the museum field.

As an aside, we should add that whatever our view of nineteenth-century museums, they helped bring science and technologies into society at a time when the impact of science was becoming evident, even if, as Hudson says, referring to the London Science Museum, 'its symbolic value was undoubtedly greater than its actual quality' (Hudson 1988: 92).

Phase 2: showing contemporary science and enhancing knowledge

Simply put, the first step towards modernity was, on the one hand, the transition from a museology oriented to the past to one focused on the present; and, on the other, asserting the value of pure science rather than the history of technologies. Jean Perrin, who conceived and created the Palais de la Découverte, wrote: 'We wanted above all to familiarise our visitors with the basic

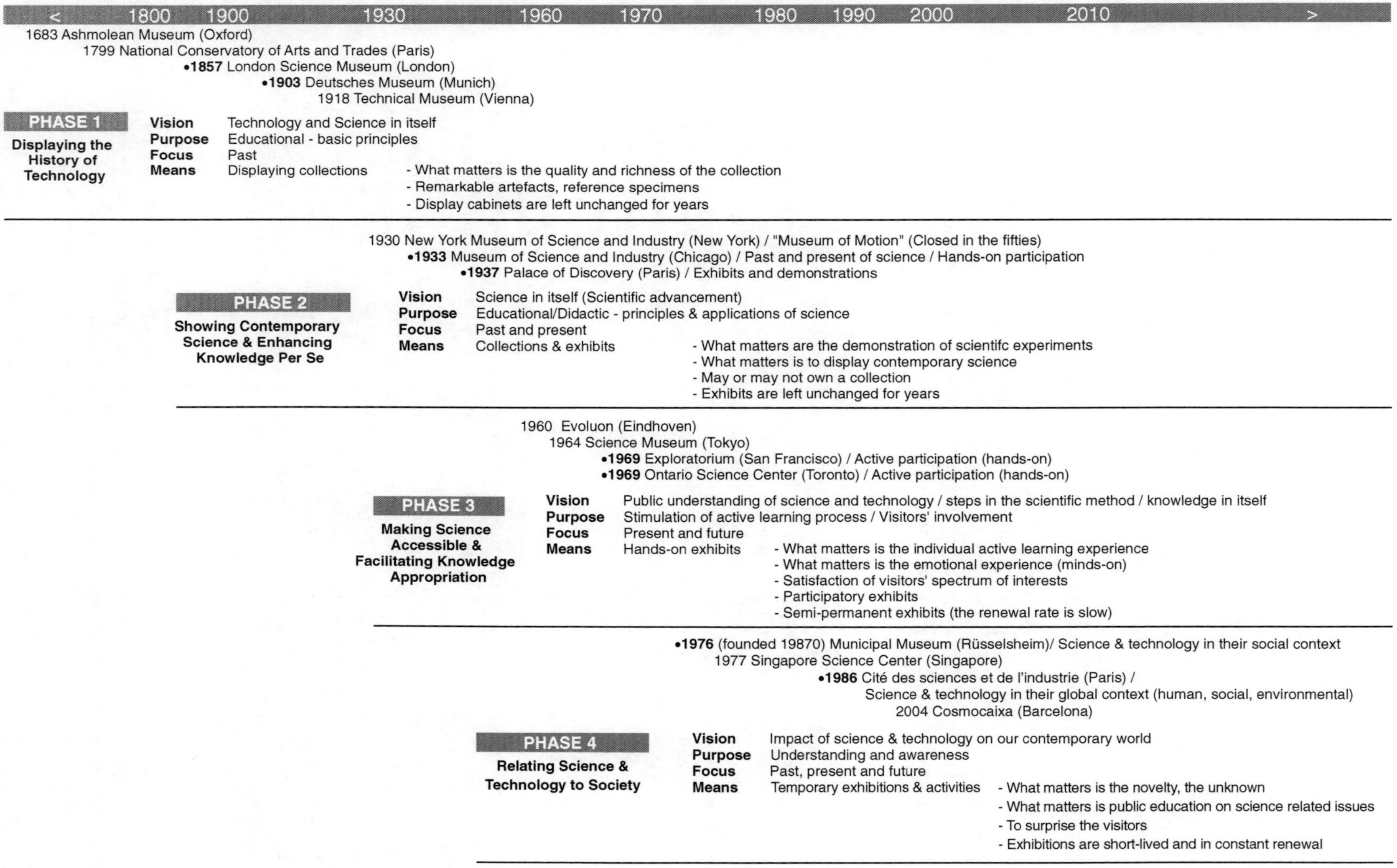

Figure 4.1 Evolution of SMCs

Graphic design by Anik Landry

research that generated science' (Perrin, quoted by Rose 1967: 206). The objective of the Palais de la Découverte was, therefore, to 'understand the decisive role of discovery in the creation of civilisation' (Roussel 1979: 2). To achieve this objective, it reproduced the great experiments of basic research for visitors on a daily basis. Between the creation of the London Science Museum and the opening of the Palais de la Découverte, the presence of science and the accompanying notion of progress were asserted and confirmed in society and in the public imagination. The theme of 'A Century of Progress' for the Chicago World's Fair of 1933–1934 bore witness to this, highlighting the interdependence of industry and scientific research with the slogan 'Science Finds, Industry Applies, Man Conforms'. It showed how much science was becoming a force for change in society on opening day by creating energy from a ray of light from the star Arcturus, captured by a photoelectric cell to create nocturnal illumination (Schroeder-Gudehus and Rasmussen 1992).

The World's Fairs helped forge the relationships SMCs would seek with their visitors. Open to all, they aimed to be educational and understandable to everyone: guided tours and gallery talks were organised in multiple languages. They also sought to entertain through spectacular presentations. These four characteristics – bridges with schools, dramatic and spectacular presentations (theatricality), guided tours, attempts to combine education and entertainment – would have a decisive impact on the future evolution of SMCs, to the point of being their principal characteristic today. Moreover, the Chicago World's Fair, like large stores, innovated in the display layout, abandoning the classification of objects into categories and instead grouping them thematically and letting visitors see everything with no special effort on their part. The SMCs quickly adopted this grouping of objects by theme rather by scientific classification. Thus the Grande Galerie de Zoologie (Paris 1889), which became the Grande Galerie de l'Evolution du Museum National d'Histoire Naturelle (reopened in 1994), today groups its specimens according to themes such as 'pelagic environments', 'coral reefs' or 'abyssal plains' rather than by animal or plant taxonomies.

Chicago's Museum of Science and Industry had already drawn inspiration from the Deutsches Museum (founded in Munich in 1903), noted for its working models that could be activated by visitors, from the Technical Museum (Vienna 1918) and, of course, from the London Science Museum, the great museums of reference of the era. In 1933, when it opened to the public for the second time, it now '[made] its visitors part of the show through ... "hands-on participation", that is, by giving them plenty of opportunity to set exhibits in motion and to follow through the results' (Hudson 1988: 104).

The Palais de la Découverte, which opened as a pavilion at the 'Arts and Technology in Modern Life' World's Fair in Paris, incorporated many of these innovations; but it opted to celebrate freewheeling curiosity, a disinterested quest of the unknown that culminates in discovery. It modelled this perspective through entertainment-exhibition, demonstrators, the invitation to touch and the push-button – these latter two foreshadowing interactivity (Eidelman 1988). It was, however, organised entirely around disciplinary knowledge and the basic sciences. Its mediation model was based on the lab-class transposed into entertainment-exhibition, animated by demonstrators who reproduced spectacular experiments and explained them to an audience. Its goal was 'to provide the perceptions the individual has looking at the outside world with a screen of concepts on which he projects and locates his perceptions' (Moles 1967: 28). The Palais de la Découverte sought to be didactic, as the preamble to the project makes clear:

> Demonstrators (with phonograph records and cinematographic films) will give the necessary explanations. Brief comments on panels will connect the experiments logically and form a logical whole for each type of science and indicate which inventions or practical applications emerged from each discovery.
>
> (quoted by Eidelman 1988: 180)

Focused on the present, it signalled its modernity by using all means of communication then available. Above all, it broke from a museology of objects to become a museology of ideas, in the dual sense of the term: by reproducing the decisive experiments, those that led to radical discoveries, and by inviting the visitors at the demonstration to retrace the intellectual path leading to this discovery.

Phase 3: making science accessible and facilitating knowledge appropriation

The opening of Exploratorium in San Francisco (1969) and the Ontario Science Centre (1969) can serve as reference points since, by focusing on visitors and interactivity, they resolutely rejected the primacy of knowledge production in favour of knowledge appropriation. From the 1970s, SMCs became more open and accessible to visitors. Western societies were being swept by winds of change and museums, summoned to adapt or disappear (Dagognet 1984), tried new ways to approach the public. New institutions emerged to meet the expectations of a public dissatisfied with the traditional cultural offering. It was a period of effervescence and experimentation that led to a diversification of the museum offering (Mairesse 2002). The advent of science centres may be seen as an adaptation of the museum field to the social situation and a response to the desire for personal freedom and satisfaction finding expression at the time. Moreover, they targeted a young public, particularly school children, and wove links with schools to do so.

Exploratorium engendered new aspirations. The new design it put forward was a complete break with the museum practices that had dominated until then, which idolised objects and kept visitors at a respectful distance. Hein, recounting the early years of the museum, recalls the prevailing state of mind: 'Exploratorium did speak a liberatory language, and Oppenheimer's personal style created an atmosphere in which people, including staff members, could enjoy maximum freedom and exploratory space' (1990: 202; Frank Oppenheimer, brother and colleague of physicist Robert Oppenheimer, was the founder of the Exploratorium). This freedom spurred them to rethink the role of visitors in the museum: it was each person's experience that became the determining factor and the main issue in designing the exhibitions. The designers and all Exploratorium personnel 'adhered to an ideal of learning as undirected and self-initiated discovery, occasioned by the experience of stimuli and advanced by opportunistic use of material and conceptual resources. They put a premium on the visitor's individual experience and saw the museum as an aid to the enrichment of that experience' (ibid.: 201). The Exploratorium's success hinged on the fact that it succeeded in 'abolishing the distance between the public and the content of the museum, reconstituting it by making it comprehensible for some yet accessible for others to enjoy for pleasure' (Desvallées 1992: 19).

The visitor-driven exhibitions were not intended to highlight the objects, but rather to offer explanations and demonstrations on scientific subjects. Describing the Exploratorium's vision, Hein (1990: 72) explains:

> The strategy is to let visitors be the laboratory subjects for their own perceptual experiments. By interacting with the museum exhibits, which provide the stimuli and the tools for observation, the subjects are able to analyze the visual process as it takes place within themselves.

What has been called the *communications turn* of museums cannot be explained without reference to the pressures exerted on museum institutions at the time. The public was no longer satisfied with a museology that confined it to a role of passive observer of a science presented 'as a set of accomplishments that have already been achieved' (ibid.: 2). Visitors had to be involved throughout their visit and become the agents of their own learning process. Instead of strolling passively from one exhibit to another, visitors would dialogue, as they wished, with these exhibits that were specifically designed to stimulate their participation.

These museums invite visitors to engage in a conversation (to use Wagensberg's metaphor) with physical and natural *reality* by offering them devices that stimulate this conversation. This conversation is situated on two levels: it is important to understand phenomena, facts, notions, concepts and theories presented by the museum, but what distinguishes SMCs from other forms of dissemination of knowledge is their obligation to display *reality* and to authenticate it. Visitors always expect to know the status of the *reality* being shown to them. Thus, the visitor 'knows he isn't being tricked nor is the victim of an illusion, nor the spectator to fiction' (Davallon 1999: 35). Furthermore, the role of the material object, the tangible asset that museums collect, was radically changing. The promotion of the object has dominated the museum project since the end of the eighteenth century – that is, since the birth of the modern museum. The SMCs of phases 1 and 2, with the exception of the Palais de la Découverte, remained attached to the object; while for the Exploratorium and the Ontario Science Centre, the object was only a stimulus intended to foster a conversation with a *reality* to be discovered.

In this context, interactive exhibits are seen as the surest way to engage and involve visitors in a dialogue that leads them to discover what the museum wants them to observe (Hein 1990) and to facilitate comprehension of the ideas it is trying to communicate, at least so they are no longer foreign to them. Oppenheimer described the then-future Exploratorium as offering an 'environment in which people can become familiar with the details of science and technology and begin to gain some understanding by controlling and watching the behaviour of laboratory apparatus and machinery; such a place can arouse their latent curiosity and can provide at least partial answers' (Oppenheimer [1968] 1990: 218). The originality of the Exploratorium and of the Ontario Science Centre following it hinged on the presentation of ideas through interactive exhibits that involved the active participation of visitors. 'An interactive exhibit is one in which the visitor has a real effect on the outcome; if the visitor is not fully engaged then the result is diminished' (Beetlestone *et al.* 1998: 7). In other words, the visitor's participation is a condition of the visit's success.

SMCs, by soliciting the visitor's active participation, revolutionised the practice of scientific and wider museology. Their success contributed greatly to the penetration of this conception, which has since become the norm. The museum has progressively become a place of communication with the visitors, thereby including their concerns within it. Their expectations and their interests were henceforth at the centre of the museum project.

Phase 4: relating science and technology to society[6]

At the beginning of the 1980s, the public was no longer satisfied with a science detached from social realities. It expected SMCs to tackle controversial subjects because the capacity of citizens to form opinions on questions they deemed important was at stake. At the same time, the deteriorating economic situation forced museums to review their objectives and rethink their ways of doing things. They would also try to reconcile two trends.

First trend

For a number of years, the relationship between society and science had been questioned, possibly more seriously in Europe than in the United States: the idea of progress, already weakened in the previous decade, was then coupled with the idea of damage and risk as the public became aware of the profound impact of science and technology on daily life, work and the environment. The sciences that were represented physically in technologies – the technosciences – were transforming society at a pace unprecedented in the history of humanity and everyone was potentially affected. Museums could no longer be limited to propagating and disseminating science culture, nor be content to merely value and celebrate science in itself, even if several of them did so enthusiastically. The presentation of science cut off from society no longer met the expectations of visitors who expected to be informed in a way that would enable them to act. This is the mindset referred to in the third report of the House of Lords Select Committee on Science and Technology (2000, Chapter 5) as a 'mood for dialogue'. Lowenthal's remarks echo those of Hudson ten years earlier: 'Today the scientific enterprise is widely feared and resented even by those who take its benefits for granted. Science is feared and resented both because its mysteries make it remote and authoritarian and because of its unintended consequences' (Lowenthal 1997: 164). SMCs were dealing with visitors who, aware of the impact of technosciences on society, wanted to foresee the consequences for themselves, their families and friends, and the community. Visitors were no longer content to merely observe changes, even less so to submit passively or adapt to them without any difficulty. These changes were too rapid, too profound and too pervasive. Visitors now expected to debate the merits of choices proposed by scientists, politicians and business people (Le Déaut 2013; Schiele 2013). They also demanded to be heard since they saw themselves as directly concerned by the stakes of the debate on the impact of technosciences on society. SMCs, thus, had to accommodate individuals with heightened awareness, who were clearly determined to be part of decisions and no longer simple observers of changes.

The Cité des Sciences et de l'Industrie (Paris), which opened in 1986, did not perceive itself as a science centre even though some of its exhibitions reflected those characteristics, such as the Cité des Enfants aimed at 3- to 12-year-olds. However, while it proposed a 'scientific stroll', as the visitors' guide explains, it was essentially 'an investment of society towards society', that referred to 'a vision of society, of its problems' and 'the way we manage our future collectively'. The Cité 'cannot be only a place of dissemination of cultural products. It must nourish reflection on the relationships between science, technology and society (and be nourished by it)' (Jantzen 1996: 6).

Obviously, the wager is not without risk. Koster, while admitting the necessity to tackle controversial questions, nevertheless issues a warning to SMCs: 'Should a museum that has yet to embark on a whole-hearted journey of relevance plunge abruptly into the realm of controversial subject matter, it will surely do so at its public relations peril' (Koster 2010: 90). This possibility was clearly demonstrated by the opposition of physicists to the Science in American Life exhibition because they saw it as betraying the ideal of a science detached from social involvement. Koster's remark invites the SMCs to recognise a necessity and challenge it at the same time since they must deal with other more constraining factors.

Second trend

Coincident with this debate, the rapid growth experienced by western economies following World War II began to stall, despite the movement towards a global economy. Traditional economic recovery methods seemed ineffective. Government indebtedness limited spending capacity, and there was a general movement of retreat or a profound reorganisation of areas of intervention in which, only a little earlier, governments saw themselves as the natural managers and guarantors.[7] Education, research, health, social and cultural programmes were sharply affected and a period of uncertainty set in for the agents of culture. In countries characterised by pronounced government support, such as France, this translated into reduced resources. Museums now had to target visitors differently and take on partners. In mixed-economy countries like Canada, or those like the United States in which the role of government is minimal, maintaining partnerships became crucial. In all cases, they had to manage declining growth over a long period. Museums had to consider redirecting their mission.

A managerial culture gradually took hold in the museum field and, more broadly, the cultural field. Performance and profitability objectives, clashing with the more traditional museum values and especially the expectations expressed by the impact of science debate, undermined their potential effect and reduced their reach in the museum field. On the other hand, museums strengthened their visitor focus, albeit in a different spirit from the access and sharing that developed in the 1970s and without the participation desired by the public. Anticipating Koster's warning, the SMCs became more circumspect in tackling controversial questions. Moreover, for science museums much more than for science centres that never had collections, this shift dissociated the traditional functions of conservation, research and education from those of communication, advertising and marketing, which were new at that time but dominant today (Tobelem 2010). It followed that the traditional functions, now lagging behind, were marginalised and the costs of maintaining and managing them considered exorbitant. In return, the management functions became professionalised. The pressure for profitability (whatever that may mean in the museum field) largely motivates the choice of programmes and activities that they offer their public today. This is clearly a more radical evolution than that of the previous decade, which did not alter the substance of the museum but instead reoriented and above all democratised it. In contrast, the shift that started in the 1980s, bolstered by a neo-liberal vision, served to redefine the museum institution. The ensuing economic difficulties, including the crisis starting in 2008, only served to reinforce the choices made at that time.

The museum field's solution: the temporary exhibition

To deal with the difficulties being faced, the museum field as a whole, and not just SMCs, invented the temporary exhibition. Of course, they also increased the number of programmes and activities. But today it is the temporary exhibition that dominates, far ahead of the arsenal of communication technologies, though SMCs invariably also have their websites, are present on Facebook and Twitter, and offer virtual tours to their visitors or at least an online overview of their exhibitions, or allow visitors to prepare for their visit or follow up on it online.

There have never been museums without visitors, nor museums without exhibitions. In science museums, as in the museum field generally, the exhibition has long been associated with the collection. It was the collection – or part of it, as in Phases 1 and 2 – that was shown to visitors. The novelty of the 1980s was making the exhibition autonomous. The exhibitions began to tackle themes quite unrelated to the collections. Moreover, the duration of these exhibitions became shorter, even very short, shrinking from several years to a few months.

From a management perspective, the temporary exhibition[8] presented undeniable benefits for SMCs always seeking visitors and hoping to see them return from one season to the next. First, the public can view a set of items (objects, specimens, devices, simulations) assembled together in one place for a limited time to which they would not otherwise have access, or only with great difficulty. This gives this kind of exhibition great appeal. Second, the temporary exhibit does away with the usual requirements of collections and scientific disciplines, such as curators or scientists to whom the institution generally entrusts the burden and responsibility of ensuring the museum mission. All sorts of groupings or associations are possible, bound by no constraint other than the theme. Third, the interpretations they offer to visitors are principally based on the theme. The recontextualisation of elements permits associations by playing on forms, colours, design, etc. This was true in the exhibition Bleu (to give an extreme example) presented at the Musée de la Civilisation (Québec) where every item exhibited was selected based on a single attribute, the colour blue. This made it possible to give an equal standing to the blue helmets worn by UN soldiers and the costumes of the Tuaregs, which are also blue. This kind of contextualisation is open to all kinds of spontaneous interpretations that the visitor will project on them. In the SMCs, the themes are generally related to science or technology, so they deal with 'new materials', 'new energies', 'climate changes', 'nanotechnologies'. Fourth, the competency of the visitors is never or rarely at issue. This in no way implies that what is shown and related to them is false. But the temporary exhibition never assumes a prior knowledge or competency in visitors that needs to be called on during the visit. Nor does it engage them in a visit that constantly mobilises their attention; the relationship is primarily entertaining. However, the aggregation of partial and fragmented information, laid out side by side, potentially induces an effect of fragmentation of meaning for the visitors (Castells 2010). Fifth, the temporary exhibition comes with a range of events such as lectures, debates, film screenings and workshops to expand the angles of approach and thereby attract more visitors. Sixth, each temporary exhibition is supported by an advertising campaign, publications, reports and interviews, which make it a unique event. The temporary exhibition constantly renews the museum offering: it invites the visitors to come and be amazed as much as to discover, and to be informed as well as to relax.

The temporary exhibition is now the traffic builder par excellence for museums, including SMCs. Science centres were its precursors precisely because they had no collections to display and therefore no need to break away from them. Certainly the SMCs still offer activities spread over longer periods, but they are careful to regularly rekindle visitor interest through new attractions. Even Montreal's Biodôme, comprised of four ecosystems that cannot be changed, enhances its offering through temporary exhibitions.

The temporary exhibition became so dominant in the museum field that, in addition to attracting visitors, it made it possible to rationalise museum activities and define performance criteria. It required the adoption of a 'logic specific to the life of media' and became a 'genre imposing its format and rules' (Jacobi 2012: 138, 139) for SMCs and other museums. The advent of temporary exhibitions 'shook up the museum world' (ibid.: 137) since it is essential to renew them frequently to sustain consumption. The museums streamlined their design and production techniques, just as the media optimised their production techniques. Thus, SMCs tried to know their publics better and design programming based on their cultural interests, expectations and habits, as shown by the extensive development of evaluation methods to survey the effectiveness of communication with the public. In this sense, they resembled the media, which need to adapt their content and above all adjust their message level to the expectations, tastes and skills of the public.

Producing temporary exhibitions requires a full array of strategies for funding, including the sale of derivative products and management of rights that could generate revenue. This leads Paquette (2009: 64) to conclude:

> In the cultural sector, management presents itself as a discourse that values a market logic and measures success in terms of profit, to the detriment of an institutional logic geared to public service and the common good …. Thus, … managerialism has the effect of … reducing public action to a delivery of services and transforming the citizen into a consumer.

Temporary exhibitions can be very expensive to produce, which places SMCs in a perpetual quest for funding. Moreover, the ever-growing number of SMCs has intensified the already existing competition among them while exacerbating competition with the other museums. All of them are vying for the resources needed to produce new exhibitions in a context of scarce funds; yet even rationalisation and standardisation in producing exhibitions has not induced them into the logic of industrial production or a broad distribution network, unlike the cultural industries (Miège 1996). Their audience will always remain limited due to the nature of the medium. Television, movies and the Internet are content to represent *reality*, to suggest or simulate it, without actually having to make it material for viewers. This is never the case with exhibitions: they create contact situations between visitors and *realities* that are physically or symbolically distant in time (dinosaurs) or in space (lunar module), but made visible and represented by real objects (Schiele 2001).

Technology today makes it possible to deploy an array of devices to enhance the experience of the visit: to incorporate virtual immersive environments, to operate within augmented reality, to use social media to exchange information and comment on the visit (Ucko 2013). All this creates a stimulating multi-sensorial environment, but does not absolve the SMCs from fostering a meeting with *reality* (Wagensberg 2006), and the public comes to the museum precisely for this encounter. If they want something else, they can go to amusement parks or arcades. For this reason, the scope of the temporary exhibition will always be limited, and its cost rarely recouped, even if visitors attend in droves as is often the case with the blockbuster shows (Ucko 2013). But like TV and movies, temporary exhibitions, with a few exceptions that boast greater originality than the others, mostly end up resembling each other. Furthermore, 'the growing number of temporary exhibitions also has a perverse trivialising effect: there are so many that it becomes difficult to attract attention at the opening of the nth temporary show of the season' (Jacobi 2013).

SMCs are therefore exploring solutions such as jointly produced travelling exhibitions. Fatal Attraction (2001), for example, an exhibition on the theme of animal communication, was designed to adapt to the different spaces of the three museums associated with this project: the Institut Royal des Sciences Naturelles de Belgique (Brussels), the Nationaal Natuurhistorisch Museum Naturalis (Leiden) and the Muséum National d'Histoire Naturelle (Paris). The effort reflects a concern to reduce costs by sharing scientific and financial resources and a determination to show originality by presenting a scientific view of the seduction and amorous rites manifested among animal species to demonstrate the similarities with human behaviour. Certain SMCs are seeking to federate in a network such as, for instance, the 'NISE Net, which involves nearly 200 member institutions in the shared development, testing, and adaptation of resources' (Ucko 2013: 24).

One wonders if Phase 4 is reaching its end, having failed to fully confront the science-society issue while trying to resolve the difficulties of innovating that museums face with their temporary exhibitions. SMCs, like the museum field overall, while not dispensing with exhibitions, are turning greater attention to events. This cannot be entirely due to the costs or the trivialisation factor of temporary exhibitions. Something else is happening here.

Towards a paradigm shift?

Society has changed radically in the past decade and museums are scarcely keeping pace. Digital technologies have spawned a profound social change whose most glaring effect is a global communicational pervasiveness. Communication is now generalised and immediately ubiquitous (Castells 1996, 2004). Interconnectivity makes interactions instantaneous, direct, unfiltered, between participants or groups of participants, wherever they are in the world. In the communicational universe, there is no longer centre or periphery, and time becomes atemporal and space, a space of flows (Castells 2010). This communicational immanence, which we can say makes the world available to everyone at all times, has sparked cultural changes whose scope we can scarcely begin to apprehend. What matters is 'what it means to be now' (Morton 1997: 169). All societies through history have sought to ward off impermanence, but our society seems to have renounced this as if only the experience lived in the present moment is worthy of attention.

Also, in adapting to this fascination with the present, SMCs progressively turn to research rather than stabilised science, since presenting current research serves to affirm the relevance of science. For Meyer (2010) representing *science* to the public is 'cold': it appears as objective, detached, and free from all ideologies; knowledge is established with certainty, all conflicts are resolved. On the other hand, a culture of *research* is 'hot': it appeals to the emotions, engages those involved in it, and mobilises passion; the as yet uncertain knowledge comes with a large measure of risk. The one presents facts established from a mainstream science, the other discloses the driving issues and the positions and relationships of the actors. So this is a double shift: from the past to the present and from the object of the knowledge to the actors engaged in producing knowledge (ibid.). These shifts are expanded in Table 4.1 below and in the following paragraphs.

Temporary exhibitions still take a long time to produce while being slow to renew and some SMCs try to innovate by designing temporary exhibitions that can be modified to keep pace with the latest scientific research. But generally the SMCs have realised that exhibitions, even temporary ones, are no longer in synch with the times since 'the boundaries between the museum and the social space of lived experience become blurred' (Cameron 2010: 60). So they are turning to the constantly renewed programming of ephemeral events: debates, forums, conferences, whose scope is further enlarged by websites and blogs. For example, the Pacific Science Center presents science in the making, by those making it. An activity entitled Scientist Spotlight brings visitors together with researchers in a live discussion and presentation of their research. The gist is to maximise direct interactions between researcher and visitors on constantly renewed themes

Table 4.1 Presentation of ready-made science versus science-in-the-making

Ready-made science	Science-in-the-making
One voice	Multiplicity of voices
Dominant view	Various contenders
Consensus	Conflicts, disagreements
Answers	Questions
Truths	Challenges
Unambiguous	Ambiguities
Linear approach	Multi-faceted approach
Physical world	Relations between people
Necessary	Contingent
Facts	Contentious themes
Results	Tentative results
Achievements	Failures, pitfalls, aberrations
Products	Processes
Stabilised knowledge	Unfinished knowledge
Closed knowledge	Open research
Secure knowledge	Engaged research
Fixed knowledge	Controversial research

Adapted from Caleb 2010; Cameron 2010; Koster 2010; Meyer 2010

(Selvakumar and Storksdieck 2013). The researchers themselves are involved in depicting their own lives as researchers and their real-life research, no longer just presenting objects or phenomena. This kind of initiative fits into the attempts to reposition SMCs as venues where researchers play a starring role in science and whose presence and descriptions demonstrate today's research. The shift outlined here is also taking effect in other forms of science communication.

Exhibitions will not disappear in the short term, nor will temporary exhibitions. This is because the material culture remains the basis of the museum, even if the immaterial heritage is now in evidence and the digital culture is coming in. However, SMCs are undergoing a genuine reconversion. Cameron (1971) pondered whether a museum should be a temple or a forum. Goaded by a public keen to participate and not content to observe, SMCs are increasingly becoming places of dialogue, exchanges, forums.

With few exceptions, the forms of presenting science that prevailed until now (phases 1, 2, 3) perpetuated the discourse on stabilised science, with contentious questions rarely discussed in public. In this perspective, science has a single, non-dissenting voice with its unambiguous answers, its implacable truths. Centered on the physical world, it presents facts, displays results, heralds achievements and products. Only that knowledge which is firmly established, fixed and hence closed is presented. The contrary approach, more prevalent in Europe than in the United States, is to present science as less sure of itself. The idea that progress entails risks is already eroding science's authority. Current cultural change, because it gives attention to all discourses circulating in the social field, is further eroding science's authority, along with all forms of 'institutional authority' (Cameron 2010: 61). Presenting science in the making opens the way to debates and controversies. It highlights the actors involved in the debates along with the views they defend. The disclosure of questions, ambiguities, disagreements and conflicts allows visitors to engage in discussions on topics that concern them, such as climate change, bioethics, sustainable development, nuclear energy, genetically modified organisms and pollution. Thus, our culture that lives in the moment revisits the question of science-society relationships. This question comes back to haunt the SMCs, while the temporary exhibition had enabled them to evade it for a time.

Key questions

- What social conditions can we identify as influencing the development of science museums and centres in the nineteenth and early twentieth centuries?
- How does the increasing emphasis on communication and the visitor's participation manifest itself in exhibitions in science museums and centres?
- What are the principal features of recently established or modified science centres that allow us to speak of a possible paradigm shift?

Notes

1 The earlier version of this chapter (Schiele 2008) reviewed the history of the development of natural science museums and science museums from their origins up to the 2000s, emphasising how the ideals of the Enlightenment helped shape both contemporary museums and science. It concluded by noting the repositioning of the museum project. This chapter, while recapitulating certain aspects covered in the previous edition, stands on its own.
2 See: www.montrealsciencecentre.com/exhibitions/sex-a-tell-all-exhibition.html (consulted May 20, 2013).
3 See: www.sciencetech.technomuses.ca/english/whatson/2012-sex-a-tell-all-exhibition.cfm (consulted May 20, 2013).
4 See: Religious groups mobilised against the exhibition on sexuality presented in Ottawa, Radio Canada, May 17, 2012; www.radio-canada.ca/regions/ottawa/2012/05/17/006-expo-sexe-evangelistes.shtml (consulted May 22, 2013).

5   The associations representing them are explicit about their areas of involvement: Ecsite (European Network of Science Centres and Museums), ASTC (Association of Science-Technology Centers, US), CASC (Canadian Association of Science Centres), ASPAC (Asia Pacific Network of Science & Technology Centres), ASMD (Association of Science Museum Directors), ANHMC (Alliance of Natural History Museums of Canada), AZAA (American Zoo and Aquarium Association), etc.
6   This part presents a considerably recast argumentation of the last part of the first version of this chapter (Schiele 2008) and of the last part of Landry and Schiele (2013).
7   Neo-liberalism advocates limiting the government role to creating and preserving a framework that permits and guarantees freedom on enterprise, private property, and free trade (Harvey 2007: 2).
8   Jacobi (2012) showed the radical change sparked by the development of the temporary exhibition. I owe to him the gist of the argument developed in this part.

References

Althins, T. (1963) 'Museums of Science and Technology', *Technology and Culture*, 4, 1: 130–147.
Beetlestone, J. G., Johnson, C. H., Quin, M. and White, H. (1998) 'The Science Center movement: contexts, practice, next challenges', *Public Understanding of Science*, 7, 1: 5–26.
Bennett, T. (1995) *The Birth of the Museum*, London: Routledge.
Caleb, W. (2010) 'The transformation of the museum into a zone of hot topicality and taboo representations: the endorsement/interrogation response syndrome', in F. Cameron and L. Kelly (eds) *Hot Topics, Public Culture, Museums*, Newcastle upon Tyne: Cambridge Scholars Publishing, 18–34.
Cameron, D. (1971) 'The museum, a temple or the forum', *Curator*, 14, 1: 11–24.
Cameron, F. (2010) 'Risk society, controversial topics and museum interventions: (re)reading controversy and the museum through a risk optic', in F. Cameron and L. Kelly (eds) *Hot Topics, Public Culture, Museums*, Newcastle upon Tyne: Cambridge Scholars Publishing, 53–75.
Castells, M. (1996) *The Rise of the Network Society, The Information Age: Economy, Society and Culture*, Vol. 1, Cambridge, MA and Oxford: Blackwell.
Castells, M. (2004) *The Network Society: A Cross-Cultural Perspective*, Cheltenham and Northampton, MA: Edward Elgar.
Castells, M. (2010) 'Museums in the information era: cultural connector of time and space', in R. Parry (ed.) *Museums in a Digital Age*, London and New York: Routledge, 427–434.
Dagognet, F. (1984) *Le musée sans fin*, Seyssel: Champ Vallon.
Danilov, V. (1982) *Science and Technology Centers*, Cambridge, MA and London: The MIT Press.
Davallon, J. (1999) *L'exposition à l'oeuvre, Stratégies de communication et médiation symbolique*, Paris: L'Harmattan.
Desvallées, A. (1992) *Vagues – une anthologie de la nouvelle muséologie*, Vol. 1, Macon: éditions W, Savigny-le-temple: M.N.E.S.
Eidelman, J. (1988) *La Création du Palais de la Découverte, Professionnalisation de la recherche et culture scientifique dans l'entre-deux guerres*, PhD Thesis, Université Paris V – René Descartes, Sciences-Humaines – Sorbonne.
Foucault, M. (1970) *The Order of Things: An Archeology of the Human Sciences*, London: Tavistock.
Foucault, M. (1977) *Discipline and Punish: The Birth of the Prison*, London: Allen Lane.
Godin, B. and Gingras, Y. (2000) 'What is scientific and technological culture and how is it measured? A multidimensional model', *Public Understanding of Science*, 9, 1: 43–58.
Harvey D. (2007) *A Brief History of Neoliberalism*, Oxford: Oxford University Press.
Hein, H. (1990) *The Exploratorium, The Museum as Laboratory*, Washington: Smithsonian Institution.
Hobsbawm, E. ([1994] 2004) *The Age of Extremes, 1914–1991*, London: Abacus.
House of Lords Select Committee on Science and Technology (2000) *Third Report, Science and Society*, London: Stationery Office; online at www.publications.parliament.uk/pa/ld199900/ldselect/ldsctech/38/3801.htm
Hudson, K. (1988) *Museums of Influence*, Cambridge, New York and Melbourne: Cambridge University Press.
Jacobi, D. (1997) 'Les musées sont-ils condamnés à séduire toujours plus de visiteurs?', *La Lettre de l'OCIM*, 49: 9–14.
Jacobi, D. (2012) 'La muséologie et la transformation des musées', in A. Meunier (ed.), *La muséologie, champ de théories et de pratiques*, Québec: Presses de l'Université du Québec, 133–150.
Jacobi, D. (2013) 'L'exposition temporaire résistera-t-elle à la montée en force de l'événementiel? Sur la fin d'un paradigme', unpublished manuscript.

Jantzen, R. (1996) *La cité des sciences et de l'industrie*, Paris: Cité des sciences et de l'industrie.
Koster, E. (2010) 'Evolution of purpose in science museums and science centres', in F. Cameron and L. Kelly (eds) *Hot Topics, Public Culture, Museums*, Newcastle upon Tyne: Cambridge Scholars Publishing, 76–94.
Landry, A. and Schiele, B. (2013) 'L'impermanence du musée', *Communication et langages*, 175: 27–46.
Le Déaut, J. -Y. (2013) 'Foreword', in P. Baranger and B. Schiele (eds) *Science Communication Today*, Paris: CNRS éditions, 7–11.
Leshner, A. (2007) 'Beyond the teachable moment', *Journal of the American Medical Association*, 298, 11: 1326–1328.
Lowenthal, D. (1997) 'Paradise and Pandora's box: why science museums must be both', in G. Farmelo and J. Carding (eds) *Here and Now: Contemporary Science and Technology in Museums and Science Centres*, London: Science Museum, 163–168.
Macdonald, S. (2001) *The Politics of Display: Museums, Science, Culture*, London and New York: Routledge.
Mairesse, F. (2002) *Le Musée Temple Spectaculaire*, Lyon: Presses Universitaires de Lyon.
Mercier, J. (2012) 'Le sexe au musée est "insultant pour les contribuables" – James Moore', *Le Droit*, Ottawa, 17 May; online at www.lapresse.ca/le-droit/arts-et-spectacles/201205/17/01-4526250-le-sexe-au-musee-est-insultant-pour-les-contribuables-james-moore.php
Meyer, M. (2010) 'From "cold" science to "hot" research: the texture of controversy', in F. Cameron and L. Kelly (eds) *Hot Topics, Public Culture, Museums*, Newcastle upon Tyne: Cambridge Scholars Publishing, 129–149.
Miège, B. (1996) *La société conquise par la communication*, Grenoble: Presses Universitaires de Grenoble.
Molella, A. (1997) 'Stormy weather: science in American life and the changing climate for technology museums', in G. Farmelo and J. Carding (eds) *Here and Now: Contemporary Science and Technology in Museums and Science Centres*, London: Science Museum, 131–137.
Moles, A. A. (1967) *Sociodynamique de la culture*, Paris and The Hague: Mouton.
Morton, O. (1997) 'Reinventing museums through the information revolution', in G. Farmelo and J. Carding (eds) *Here and Now: Contemporary Science and Technology in Museums and Science Centers*, London: Science Museum, 169–171.
Oppenheimer, F. ([1968] 1990) 'A rationale for a science museum', Appendix 1 in H. Hein, *The Exploratorium, The Museum as a Laboratory*, Washington: Smithsonian Institution, 217–221.
Paquette, J. (2009) 'Communiquer la science: métier, conflit de normes et harcèlement social', *Éthique publique*, 11, 2: 61–71.
Persson, P. -E. (2000) 'Science centers are thriving and going strong!', *Public Understanding of Science*, 9, 4: 449–460.
Rose, A. J. (1967) 'Le Palais de la Découverte', *Museum*, 20, 3: 204–207.
Roussel, M. (1979) *Le public adulte au Palais de la Découverte*, Paris: Palais de la Découverte, Manuscript.
Schiele, B. (2001) *Le musée de sciences*, Paris: L'Harmattan.
Schiele, B. (2008) 'Science museums and science centres', in M. Bucchi and B. Trench (eds) *Handbook of Public Communication of Science and Technology*, London and New York: Routledge, 27–39.
Schiele, B. (2011) 'La participation en science à l'ère des enjeux globaux', *Communication et langages*, 169: 3–14.
Schiele, B. (2013) 'Five things we must keep in mind when talking about the mediation of science', in P. Baranger and B. Schiele (eds) *Science Communication Today*, Paris: CNRS Éditions, 305–318.
Schroeder-Gudehus, B. and Rasmussen, A. (1992) *Les fastes du progrès: le guide des expositions universelles 1851–1992*, Paris: Flammarion.
Selvakumar, M. and Storksdieck, M. (2013) 'Portal to the public: museum educators collaborating with scientists to engage museum visitors with current science', *Curator – The Museum Journal*, 56, 1: 69–78.
Tobelem, J. -M. (2010) *Le nouvel âge des musées – Les institutions culturelles au défi de la gestion*, Paris: Armand Colin.
Ucko, D. A. (2013) 'Science centers in a new world of learning', *Curator – The Museum Journal*, 56, 1: 21–30.
Wagensberg, J. (2006) 'Toward a total museology through conversation between audience, museologists, architects and builders', in R. Terradas, E. Terradas, M. Arnal and K. -J. Van Gorsel (eds) *The Total Museum*, Barcelona: Sacyr, 11–103.

5
Public relations in science
Managing the trust portfolio

Rick E. Borchelt and Kristian H. Nielsen

Introduction

Public relations (PR) have come to play an important and inextricable part in science communication. Whether they acknowledge it or not, scientific organisations use PR in a variety of ways. Higher education organisations advertise the quality of their programmes to recruit new students and faculty. Research-performing organisations (some of which also have higher education programmes) release information on new results to publicise work being funded by government and private foundations in order to display accountability, but also in hopes of gaining public visibility and thus attracting additional funding. Non-profit advocacy organisations publicise their work to appeal to new donors or call attention to their issues and how they have been able to effect legislative or policy change. Corporate scientific institutions attract and retain customers or investors or seek to change consumer behaviour toward their, or a similar, product.

Recent decades have seen an increased professionalisation and institutionalisation of PR in scientific organisations. Yet, the role of PR in science communication has escaped much of the dedicated research effort that has helped scholars understand other variables, such as media coverage of science, public attitudes toward science and the impact of the Internet on acquiring scientific knowledge. Notable exceptions include the singular role of the science Public Information Officer (PIO) as actor or gatekeeper in communication (Ankney and Curtin 2002; Kallfass 2009), the relationship between PR and science journalism (Bauer and Bucchi 2007; Göpfert 2007; Friedman *et al*. 1986; Müller 2004; Nelkin 1995), PR in risk and health communication (Hamilton 2003; McComas 2004; Palenchar and Heath 2007; Springston and Lariscy 2003), and the implications of digital media for science PR (Duke 2002; Lederbogen and Trebbe 2003).

This chapter[1] discusses these issues and some of the various approaches to PR that scientific organisations employ; the goals these organisations have for PR; the levels in scientific organisations at which PR is or can be employed; management of PR in terms of organisational behaviour; and ways to evaluate PR practice as both a process and an outcome. PR is the term most often used by academic researchers in communications and in the corporate world to refer to the communication management function of an organisation, although in many situations in the scientific world, PR has come to denote a less than savoury bag of tricks to confuse or dupe potential customers or citizens. As we use it, PR is the art and science of developing meaningful relationships with the public necessary for continuing the work of an organisation. It is not intended as a synonym for marketing, although marketing may be a component of PR practice in some scientific organisations.

The approach taken is to view PR in a scientific organisation as *managing the trust portfolio* – both for the organisation and for the scientific enterprise more generally, and as a unifying concept for future scholarship. The trust portfolio has several components: accountability, competence, credibility, dependability, integrity, legitimacy and productivity. Managing the trust portfolio means planning and managing a wide variety of strategic communication programmes building diverse relationships between science and different publics. Paraphrasing PR scholars Grunig and Hunt (1984), the trust portfolio depends on what scientists, information officers and communications managers do with what they know and what others think about what they say.

PR in science communication: a historical outlook

PR is a relatively young field, practiced by the corporate and scientific world for about a century and studied by academic scholars for even fewer years. In some form, marketing and publicity have been with us throughout the history of science. From the seventeenth century up to and including most of the nineteenth century, members of the learned communities used natural theology to establish trust in their pursuit of knowledge. Robert K. Merton ([1938] 2002) showed how natural philosophers in seventeenth-century England achieved legitimacy and credibility by stressing intimate relations between their cultivation of early experimental science and Puritanism. The development of effective means of mass communication in the nineteenth century created an entirely new field of play for scientists and others as they were taking science to the *marketplace* of public education, entertainment, participation and propaganda (Fyfe 2004; Fyfe and Lightman 2007; Secord 2000).

By the start of the twentieth century, science had become a professional and highly specialised domain in its own right. The notion that *real* science and the public were separate spheres demarcated by social as well as epistemological boundaries was gaining ground, partly in consequence of the many attempts to make science popular with mass audiences (Shapin 1990). In the United States, partly in acknowledgment of the differentiation of science and the public, scientific associations and universities from the late 1910s onwards began hiring public relations directors and setting up news services (LaFollette 1990). The Science Service, an institution for the dissemination of science news in Washington, DC, was founded in 1921 by scientists concerned about both soliciting funds and nurturing the public image of science, along with wealthy newspaper publisher E.W. Scripps. For Scripps, public trust of science depended on its ability to secure the advance of democracy. Science news was the most efficient way to 'democratise' science by providing 'the 95 per cent', as he called it, 'with the basis for forming intelligent opinions on matters of national importance' (cited in Rhees 1979).

In Europe, scientists engaged in the social relations of science movement of the 1930s and 1940s also promoted science as a key to democracy. One of the leading proponents in Britain, the crystallographer and public scientist J. D. Bernal, expounded the view that dissemination of (proper) scientific information and world views to the public was necessary to combat fascism. Like Scripps, he lamented the lack of serious science coverage in the press. Science journalism, according to Bernal (1944: 304), had to make 'a real understanding of science become part of the common life of today' by giving 'adult minds … the opportunity of appreciating what science is doing and how it is likely to affect human life'. Some science journalists, such as J. G. Crowther in Britain and B. Michelsen in Denmark, used Bernal's and other scientists' ideas in their attempts to enhance public understanding and appreciation of science. After World War II, Michelsen led UNESCO's Division of Science and Its Popularisation for a short period, cultivating science journalism and PR all across the globe (Nielsen 2008).

Institutions such as Science Service and the UNESCO Division of Science and Its Popularisation combined two ideal types of PR, which today still feature prominently in science PR: *publicity* and *public information* (Grunig and Hunt 1984). Publicity aims at maximising public awareness of science, benefitting from mass communication technologies to reach large audiences more quickly, with more targeted messages. The number of press releases produced, how many were used by reporters and how many stories resulted from the press releases are typical measures of success. An early study of 97 American newspapers from 1939 to 1941 revealed that about 5 per cent of non-advertising space was devoted to science news (Krieghbaum 1941). Public information, conversely, internalises scientists' concern for the quality of information disseminated to the public. It strives for scientific credibility and public education. For science journalists and others engaged in science PR in the first part of the mid-twentieth century, balancing the need for making the news with the emphasis on disseminating correct information was an important aspect of PR in science, and it continues to be so to this day.

Following World War II, scientists and scientific organisations adopted a third approach to PR: the two-way asymmetric model (Grunig and Hunt 1984). This PR model is based on a (semi-)scientific approach to studying public opinion, aiming to achieve maximum change in understanding, attitudes and behaviour rather than maximum coverage in the mass media or maximum credibility of science. In 1957, supported by Rockefeller Foundation, the US National Association of Science Writers (NASW) commissioned what is probably the first research-based survey of the public impact of science in the mass media (Survey Research Center, University of Michigan, and National Association of Science Writers 1958). It found that only a small segment of newspaper readers took an interest in science news, and that 'the science consumer tends to view the world from a perspective similar to that of science' (225).

Beginning in 1972, the National Science Board commenced its biennial surveys of public attitudes toward science and technology, which now form part of the annual *Science and Engineering Indicators* (National Science Board 2013; see also Miller 1987). Today, similar surveys are being carried out in most developed and in some developing countries, showing public level of knowledge, interest in, and attitudes toward science and technology in relation to various socio-economic indicators (see Bauer and Falade in this volume and, e.g. China Research Institute for Science Popularisation (CRISP) 2008; European Commission 2010; Korea Foundation for the Advancement of Science and Creativity (KOFAC) 2009; Lamberts *et al.* 2010; National Institute of Science and Technology Policy (NISTEP) 2002; Shukla 2005). The results are being used to varying degrees by decision- and policy-makers as they identify the policies and procedures of scientific organisations with public interests or deficits.

The two-way symmetric model focuses on dialogue, participation and engagement in a response to *active publics* (Grunig and Hunt 1984). In the late 1950s and 1960s, the rise of environmental and consumer movements in conjunction with increasing public expenditures on science placed new demands on scientific organisations. Responding to public concerns about the accountability and responsibility of science, some scientific organisations began developing special PR programmes aimed at facilitating interactive relations with concerned citizens. A few prominent examples include: the science shop system, launched by Dutch universities in the 1970s to enable citizens or NGOs to commission research; the guidelines resulting from the Asilomar Conference on Recombinant DNA in 1975, which included calls for devising mechanisms and models for bringing science into the public eye; and the consensus conferences pioneered in the late 1980s by the Danish Board of Technology to communicate citizens' views and attitudes on potentially controversial technologies. From the 1990s and onwards, governments, scientific organisations and NGOs all around the world have devoted significant efforts to developing forums for public participation in science, in particular with regard to controversial science and technology issues (Bucchi and Neresini 2008).

Online challenges: promoting and debating science on the web

The turn to symmetric PR pervades the field of science communication, but so do the commercialisation of scientific research and the ensuing corporatisation of scientific organisations (Bauer and Bucchi 2007). While many organisations and governments are making genuine efforts to enable dialogue between science and its publics, there are also attempts to turn science PR into pure marketing and branding. Public dialogue and engagement has been shown to form part of a new type of scientific governance that emerged in the late 1990s in Europe, building on transparency and openness to win over active publics who have grown sceptical of scientific organisations' and governments' handling of risk (Irwin 2006). The tension between, on the one hand, debating science and its societal implications and, on the other, *selling science* has been identified within US science journalism since the 1980s (Nelkin 1995). Today, it is probably most strongly felt within the domain of online science communication.

Two surveys of members of NASW, conducted in 1994 and 1999, showed that most science journalists use email regularly and 'have relatively little shyness for the use of the Web as an information source in news making' (Trumbo *et al.* 2001: 361). The fact that today this result appears to be a truism testifies to the rapid development of online communication. About half of the nearly 2,000 NASW members registered in the early 2000s were listed as associate members, most of which were employed as information officers in scientific organisations or research-intensive corporations. They were not included in the two surveys just mentioned, which only addressed active science writers. Another survey, conducted in 2000, specifically sampled NASW members identified as PR practitioners, and they reported that email and the Web had become an integral part of science PR, especially media relations work (Duke 2002).

Access to the Internet has increased dramatically since then, opening up many aspects of science to public scrutiny and comment and providing new modes of interaction between science and its publics. At the same time, the advent of online science has deepened certain inequities in the access to knowledge, often described as 'the digital divide' (Montgomery 2009: 93). Moreover, discerning reliable and trustworthy information has become even harder, as science journalists report (Dumlao and Duke 2003). The science PIO has to find ways of ensuring the online credibility of the organisation, while the science journalist needs to make critical assessments of the sources of information provided online. As Trench (2007) puts it, scientific organisations have to do trust management to reduce the increased uncertainties resulting from the overabundance of Internet-based media for the dissemination of scientific information. He suggests that organisations support their users by providing context for all news items, providing information in multiple layers for groups of users, categorising documents – e.g. as peer-reviewed papers, self-published research reports, corporate press releases or advocacy groups' statements – and rating the relevance of information based on editorial judgement rather than user popularity.

Take a current example of high importance: online communication has played a highly significant role for the public debate about climate change and the appropriate responses to it. Despite the centrality of climate science in defining and assessing climate change, it is found that 'climate scientists and scientific institutions from the field do not seem to be the major players in online climate communication' (Schafer 2012: 529). Even though academic and non-academic research institutions, including institutions engaged in climate research, have expanded and professionalised their PR efforts online, their influence in online climate communication remains limited. This may have to do with the fact that the websites of many scientific organisations still cater primarily to the interests of the scientific community and 'do not fulfil the user's demand for up-to-date and easy-to-understand contents', as it was found in a survey of the websites of 22 German universities and non-university research institutions performed in 2000 (Lederbogen and Trebbe 2003: 350). Regardless of whether the online presence of scientific organisations is to blame or not, it seems to be the case that the general public mostly consults other online sources for information about climate change and that the impact of research-based and institutional online climate communication on specialised audiences such as journalists, scientists and politicians remains uncertain (Schafer 2012). Trust management of online climate communication has turned out to be almost impossible.

The development of online communication holds promises for more interactive, symmetric approaches to science PR. Yet, it seems as if asymmetrical communication models are still the preferred mode of PR practice for many scientific organisations, with the public information model probably most widely practised in the scientific world today. This practice, to a high degree, reflects scientists' own interests in science PR and is probably best demonstrated by the increasing numbers of science writers being recruited by universities and research agencies, often to write lay-language brochures, news materials, website material and annual reports. While such practices have been a staple of US science PR activity for quite some time, public information work like this is relatively new in much of Europe and the developing countries (Kallfass 2009; Moore 2000; Schiele *et al.* 2012).

Managing the trust portfolio

Two of the key concerns for scientists with respect to science PR are to keep the public informed about science topics and maintain the trustworthiness of the scientific enterprise (Besley and Nisbet 2013). For scientific organisations, therefore, the PR function might usefully be thought of as *managing the trust portfolio* (Borchelt 2008). By the trust portfolio, we mean the principal relationships that exist between the organisation and its many stakeholders. Science PR, done effectively and strategically, helps the other parts of the organisation to do their job more effectively by cultivating or maintaining trust in the ability of the organisation to do science, advocacy or science policy.

For example, public affairs officers (PAOs; yet another term for PR practitioners, most often used in government) at a government-funded research institution may have a number of stakeholders for whom science communication would be helpful in establishing and maintaining trust. First and foremost, the organisation is probably concerned about its funding stream, and the appropriate kinds of PR can help the agency or laboratory convince legislators or agency heads that money sent to this organisation is money well spent and that its research is top quality and worth supporting. Second, the organisation probably has a need to make sure that other scientists and researchers elsewhere know about the range of research being conducted there, in order to facilitate collaboration and keep abreast of scientific research conducted by other organisations, and to position the organisation as a credible and reliable scientific collaborator (or future employer). Third, the organisation may need to have a good relationship with the people in the area surrounding the facility: government-run laboratories increasingly face the need to maintain the trust and support of their local communities in order to do research in community settings.

Of course, media are an important public, but in reality the organisation is seldom interested in the media for the media's sake. The scientific organisation is interested in media because they are able to reach other primary stakeholders whose actions directly affect the ability of the organisation to stay open and conduct research. Media in this context are third-party validators, as Dorothy Nelkin (1995: 124) posits: '[s]cientists ventriloquate through the media to those who control their funds'. Bad press certainly can affect the disposition of key stakeholders toward the organisation, and, conversely, good press can validate the work and integrity of the organisation among groups that materially affect the organisation's ability to do its research. Good PR practitioners, however, never confuse the route they use to get to strategic publics with the publics themselves. Similarly, good communications researchers should not focus unduly on the role of media in science communications. Media content studies are low-hanging fruit from many perspectives – discrete, quantitative, archivable – yet they give only a very incomplete picture of trust relationships between an organisation and its true stakeholders.

Trust generally is understood as a means for reducing social complexity. Thus, trust in science allows us, in complex situations, to solve problems and make decisions based on scientific research without having to scrutinise the scientific evidence. Sociologists like Giddens and Beck, since the late 1980s, have alerted us to the fact that scientific authority no longer can be taken for granted but, increasingly, has to be actively earned and supported by active investment'. The emergence of global risks such as climate change, Beck argues, means that key institutions, such as science, business and politics, 'are no longer seen as managers of risks, but also as sources of risk' (2009: 54). We might add that science cannot simply be seen as a source of trust, but increasingly as a manager of trust.

Specifically, important components of managing the trust portfolio of scientific organisations are:

- *Accountability*: organisations need to acknowledge and assume responsibility for their actions, products, decisions and policies. To what or to whom is the organisation accountable? What are the measures of accountability?
- *Competence*: normally, the production of scientific knowledge is associated with high levels of competence. Can the organisation do the work that is expected of it? Do the researchers who work there have the right credentials, and are they considered pre-eminent in their field?
- *Credibility*: according to public opinion polls, science generally has a high degree of credibility. How credible is the organisation? Can the organisation maintain credibility when dealing with controversial topics?
- *Integrity*: dependent on external funding and often engaged in collaborations with other public institutions and businesses, scientific organisations must still maintain integrity. Do outside observers believe that researchers and management at the organisation know the difference between right and wrong in a science setting? Do they know the safety and ethical codes of conduct for research?
- *Legitimacy*: science is an authority on almost all matters. How does the organisation promote the kinds of values whereby science is recognised and accepted as a right and proper way of influencing governments as well as the lives of individual citizens? What is the basis for the legitimacy of science and how does it translate into PR for the organisation?
- *Productivity*: science must pay heed to its integrity, but also make concessions to the demand for useful and relevant products (knowledge and graduates). Does the organisation respect outside requests for productivity? Do policymakers, industrial leaders and others praise the organisation for delivering on time, within budget and with no bugs?

Science PR in practice: trends and contexts

Managing the trust portfolio of science is a difficult and complex task. How do science PIOs, PAOs and communications managers go about their business in practice? An exploratory, interview-based study of 45 organisations in Germany, Britain and France, conducted under the auspices of the German research project INWEDIS (Integrating Scientific Expertise in Public Media Discourse), provides some answers (Kallfass 2009). The organisations included universities, university hospitals, non-university research institutions, and research support and science communication organisations, most of which employed a relatively small staff (often just two persons) for PR and communications. Unsurprisingly, in all types of organisations, the persons in charge spent a great deal of time cultivating their contacts to the media. This included reactive and proactive communications. Describing the full scope of their media relations, PIOs mentioned feeding press releases and background material (including embargoed information) to journalists and editors, promoting the brand of the organisation, maintaining a high degree of media presence and visibility, directing communications to different target groups (different types of media), mediating relationships between researchers and journalists, training researchers in communications, handling public controversies and developing uniform patterns of media relations.

Internal communications also was seen as important to organisational PR. As one PIO at a British university explained: '[t]he first thing, we need to have a good internal communication before we can effectively talk to the media. So we need to improve communications between academics and the Press Office, that's a big on-going job' (Kallfass 2009: 113).

In facilitating researchers' contact to the media, some PIOs found it useful to distinguish between three types of researchers: the media stars, who are much in demand and used to dealing with journalists; those who take PR seriously enough to regularly consult the press office or communications department for advice and support; and those who think that PR is inferior to research and prefer not to have anything to do with it. Most of the PIOs expressed high respect for the researchers' passion and took great pains to manage their expectations and to explain to them how the news system works (Kallfass 2009).

An important finding relates to the contextual nature of PR. All of the organisations surveyed were found to give relatively low priority to PR compared with other types of organisations such as private corporations and governments. PR in science has to struggle for legitimacy and visibility. Moreover, there were marked differences between the national contexts. In France, scientific research for the most part takes place in national laboratories, which may account for the fact that few universities have integrated PR into the organisation. Science PR in France was found to be in embryo. In Great Britain, science PR often is split into media relations and higher education, the latter of which deals with the promotion of educational programmes. Media relations in Britain have a strong element of co-operation, since universities and private foundations often co-operate to communicate co-funded research projects. Finally, for linguistic reasons, German and French PIOs typically focus on national media, whereas British ones spend more time attracting international media attention (Kallfass 2009).

In other cultural contexts, science PR and science communication again take on other dimensions (Bauer et al. 2012; Cheng et al. 2008; Schiele et al. 2012). Scholars have shown how global trends of public participation in science and dialogue-based science communication have been translated into national practices. Latin American countries such as Brazil, Argentina and Mexico, all of which have experienced rapid economic growth in past decades, have witnessed many new initiatives in science communication with a strong link to national traditions for top-down policymaking, but also with a high degree of cultural diversity across the countries and an important role for social movements such as indigenous rights and environmental protection (Polino and Castelfranchi 2012; see also Trench et al. in this volume). In Japan, too, the national government, since the mid-2000s, has become an influential player in science communication. Many activities, including science cafés, based on dialogue and informal settings for science communication, have been embedded in the institutional culture of Japanese universities and research institutions, often with a distinctive *promotional* flavour (Nakamura 2010).

The increasing institutionalisation of science PR, which formed the basis for Kallfass' (2009) survey of scientific organisations in Europe, also appears in many other parts of the world, albeit in a form shaped by other circumstances. Australia, since the 1990s, has seen a professionalisation of the field of science communication with the establishment of a professional organisation, Australian Science Communicators, and the consolidation of three centres for training science communicators. The affiliation of science communicators with universities and Australia's national science agency, the Commonwealth Scientific and Industrial Research Organisation, means that there are intimate connections between science communication practice and research (Metcalfe and Gascoigne 2012).

China, for its part, is characterised by strong emphasis on science popularisation – a term which has many connotations in the Chinese language – and public science literacy, defined as basic knowledge of science and technology, including methods and ethos of science, and 'the ability to apply [this knowledge] to resolve practical problems and participate in public affairs' (Ren et al. 2012: 73). In 2002, the Chinese government enacted its Law of the People's Republic of China on Popularisation of Science and Technology, laying out national terms for the organisation and administration of popularisation of science and technology as 'a common task for the society as a whole' (People's Republic of China 2002, article 13). Interestingly, the western debate about appropriate methods for research into public understanding of science is being played out by Chinese scholars who dispute the applicability of western indicators in developing countries like China with different cultural and ideological institutions (Bauer et al. 2007; Ren et al. 2012).

Future research on PR in science

What do we know about PR in science? Not a lot, it turns out. There is substantial information available about science in the media: how journalists portray science and how they interact with scientists. Some of this information is pertinent to science PR as we have good indications that many scientific organisations see media relations as an important, if not the most important, part of their PR portfolio. We also know that science communication including science PR depends on context.

In developing future research strategies for PR in science, it will certainly be necessary to take into account cultural differences, but it may also be useful to take a broader approach to PR rather than to think of it as something that one unit in an organisation does. Scientific institutions that are effective in their job of engendering trust and developing satisfying relationships with their publics must be effective at four different levels of organisational management. Certainly, communications success at the lower levels contributes to success at the higher levels but an institution has little hope of using PR effectively until and unless it also is committed to, and attains, success at higher organisational levels.

The programme level is the individual component of an overall PR programme, such as media relations, publications, events planning and so on. The effectiveness of these individual programmes usually can be deduced by whether they meet specific objectives: do they change knowledge, attitudes or behaviours of the publics to which they are targeted? But success at the programme level of organisation does not guarantee, or necessarily even contribute to, success of the institution as a whole unless the publics reached are the ones that really matter to the health and survival of the institution and the programme contributes to cultivation of mutually satisfactory relationships with these strategic publics. A media relations programme that consistently produces reams of coverage in national newspapers about an institution's research is disconnected from the overall effectiveness of the institution if the strategic publics – policymakers, say, or government funding agencies – do not read those newspapers, or if these publics use other criteria to judge an organisation's success.

The functional level is the overall communications or PR function of the institution, typically including all of the individual programme-level units discussed above. While the PR function of an institution may appear to be effective – communications teams may regularly win awards for excellence in news writing or campaign planning – its effectiveness is only as good as its relationship to overall institutional management objectives. For example, if the PR function of a biomedical research institute does not understand that donors and foundations are also a critical element of the PR portfolio, the PR function cannot be considered successful.

At the organisational level, PR must contribute in some way to the organisation's bottom line – financially, in the case of a corporate entity; by attracting new students or helping retain world-class staff at a university; or by attracting membership or donor support for non-profit advocacy organisations. PR is most effective at the organisational level when it helps the organisation to identify its strategic publics, how best to interact with them and their expectations in return. This is a management function of PR, and requires that PR has a place at the table among senior organisation executives to be truly successful.

PR is a strategic function of a successful organisation as well as a tactical one. Too many scientific organisations see only the tactical value of PR and are content to manage it through the human resources department, laboratory administration or some other programme not at all connected functionally to PR. This has important ramifications for the qualifications of PR managers in scientific organisations. They really must understand the scientific issues that come before the senior management and speak with authority to the scientists and researchers at the lab bench. They must be seen as independent experts in their own right – there should be no temptation on the part of the CEO or other senior staff to make PR decisions without them – and they must have the full and absolute backing of senior management in implementing practices of engagement with stakeholder publics but also be seen as credible by those stakeholders. These internal relationships have been very poorly studied in science organisations, although they undoubtedly have great impact on science communications as an enterprise.

Finally, at the societal level, PR professionals can help their organisations understand what it means to be socially responsible and help contribute to the ethical behaviour and social commitment of the organisation. At this level, management of the trust portfolio goes beyond the trust engendered between the organisation per se and its publics; it helps the organisation manage the trust portfolio for the entire scientific enterprise. Socially responsible scientific organisations help cultivate public trust in science and technology. PR – if empowered by management – can play a vital role in articulating social responsibility and finding ways for an organisation to allay public mistrust in, and wariness of, science and scientists.

Understanding the role of PR in science communications will require a much more robust research effort aimed at elucidating these organisational and societal level PR activities. Scholars need to examine, for example, the degree of disconnection between PR practices at the programme or functional level and the organisational level. Activities at higher management levels frequently undermine programme-level attempts at two-way dialogue and participation by putting in place rigid information control policies for corporate managers. Because PR is a function of entire organisations, not just science communicators or scientific officers, science communications researchers need to turn their attention to the relationship between senior management and PR departments and to its impact on the trust portfolio at the societal level. Otherwise, we will have great difficulty in comprehending the contributions of PR to the public understanding of science and technology. The ongoing institutionalisation and professionalisation of different types of PR at the level of scientific organisations and government provides an opportune moment for the establishment of sustained and diverse research programmes on PR in science.

Questions for discussion

- What types of relationship exist between PR and science journalism? Give examples of the ways in which the two practices are mutually beneficial or conflicting. How should this relationship be developed in the future?
- How does the role of PR in a science organisation relate to other functions such as senior management, research and teaching (if applicable)?
- What are the most important tasks for science public information officers (PIOs) in routine operations and when controversy arises?
- How does the role of science PR differ in various parts of the world? Is science PR just a western thing, or can it be exported in a meaningful way to other cultural contexts? If so, how?

Note

1 In the revision of this chapter, as compared with Borchelt (2008), the section on the historical outlook on PR in science communication has been rewritten to include examples from the history of science communication. Two new sections on online challenges to PR in science and on science PR in practice have been added. Special attention has been given to covering developments across the globe. Moreover, the notion of *trust portfolio* has been expanded to include more elements.

References

Ankney, R. N. and Curtin, P. A. (2002) 'Delineating (and delimiting) the boundary spanning role of the medical public information officer', *Public Relations Review*, 28, 3: 229–241.
Bauer, M. W. and Bucchi, M. (eds) (2007) *Journalism, Science and Society: Science Communication between News and Public Relations*, New York and London: Routledge.
Bauer, M. W., N. Allum and S. Miller (2007) 'What can we learn from 25 years of PUS survey research? Liberating and expanding the agenda', *Public Understanding of Science*, 16, 1: 79–95.
Bauer, M. W., Shukla, R. and Allum, N. (eds) (2012) *The Culture of Science: How the Public Relates to Science Across the Globe*, New York and London: Routledge.
Beck, U. (2009) *World at Risk*, Cambridge: Polity.
Bernal, J. D. (1944) *The Social Function of Science*, London: George Routledge.
Besley, J. C. and Nisbet, M. (2013) 'How scientists view the public, the media and the political process', *Public Understanding of Science*, 22, 6: 644–659.
Borchelt, R. (2008) 'Public relations in science: managing the trust portfolio', in M. Bucchi and B. Trench (eds) *Handbook of Public Communication of Science and Technology*, London and New York: Routledge, 147–157.
Bucchi, M. and Neresini, F. (2008) 'Science and public participation', in E. J. Hackett, O. Amsterdamska, M. Lynch and J. Wajcman (eds) *Handbook of Science and Technology Studies*, Cambridge, MA and London: MIT Press, 449–472.
Cheng, D., Claessens, M., Gasgoigne, T., Metcalfe, J., Schiele, B. and Shi, S. (eds) (2008) *Communicating Science in Social Contexts: New Models, New Practices*, Dordrecht: Springer.

China Research Institute for Science Popularization (CRISP) (2008) *Chinese Public Understanding of Science and Attitudes Towards Science and Technology*, Beijing: CRISP.

Duke, S. (2002) 'Wired science: use of World Wide Web and e-mail in science public relations', *Public Relations Review*, 28, 3: 311–324.

Dumlao, R. and Duke, S. (2003) The Web and E-Mail in Science Communication, *Science Communication*, 24, 3: 283–308.

European Commission (2010) *Science and Technology: Special Eurobarometer 340*, European Commission Directorate-General for Research. Online at: http://ec.europa.eu/public_opinion/archives/ebs/ebs_340_en.pdf (accessed 3 July 2013).

Friedman, S. M., Dunwoody, S. and Rogers, C. L. (eds) (1986) *Scientists and Journalists: Reporting Science as News*, New York: Free Press.

Fyfe, A. (2004) *Science and Salvation: Evangelical Popular Science Publishing in Victorian Britain*, Chicago and London: University of Chicago Press.

Fyfe, A. and Lightman, B.V. (eds) (2007) *Science in the Marketplace: Nineteenth-Century Sites and Experiences*, Chicago and London: University of Chicago Press.

Göpfert, W. (2007) 'The strength of PR and the weakness of science journalism', in M. W. Bauer and M. Bucchi (eds) *Journalism, Science and Society*, New York and London: Routledge, 215–226.

Grunig, J. E. and Hunt, T. (1984) *Managing Public Relations*, New York: Holt, Rinehart and Winston.

Hamilton, J. D. (2003) 'Exploring technical and cultural appeals in strategic risk communication: the Fernald radium case', *Risk Analysis*, 23, 2: 291–302.

Irwin, A. (2006) 'The politics of talk: coming to terms with the "new" scientific governance', *Social Studies of Science*, 36, 2: 299–320.

Kallfass, M. (2009) Public Relations von Wissenschaftseinrichtungen – explorative Studie in Deutschland, Frankreich und Großbritannien, *Medienorientierung biomedizinischer Forscher im internationalen Vergleich. Die Schnittstelle von Wissenschaft & Journalismus und ihre politische Relevanz*, Jülich: Forschungszentrum Jülich, 101–175. Available online http://wwwzb1.fz-juelich.de/contentenrichment/onlinepublikationen/Gesundheit_18.pdf (accessed 4 July 2013).

Korea Foundation for the Advancement of Science and Creativity (KOFAC) (2009) *Survey of Public Attitudes Towards and Understanding of Science and Technology 2008*, Seoul: KOFAC.

Krieghbaum, H. (1941) 'American newspaper reporting of science news', *Kansas State College Bulletin*, XXV, 5: 1–73.

LaFollette, M. C. (1990) *Making Science Our Own: Public Images of Science 1910–1955*, Chicago: University of Chicago Press.

Lamberts, R., Grant, W. J. and Martin, A. (2010) *ANU Poll: Public Opinion about Science*, The Australia National University. Available online: http://lyceum.anu.edu.au/wp-content/blogs/3/uploads//ANUpoll%20on%20science1.pdf (accessed 4 July 2013).

Lederbogen, U. and Trebbe, J. (2003) Promoting science on the Web: public relations for scientific organizations – results of a content analysis, *Science Communication*, 24, 3: 333–352.

McComas, K. (2004) 'When even the "best-laid" plans go wrong – Strategic risk communication for new and emerging risks', *EMBO Reports*, 5, Supplement 1: S61–S65.

Merton, R. K. ([1938] 2002) *Science, Technology and Society in Seventeenth Century England*, New York: Howard Fertig.

Metcalfe, J. and Gascoigne, T. (2012) 'The evolution of science communication research in Australia', in B. Schiele, M. Claessens and S. Shi (eds) *Science Communication in the World: Practices, Theories and Trends*, Dordrecht: Springer, 19–32.

Miller, J. D. (1987) 'Scientific literacy in the United States', in D. Evered and M. O'Connor (eds) *Communicating Science to the Public*, Chichester, New York, Brisbane, Toronto and Singapore: Wiley, 19–40.

Montgomery, S. L. (2009) 'Science and the online world: realities and issues for discussion', in R. Holliman, J. Thomas, S. Smidt, E. Scanlon and E. Whitelegg (eds) *Practising Science Communication in the Information Age: Theorising Professional Practices*, Oxford: Oxford University Press, 83–97.

Moore, A. (2000) 'Would you buy a tomato from this man? How to overcome public mistrust in scientific advances', *EMBO Reports*, 1, 3: 210–212.

Müller, C. (ed.) (2004) *SciencePop: Wissenschaftsjournalismus zwischen PR und Forschungskritik*, Graz: Nausner und Nausner.

Nakamura, M. (2010) 'STS in Japan in light of the science café movement', *East Asian Science, Technology and Society: An International Journal*, 4, 1:145–151.

National Institute of Science and Technology Policy (NISTEP) (2002) *The 2001 Survey of Public Attitudes Toward and Understanding of Science and Technology in Japan*, NISTEP. Available online http://data.nistep.go.jp/dspace/bitstream/11035/612/2/NISTEP-NR072-SummaryE.pdf (accessed 4 July 2013).

National Science Board (2013) *Science and Engineering Indicators*, Arlington, VA: National Science Foundation. Available online http://www.nsf.gov/statistics/indicators/ (accessed 3 July 2013).

Nelkin, D. (1995) *Selling Science: How the Press Covers Science and Technology*, revised edition, New York: W. H. Freeman.

Nielsen, K. H. (2008) 'Enacting the social relations of science: historical (anti-)boundary-work of Danish science journalist Børge Michelsen', *Public Understanding of Science*, 17, 2: 171–188.

Palenchar, M. J. and Heath, R. L. (2007) 'Strategic risk communication: adding value to society', *Public Relations Review*, 33, 2: 120–129.

People's Republic of China (2002) *Law of the People's Republic of China on Popularization of Science and Technology*, Ministry of Science and Technology of the People's Republic of China. Available online www.most.gov.cn/eng/policies/regulations/200501/t20050112_18584.htm (accessed 3 July 2013).

Polino, C. and Castelfranchi, Y. (2012) 'The "communicative turn" in contemporary techno-science: Latin American approaches and global tendencies', in B. Schiele, M. Claessens and S. Shi (eds) *Science Communication in the World: Practices, Theories and Trends*, Dordrecht: Springer, 3–17.

Ren, F., Yin, L. and Li, H. (2012) 'Science popularization studies in China', in B. Schiele, M. Claessens and S. Shi (eds) *Science Communication in the World: Practices, Theories and Trends*, Dordrecht: Springer, 65–79.

Rhees, D. J. (1979) *A New Voice for Science: Science Service under Edwin J. Slosson 1921–29*, Master's Thesis, University of North Carolina. Available online: http://scienceservice.si.edu/thesis/ (accessed 3 July 2013).

Schafer, M. (2012) 'Online communication on climate change and climate politics: a literature review', *Wiley Interdisciplinary Reviews – Climate Change*, 3, 6: 527–543.

Schiele, B., Claessens, M. and Shi, S. (eds) (2012) *Science Communication in the World: Practices, Theories and Trends*, Dordrecht: Springer.

Secord, J. A. (2000) *Victorian Sensation: The Extraordinary Publication, Reception, and Secret Authorship of Vestiges of the Natural History of Creation*, Chicago: University of Chicago Press.

Shapin, S. (1990) 'Science and the public', in R. C. Olby, G. N. Cantor, J. R. R. Christie and M. J. S. Hodge (eds) *Companion to the History of Modern Science*, London and New York: Routledge, 990–1007.

Shukla, R. (2005) *India Science Report: Science Education, Human Resources and Public Attitude towards Science and Technology*, New Delhi: National Council of Applied Economic Research. Available online: www.eaber.org/sites/default/files/documents/NCAER_Shukla_2005.pdf (accessed 4 July 2013).

Springston, J. K. and Lariscy, R. A. W. (2003) 'Health as profit: public relations in health communication', in T. L. Thompson, A. Dorsey, K. I. Miller and R. Parrot (eds) *Handbook of Health Communication*, Mahwah, NJ: Lawrence Erlbaum, 537–556.

Survey Research Center, Universty of Michigan, and National Association of Science Writers (1958) *The Public Impact of Science in the Mass Media: A Report on a Nation-Wide Survey for the National Association of Science Writers*, Ann Arbor: University of Michigan.

Trench, B. (2007) 'How the Internet changed science journalism', in M. W. Bauer and M. Bucchi (eds) *Journalism, Science and Society*, New York and London: Routledge, 132–141.

Trumbo, C., Sprecker, K., Dumlao, R., Yun, G. and Duke, S. (2001) 'Use of e-mail and the Web by science writers', *Science Communication*, 22, 4: 347–378.

6
Scientists as public experts
Expectations and responsibilities

Hans Peter Peters

Introduction

Through the mass media we frequently encounter experts who analyse the current economic situation, give advice on health problems, warn against global warming or comment on the chances and risks of technologies. These experts are affiliated with government agencies, companies, hospitals, NGOs, universties or other scientific organisations. This chapter focuses on scientists as public experts and explores their role when they not only talk about their research in public but use their special knowledge to provide orientation and advice relevant to individual or political *problems* of a lay audience.[1] The expert role is particularly challenging for scientists because most are primarily concerned with the creation of knowledge, not its practical application, and scientific knowledge is usually insufficient as the sole base for giving advice. Furthermore, as information sources for journalists, or as contributors to blogs or websites, scientists face the challenge of being relevant and comprehensible for a lay audience. Finally, because of the practical implications of expert advice, scientists as public experts often find themselves intertwined with political and commercial interests – an experience not cherished by all scientists.

It is important to distinguish the role of scientists as *public experts* from other possible roles scientists may take in public. Besides the communication of scientific expertise, i.e. the use of scientific knowledge in the public reconstruction of non-scientific *problems* (e.g. climate change), there are two other types of science communication in which scientists are involved: popularisation of research as the public reconstruction of scientific projects, discoveries, achievements and theories from a science-focused point of view; and meta-discourses about science and technology and the science-society relationship such as disputes about science funding and science policy, and conflicts between science and social values, e.g. in the case of animal experimentation and research with human embryonic stem cells. In popularising their research, they act in a teacher role; in the meta-discourses they take the role of a stakeholder. Of course, in actual communication practice these roles are often not clearly distinguished but mixed.

Looking at scientists as *public experts* combines two interesting aspects that have been extensively studied by scholars: scientists as (policy) advisors (e.g. Jasanoff 1990; Maasen and Weingart 2005) and scientists as public communicators (e.g. Friedman *et al.* 1986; Peters 2013). Both policy advice and public communication challenge scientific norms and present dilemmas to scientists (Sarkki *et al.* 2013). Public communication of scientific expertise often has political impacts and – in response – political organisations and groups with political goals try to govern the production and use of scientific expertise (Stehr 2005). Because of its practical relevance, scientific expertise is attractive for journalism to report. It is frequently covered by the media – including outside, specialised science sections. Surveys suggest that roughly one-third of the oral communications between journalists and researchers in the *hard* sciences focus on general expertise rather than research results; this proportion is considerably higher for researchers from the humanities and social sciences (Peters 2013).

In the case of global climate change, for example, studies in several countries have shown that scientific sources are prominently represented in media coverage (Bell 1994; Wilkins 1993; Peters and Heinrichs 2005). The same has been found for the reporting of biotechnology (Kohring and Matthes 2002; Bauer *et al.* 2001). Food biotechnology, stem cell research, bird flu and nuclear safety are just a few examples of issues in which scientists are actively involved in the construction of a social reality by means of public communication. Because of its public character, that *reality* cannot be ignored by policymakers. Through the public sphere, scientific expertise – transformed by the logic of mass media, however – enters the realm of policymaking (Petersen *et al.* 2010).

The expert role of scientists in public communication

Scholars of science make a clear distinction between scientific knowledge per se and scientific expertise (e.g. Horlick-Jones and De Marchi 1995). Scientific knowledge per se is essentially concerned with the understanding of cause-effect relations. Its concepts and theories tend to be general, abstracting as far as possible from specific situations, observations and experiments. Expert knowledge, in contrast, is concerned with the analysis and solution of practical problems in specific situations. It relates to the provision of concrete advice in specific situations to decision-makers. Successful action undoubtedly requires the anticipation of the consequences of this action and, to this extent, causal knowledge is a necessary component of expertise. But expertise goes beyond scientific understanding; it is concerned with the explanation of practical problems and giving advice to *clients* responsible for the solution of these problems. It should be noted, however, that science is not the exclusive source of expertise and that scientific knowledge, even if it is relevant, is usually not a sufficient source of expertise (Collins and Evans 2002). Two aspects of the basic definition of scientific expertise as advice to decision-makers based on scientific inquiries and knowledge require our attention: its relation to decision-making processes and its provision within a social expert-client relationship.

Scientific knowledge and decision-making

Expertise, by definition, is provided in the context of decision-problems, i.e. problems of orientation, forming opinions and acting. The decision-problems may either be individual problems regarding, for example, understanding a disease and choosing a medical therapy, or policy problems such as regulation of food biotechnology or improvement of coastal protection in order to adapt to rising sea levels. Normative decision models distinguish between decision options (and their expected consequences) and preferences. In these models, the expert as advisor to the decision-maker has three functions (cf. Jungermann and Fischer 2005): making the client's implicit preferences explicit, developing possible decision options, and determining and assessing consequences using the clients' preferences. Setting the preferences, however, remains the privilege and responsibility of the decision-maker.

Applied to politics, this normative decision-model resembles Weber's (1919) concept of the relationship between civil service and politics, according to which the professional civil service, including scientific experts, is responsible for consulting and for implementation of political decisions, and politics for setting the goals and decision-making. Habermas (1971) has called this concept the 'decisionistic model' and distinguishes it from the 'technocratic model' (63), in which science *de facto* makes the decisions and politics accepts them, and a 'pragmatistic model' (66), preferred by Habermas, which proposes a critical interrelationship of science and politics rather than a strict division of functions.

There is a broad consensus among scholars that the decisionistic model does not properly describe the empirical reality of science-policy interactions. However, this model seems to fit best to *mainstream* self-descriptions of science in this context. Empirical analyses of scientific expertise in controversies and policy issues clearly show that the expertise delivered by scientists is not value-free (e.g. Mazur 1985) and that scientific experts in public not only provide knowledge or comment on knowledge claims but also evaluate options, decisions and policies and demand strategies of political or personal action based on their own values (e.g. Nowotny 1980).

Besides the value problem, science as expertise faces the challenge of how to deal with uncertainty. In principle, uncertainty – even controversy – is not a problem for science. Research will continue until the uncertainties are resolved and a consensus is reached. This may take quite a while but in the end – so scientists are convinced – the research process will lead to unambiguous knowledge. In the case of expertise, however, decisions may be urgent and cannot be postponed until all uncertainty is resolved, and this pressing situation may be the rule rather than the exception.

Funtowicz and Ravetz (1991) distinguish several realms of science according to the degree of uncertainty involved. They develop the concept of a *second-order science* where 'facts are uncertain, values in dispute, stakes high and decisions urgent' (ibid: 137). In their view, a central task for experts in this field is the management of uncertainties rather than the provision of unequivocal certainties. Böschen and Wehling (2004) argue that the ways of dealing with *non-knowledge* are important characteristics of scientific epistemic cultures. Explicit and *rational* ways of dealing with uncertainties are thus central elements of scientific expertise.

The media implicitly or explicitly construct the certainty or uncertainty of expert knowledge in several forms: by including or omitting explicit uncertainty reservations when referring to expert knowledge, by challenging expert knowledge with non-scientific knowledge such as common sense, or by quoting several expert sources that either confirm or contradict each other. An important journalistic strategy for communicating scientific uncertainty in the mass media is framing a topic as an expert controversy (Boykoff and Boykoff 2004; Ren *et al.* 2012).

Expert-client relationship and the responsibility of public experts

In psychological expert research, experts are defined by having particularly high competence in a certain field. The counterpart to an expert is the novice or layperson. For sociologists, however, the expert role is defined not only by possession of special knowledge, i.e. knowledge that not everyone can be expected to have, but also by the function of giving advice (or a knowledge-based service) to a client (Peters 1994). The social complement to the *expert* thus is the *client*. In the case of scientists as public experts, the expert-client relationship is largely implied. Sometimes scientists who are quoted in the media may indeed have the intention of giving advice to specific clients such as policymakers, citizens, patients or consumers. More often, however, journalists put scientists in the expert role, relating their knowledge to policy issues or individual problems – sometimes to their surprise.

The task of experts is not only to provide general knowledge and then to leave it to the people to make sense of it; their job is to provide the best available advice in a meaningful way in order to enable rational decision-making. What this means in terms of normative expectations towards experts and their responsibility is well illustrated by the tragic case of the earthquake in L'Aquila, Italy on 6 April 2009 which killed 309 inhabitants of the town. Six days before the earthquake, seven members of the National Commission for the Forecast and Prevention of Major Risks had met, and the Chairman of that meeting (a public official) had issued a reassuring summary of the proceedings in the name of all members. After the reassuring message was disproved by the actual disaster, the seven committee members were charged, tried, convicted and sentenced to six years in prison for manslaughter (Hall 2011; Nosengo 2012; Cartlidge 2013). The judge assumed in his verdict that several victims had returned to their houses as a consequence of the reassuring message of the Commission, while the series of minor seismic shocks preceding the major earthquake might otherwise have motivated them to take precautionary measures.

The judge acknowledged that it is impossible to predict an earthquake, but he held the committee members responsible for 'their complete failure to properly analyse, and to explain, the threat' (Cartlidge 2013). A resident who lost his daughter and wife was quoted as saying: '[e]ither they [the experts] didn't know certain things, which is a problem, or they didn't know how to communicate what they did know, which is also a problem' (Hall 2011: 266). The case is complex and the actual responsibilities of the scientists are controversial (cf. Nosengo 2012). The verdicts and sentences have been appealed. However, the issue relevant for this chapter is not whether the jail sentence is justified but that the public – and the law – holds normative expectations of experts who publish assessments and advice.

Scientists who present themselves as public experts are responsible, first, for acquiring und using the full available knowledge relevant to the problem; second, for making a systematic and comprehensive assessment; and, third, for communicating it in a way that supports the decision-making of members of the public. Publishing a piece of general scientific knowledge, for instance about the causes and effects of earthquakes, is usually not sufficient action for an expert in a risky situation. In addition to general knowledge, familiarity with relevant characteristics of the specific situation – the geology of the region, the buildings, even the psychology of the people – is required as well as good judgment and skills to communicate with the public. Of course, L'Aquila is an extreme case and scientists talking to journalists usually do not adopt a formal role as experts that would make them liable in a legal sense, as the committee members in L'Aquila were judged to be. But the question of responsibility as an expert is always present when scientists provide information that might affect the decisions and behaviour of members of the public with serious consequences.

Scientific authority and trust

Turner (2001: 124) argues that the state accepts the authority of science and expertise by 'requiring that regulations be based on the findings of science or on scientific consensus'. Moreover, public opinion surveys in Europe as well as in the US show that scientific institutions are among the most trusted in society and that, compared with other professions, scientists possess high credibility (e.g. European Commission 2005; National Science Board 2004). In a content analysis of media coverage of climate change we found that scientific sources, on average, were rated more positively than non-scientific sources (Peters and Heinrichs 2005).

Several authors nevertheless diagnose a credibility crisis of scientific expertise (e.g. Horlick-Jones and De Marchi 1995). The high general trust in scientific expertise is indeed challenged in contexts in which scientists are perceived as an interest group or as advocates for a technology (e.g. Peters 1999) or if experts publicly disagree (Rothman 1990). Media reception studies show that recipients of media stories have critical thoughts about experts who are quoted in these stories if the experts express opinions that contradict the recipients' previously held attitudes (Peters 2000). Trust is not unconditionally assigned to science and scientists but rather it is modified and specified depending on the context – taking into account, for example, organisational affiliation (e.g. academia vs. industry) and advocacy relationships of scientists.

Giddens (1991) emphasises the ubiquity of trust in *expert systems* and in the experts that operate them. Trust in science and scientific expertise is still the default, unless – in a few selected domains – its authority is purposely dismantled by distrust and counter-expertise. For example, as strategies against the *risk technocracy*, Beck (1988) recommends challenging the monopoly of technical risk experts in the discussion of safety issues and changing the burden of proof for safety claims. The authority of science is so strong that it usually can be neutralised only by counter-expertise. Critics of established scientific expertise have to refer to science in order to effectively support their claims. In the US, for example, the opponents of Darwin's theory of evolution formulate their alternative, motivated largely by religious beliefs, as a quasi-scientific theory in order to increase its persuasiveness (Park 2001).

Many public controversies refer to science and technology, and scientists participate as advocates on both sides of the controversy (e.g. Mazur 1981; Frankena 1992; Nelkin 1992). In controversies over science and technology, science can be involved in different ways: scientific research or science-based technologies (e.g. nuclear power or genetic engineering) may be the controversial issue, or problem definition and the assessment of pros and cons of policy options may depend on scientific expertise. In both cases the control over the definition (Who is a legitimate expert?) and content of public expertise (Which experts are prepared to *testify* in the media?) is crucial for the opening and closure of controversies. Nowotny (1980) and others have found that the readiness of scientists to serve as public experts partly depends on their position in the controversy. Experts challenging the establishment side are usually more prepared to participate in public debates than those defending a technology or a government policy. Experts are therefore an important power resource for NGOs. As legitimate journalistic sources, they help in securing access to the public sphere, they increase the rationality of the claims, they provide legitimacy drawing from scientific authority and they contribute to the construction of a social reality that favours the acceptance of their clients' claims. In order to have better access to and control over the content of expertise, NGOs have developed an infrastructure of research institutes and of networks of sympathising experts providing *critical expertise* or *counter expertise*.

A common problem of all forms of science communication is relating it to the relevance structure of the audience, i.e. giving the audience a good reason to *listen* to the communication offer. The esoteric character of modern science, its incomprehensibility and detachment from everyday culture, makes it particularly difficult to connect scientific knowledge to everyday discourses and common sense. The 'mad scientist' scheme (e.g. Haynes 2003), and scientific 'miracles' and practical applications of science (Fahnestock 1986) are examples of semantic structures used by journalism to construct connections between science and the everyday world. Scientific expertise relates to the everyday world particularly well by addressing well-known and relevant problems.

The relatively easy solution of the problem of relevance makes the reporting of scientific expertise attractive for journalism. Furthermore, unlike popularisation the reporting about scientific expertise does not only address a special science-attentive audience but also larger audiences interested in the practical problems of, for example, health, environment and technical risk. Scientific expertise is, thus, not confined to science sections or science programmes but often included in the general news coverage.

Scientific expertise and other forms of knowledge

By definition, scientific expertise is scientific knowledge applied to the understanding and solution of practical problems. Sometimes these problems are only known because of science. Without science, for example, we would not know about the ozone hole or global climate change. More often, however, the problems are obvious or already well known by experience, and more or less successful strategies to cope with them are already implemented. When scientists offer their expertise, their knowledge frequently meets competing knowledge sources: everyday knowledge, special knowledge based on practical experience, or traditional knowledge stemming, for example, from religion, folk wisdom or indigenous culture.

The competition of scientific knowledge with other knowledge forms causes two kinds of problems. First, prior knowledge about a subject may hinder the understanding and acceptance of new knowledge, leading to problems of explanation. Rowan (1999) has analysed such cognitive problems in the acquisition of scientific knowledge by lay people. As explanatory strategy to overcome that problem, she recommends making the contradiction between everyday knowledge and scientific knowledge explicit. Second, the question arises which of the competing knowledge forms – for example, scientific knowledge and everyday knowledge – is more valid and suitable to solve the problem. Like many others, Wynne (1996) has challenged the assumption that scientific knowledge is per se superior to *local knowledge* based on experience.

It is quite obvious that the solution of practical problems requires several kinds of knowledge. Even if valid and relevant scientific theories are available, the conditions and constraints of the specific case have to be known in order to *calculate* the conditions and effects of interventions. Intimate knowledge about available resources, legal, political and psychological constraints and implementation barriers, for example, are necessary to devise efficient action strategies. A climate researcher in a media interview recommending the reduction of energy consumption by 50 per cent in the next ten years will immediately provoke the journalist's next question: 'How?'.

In complex problems we cannot expect a single expert to be competent in all aspects. The need to use different sources of knowledge and to integrate them in order to solve a practical problem is obvious. This demand for integrated expertise has consequences for knowledge management, including journalistic strategies of inquiry, but also for knowledge production. In the current debate about research strategies to deal with, for example, ecological problems, there is an almost ubiquitous call of more *interdisciplinary* or *transdisciplinary* research, i.e. of research that exceeds the boundaries of scientific disciplines or even the classical framework of science at large (e.g. Somerville and Rapport 2003). Gibbons *et al.* (1994) claim the emergence of a new way of knowledge production (mode 2) which involves a broader spectrum of actors such as practitioners and users of knowledge, takes place outside academic institutions and leads to more contextualised and *socially robust* knowledge (Nowotny *et al.* 2001). Grunwald (2003) even discusses the possibility of *transsubjective* normative advice as part of scientific expertise.

Scientists as experts in the mass media

Scientific expertise reaches the public through many channels. For example, scientists may give public talks about climate change or Alzheimer's disease, or contribute to exhibitions or other public events focusing on these topics. But under the conditions of the *media society*, the mass media – television, newspapers (online or print), websites and blogs – are particularly formative for the public sphere. Additionally, scientists may have their own personal websites or blogs, or they may contribute to information resources provided by universities, government agencies or NGOs. Professional health communication platforms provided by research institutes, government agencies or pharmaceutical companies play an important role as a source of medical expert information for patients and their carers, with implications for the relationship of doctors and their patients. YouTube and other video-sharing platforms are also full of videos presenting expertise on problems such as climate change or prevention of heart attack – provided by scientific experts and those who claim to be experts. Online resources are used by members of the public who actively search specific information, using search engines or following links distributed in social networks such as Twitter or Facebook.

Despite the growing importance of direct communication by scientists, scientific organisations and scientific media, journalism is still an important public mediator of scientific expertise. The role of the journalistic mass media in the communication of science and scientific expertise has received much attention. Besides the media coverage itself, scholars have studied the role of scientists as information sources for journalists. In a case study of reporting of the marijuana controversy, Shepherd (1981) found that the media did not primarily quote the most relevant and experienced researchers as experts but rather health administrators and highly prominent scientists, regardless of their specific field of expertise. From the perspective of journalists, it is not research productivity but other qualities that define a good public expert. Practitioners as well as senior scientists with overview knowledge and general experience may be better suited than the actual researchers to relate research to decision-problems, to integrate different knowledge sources and to provide contextualised expertise. Rothman (1990) concludes from several case studies of expert controversies that the journalistic selection of experts is biased: experts representing minority positions are usually over-represented in the coverage. Kepplinger *et al.* (1991) argue that media tend to select expert sources that support their editorial policies. Goodell (1977: 4) concludes that the media focus on few *visible scientists* and select scientific sources 'not for discoveries, for popularising, or for leading the scientific community, but for activities in the tumultuous world of politics and controversy'.

Selecting expert sources for journalists is a complex process in which scientific productivity and reputation is but one factor. The main journalistic criterion in the selection of sources is whether a source makes a good story or improves a story. What makes a good story, however, differs between media, between sections and programmes, and between different topics. Relevance is one of the more important factors that influence the likelihood of scientists appearing in the media: a scientific source must be able to comment on something *relevant* for the audience. In some cases, the relevance is quite obvious, as, for example, with medical therapies, environmental risks or government decisions. The *news values* concept (e.g. Badenschier and Wormer 2012) describes other criteria journalists use to assess public relevance, among them visibility, accessibility and media appropriateness.

Scientists become visible to journalists by their involvement in events and debates outside science (e.g. as members of a policy advisory board or as authors of expert opinions commissioned by the government). They are more visible if they publish in journals (e.g. *Science* or *Nature*) or talk at conferences (e.g. AAAS [American Association for the Advancement of Science] annual meetings) that are regularly monitored by journalists. Furthermore, the public relations efforts of scientific organisations, journals, associations and congresses strongly influence the visibility of scientists. Finally, prior media coverage makes scientists more visible. This leads to the feedback loops of media attention resulting in the *visible scientists* described by Goodell (1977).

Journalists work with limited resources and in a narrow time frame. The anticipated effort required to deal with a scientist – thus, the accessibility – is an important selection criterion. Journalists prefer scientists who are able and willing to speak crisply and concisely, to answer the questions asked and to explain complicated matters using comparisons and metaphors, and those who draw bold conclusions. Furthermore, journalists prefer scientists with high organisational rank and public reputation and who are, in that sense, media appropriate.

Interactions of scientific experts and journalists

There are several reasons to expect that the relationship between scientists and journalists is tense and the interactions difficult. A number of surveys have studied scientists' attitudes towards the media and their contacts with journalists (e.g. Dunwoody and Ryan 1985; Hansen and Dickinson 1992; Peters *et al.* 2008; Bentley and Kyvik 2011; Kreimer *et al.* 2011). In two studies in which scientists and journalists were surveyed with matching questionnaires, we found several systematic differences in the mutual expectations of each group (Peters 1995; Peters and Heinrichs 2005). These revolve around conceptions of communication norms, of the model of journalism and of control of communication.

Scientists tend to apply scientific communication norms also to public communication. They prefer to focus on knowledge in their specialist field and – compared to journalists – they like a serious, matter-of-fact, cautious and educational style of communication. Those they interact with in journalism do not completely disagree but look for overview knowledge, prefer clear messages, evaluative comments and an entertaining style.

Scientists favour a kind of *service model* of journalism, expecting journalists to help them promote scientific goals and interests. Based on their professional norms, journalists, at least verbally, insist on distance from the objects they report, on their independence and on a watchdog perspective.

A clear-cut disagreement exists on the issue of control: who should control the communication with the public and the media content? Similar studies conducted in Britain (Gunter *et al.* 1999) and Taiwan (Chen 2011) replicated this finding. Journalists consider themselves as responsible authors and scientists as their *sources*, i.e. as a resource for their task of writing a story. According to their own norms, journalists owe sources a fair treatment (e.g. correct quotation) but nothing more. In particular, they are very critical of demands from sources that may be viewed as censorship. Scientists, however, think that they are the real authors and should control the communication process because they are the originators of the message to be conveyed to the public. In accordance with their service model of journalism, they tend to assign journalists a role as disseminator only.

Despite the disagreement in some aspects, the general conclusion from the two German surveys (Peters 1995; Peters and Heinrichs 2005) is that of a surprisingly strong co-orientation of scientific experts and journalists. In many respects the expectations of both groups are actually congruent and in most other respects there are moderate discrepancies. It is only in relation to the control of communication that there is outright disagreement. Scientists and journalists seem to possess effective strategies to overcome the problems arising from the discrepant expectations. Even more, the surveys show that scientists, to a large degree, not only accept the journalistic treatment of science but anticipate it in their statements. Surveys in several countries show surprisingly high satisfaction of scientists with their own encounters with the media (Peters *et al.* 2008; Peters 2013).

Public (re-)construction of scientific expertise

Public communication of science cannot be understood as *translation*. Translation would require a structural equivalence of source and target language and a shared reality serving as background for making sense of information. Neither is there an equivalence of scientific and everyday language, nor a shared reality. The worlds of modern science are esoteric and rather inaccessible by everyday reasoning.

Scientists and the media therefore have to construct public images of the esoteric worlds of science and their events using terms, metaphors, comparisons and concepts from the everyday world. With respect to public expertise, this means that while expert advice itself can be expressed in everyday language and refers to the objects and events of the everyday world, the scientific justification of that advice will often be incomprehensible. The validity of scientific expert advice can therefore not be proven to clients but only made plausible. In the end clients must trust their experts. To inform trust, however, the social context of expertise (e.g. interests, neutrality, independence) becomes crucial and a legitimate field of journalism inquiry and reporting (Kohring 2004).

The analysis of how scientific expertise in the media is constructed is an urgent field for research. Some general features of public constructs of science and scientific expertise can be mentioned, however. The neglect of scientific detail and accuracy, demonstrated by numerous studies (e.g. Singer 1990; Bell 1994), is not an indicator of unprofessional journalism but, on the contrary, the consequence of journalistic professionalism. Journalists do not adopt the quality criteria of science, such as accuracy, but follow their own criteria (Salomone *et al.* 1990). Scientific errors and inaccuracies are only the *tip of the iceberg* of semantic discrepancies between scientific and public constructs of science.

Dunwoody (1992) and others use the concept of *story frames* to explain which kind of information journalists include or exclude from their stories. These frames guide the organisation of information in the story, but also journalistic inquiry. Two-thirds of the journalists interviewed in the survey of climate change experts and journalists by Peters and Heinrichs (2005) said that they had the outline of the story in mind when contacting a scientist, and one-third of the scientists said that they had the impression the interviewing journalist wanted to hear something particular. Journalists often have pronounced expectations not only with respect to the topic they ask their sources to comment about but also regarding the stances their sources take.

Comparing popular science stories written by scientists and by journalists about the same research, Fahnestock (1986) found several systematic differences that illustrate some of the journalistic rules of constructing public science. She found, for example, that journalists focused more than scientists on explaining the purpose of research, and liked to focus on practical applications. She also found that journalists did not focus as much on the scientific results of research as the scientists but rather looked at its consequences for evaluation and action. Peters and Heinrichs (2005) analysed journalists' use of information provided by climate change experts in media interviews and concluded that journalists expected problem-oriented, interpreted knowledge rather than pure scientific research results. Journalists contextualised information received from scientists by relating it to concrete events or problems. Furthermore, as Fahnestock (1986) observed, journalists omitted reservations and thus let conclusions appear less uncertain and more general than the scientists had intended. Dealing with scientific uncertainty is a particular challenge for journalism and especially important in the coverage of risk issues (Friedman *et al.* 1999; Singer and Endreny 1993).

In many cases journalism not only makes scientific expertise public and emphasises the expertise-based character of research, but actively contributes to the creation of *public expertise*. Journalists confront their scientific sources with specific events (e.g. weather anomalies), asking them to explain these, and they urge them to comment on policy problems. Contextualising scientific knowledge and complementing it with non-scientific forms of knowledge in such ways may be an important task of (science) journalism. Discussing two different forms of rationality and their respective advantages and disadvantages for problem-solving, Spinner (1988) calls journalists *agents of occasional reason* and assigns them the task of challenging principles-based scientific-technical rationality with *occasional rationality*, i.e. with a cunning orientation to specific, local and temporary aspects of events and situations.

Scientists may be increasingly ready to comply with the expectation to link their knowledge to non-scientific issues. For example, many climate researchers accept a political role and go beyond their facts in public communication with the goal of urging governments to promote a policy of climate change mitigation (Peters and Heinrichs 2005). The strong policy orientation of many climate researchers may imply risks for the credibility of climate research, however. By means of hermeneutical analysis of journalistic coverage of epidemiology, Jung (2012) has demonstrated that journalists often challenge the credibility of science when portraying it as closely linked to politics or economy.

Concluding remarks

As compared with pure scientific knowledge, expertise is defined by its reference to social *problems*, to decision-making and to action. Being framed as *public experts* thus implies the expectation that scientists apply their knowledge to the explanation and solution of non-scientific problems. On the one hand, providing expertise is rewarding for scientists because, in contrast to esoteric scientific discoveries or theories, expertise usually connects rather easily to what the media and their audiences consider relevant. On the other hand, being an expert means crossing the boundary of science, entering society as an actor, and exposing oneself to internal and external criticism. As experts, scientists do not possess a monopoly of relevant knowledge; values and interests will come into play, public controversies may evolve and the credibility of science may be challenged.

While some scientists are prepared to become involved in issues of policy or public health, other scientists may be reluctant to enter the 'tumultuous world of politics and controversy' (Goodell 1977: 4). Journalists tend to focus on the connections of scientific knowledge with the non-scientific world, however, and will often push scientists in interviews to the limit (or even beyond) of what they are prepared to offer about the practical implications of their knowledge. Journalism, thus, has an important function not only in the public communication of scientific expertise but also in its creation.

Key questions

- What are the specific features of scientists' role as experts, compared with other public roles they may take?
- How are public expectations towards experts, including scientific experts, changing in contemporary society?
- What roles does journalism play in constructing scientific expertise for society?
- What types of expertise, other than scientific, come into play in public discussion of science-based issues?

Note

1 The revision of this chapter has included reference to more recent relevant research, the addition of a recent case study concerning the fate of experts who advised on the L'Aquila earthquake and expansion of the discussion on use of Internet media by experts.

References

Badenschier, F. and Wormer, H. (2012) 'Issue selection in science journalism: towards a special theory of news values for science news?', in S. Rödder, M. Franzen and P. Weingart (eds) *The Sciences' Media Connection – Public Communication and its Repercussions,* Dordrecht: Springer, 59–85.

Bauer, M., Kohring, M., Allansdottir, A. and Gutteling, J. (2001) 'The dramatisation of biotechnology in elite mass media', in G. Gaskell and M.W. Bauer (eds) *Biotechnology 1996–2000: The Years of Controversy,* London: Science Museum, 35–52.

Beck, U. (1988) *Gegengifte. Die organisierte Unverantwortlichkeit,* Frankfurt am Main: Suhrkamp.

Bell, A. (1994) 'Media (mis)communication on the science of climate change', *Public Understanding of Science,* 3, 3: 259–275.

Bentley, P. and Kyvik, S. (2011) 'Academic staff and public communication: a survey of popular science publishing across 13 countries', *Public Understanding of Science,* 20, 1: 48–63.

Böschen, S. and Wehling, P. (2004) *Wissenschaft zwischen Folgenverantwortung und Nichtwissen,* Wiesbaden: Verlag für Sozialwissenschaften.

Boykoff, M.T. and Boykoff, J. M. (2004) 'Balance as bias: global warming and the US prestige press', *Global Environmental Change – Human and Policy Dimensions,* 14, 2: 125–136.

Cartlidge, E. (2013) 'Judge in L'Aquila earthquake trial explains his verdict', *Science News,* 21 January; online at http://news.sciencemag.org/earth/2013/01/judge-laquila-earthquake-trial-explains-his-verdict.

Chen, Y. -N. K. (2011) 'An explorative study on the differences of the two professions' perceptions of science news', *Chinese Journal of Communication Research,* 19: 147–187.

Collins, H. M. and Evans, R. (2002) 'The third wave of science studies: studies of expertise and experience', *Social Studies of Science,* 32, 2: 235–296.

Dunwoody, S. (1992) 'The media and public perceptions of risk: how journalists frame risk stories', in D.W. Bromley and K. Segerson (eds) *The Social Response to Environmental Risk,* Boston: Kluwer, 75100.

Dunwoody, S. and Ryan, M. (1985) 'Scientific barriers to the popularization of science in the mass media', *Journal of Communication,* 35, 1: 26–42.

European Commission (2005) *Social Values, Science and Technology: Eurobarometer 2005,* Brussels: European Commission.

Fahnestock, J. (1986) 'Accommodating science: the rhetorical life of scientific facts', *Written Communication,* 3, 3: 275–296.

Frankena, F. (1992) *Strategies of Expertise in Technical Controversies: A Study of Wood Energy Development,* Bethlehem, PA: Lehigh University Press.

Friedman, S. M., Dunwoody, S. and Rogers, C. L. (eds) (1986) *Scientists and Journalists: Reporting Science as News,* New York and London: The Free Press and Macmillan.

Friedman, S. M., Dunwoody, S. and Rogers, C. L. (eds.) (1999) *Communicating Uncertainty,* Mahwah, NJ: Lawrence Erlbaum.

Funtowicz, S. O. and Ravetz, J. R. (1991) 'A new scientific methodology for global environmental issues', in R. Costanza (ed.) *Ecological Economics: The Science and Management of Sustainability,* New York: Columbia University Press, 137–152.

Gibbons, M., Limoges, C., Nowotny, H., Schwartzman, S., Scott P. and Trow, M. (1994) *The New Production of Knowledge,* London: Sage.

Giddens, A. (1991) *The Consequences of Modernity,* Cambridge: Polity Press.

Goodell, R. (1977) *The Visible Scientists,* Boston: Little, Brown and Company.

Grunwald, A. (2003) 'Methodological reconstruction of ethical advices' in G. Bechmann and I. Hronszky (eds) *Expertise and Its Interfaces: The Tense Relationship of Science and Politics,* Berlin: edition sigma, 103–124.

Gunter, B., Kinderlerer, J. and Beyleveld, D. (1999) 'The media and public understanding of biotechnology: a survey of scientists and journalists', *Science Communication,* 20, 4: 373–394.

Habermas, J. (1971) 'The scientization of politics and public opinion', in J. Habermas, *Toward a Rational Society,* London: Heinemann, 60–80.

Hall, S. S. (2011) 'Scientists on trial: at fault?', *Nature,* 477: 264–269.

Hansen, A. and Dickinson, R. (1992) 'Science coverage in the British mass media: media output and source input', *Communications*, 17, 3: 365–377.

Haynes, R. (2003) 'From alchemy to artificial intelligence: stereotypes of the scientist in Western literature', *Public Understanding of Science* 12, 3: 243253.

Horlick-Jones, T. and De Marchi, B. (1995) 'The crisis of scientific expertise in fin de siècle Europe', *Science and Public Policy*, 22, 3: 139–145.

Jasanoff, S. (1990) *The Fifth Branch: Science Advisors as Policymakers*, Cambridge, MA: Harvard University Press.

Jung, A. (2012) 'Medialization and credibility: paradoxical effect or (re)-stabilization of boundaries? Epidemiology and stem cell research in the press', in S. Rödder, M. Franzen and P. Weingart (eds) *The Sciences' Media Connection: Public Communication and its Repercussions*, Dordrecht: Springer, 107–130.

Jungermann, H. and Fischer, K. (2005) 'Using expertise and experience for giving and taking advice', in T. Betsch and S. Haberstroh (eds) *The Routines of Decision Making*, Mahwah, NJ: Lawrence Erlbaum, 157–173.

Kepplinger, H. M., Brosius, H. -B. and Staab, J. F. (1991) 'Instrumental actualization: a theory of mediated conflicts', *European Journal of Communication*, 6, 3: 263–290.

Kohring, M. (2004) *Vertrauen in Journalismus. Theorie und Empirie*, Konstanz: UVK.

Kohring, M. and Matthes, J. (2002) 'The face(t)s of biotech in the nineties: how the German press framed modern biotechnology', *Public Understanding of Science*, 11, 2: 143–154.

Kreimer, P., Levin, L. and Jensen, P. (2011) 'Popularization by Argentine researchers: the activities and motivations of CONICET scientists', *Public Understanding of Science*, 20, 1: 37–47.

Maasen, S. and Weingart, P. (eds) (2005) *Democratization of Expertise? Exploring Novel Forms of Scientific Advice in Political Decision-Making*, Dordrecht: Springer.

Mazur, A. (1981) *The Dynamics of Technical Controversy*, Washington, DC: Communications Press.

Mazur, A. (1985) 'Bias in risk-benefit analysis', *Technology in Society*, 7, 1: 25–30.

National Science Board (2004) *Science and Engineering Indicators 2004*, Arlington, VA: National Science Foundation.

Nelkin, D. (ed.) (1992) *Controversy: Politics of Technical Decisions*, third edition, Newbury Park, CA: Sage.

Nosengo, N. (2012) 'Wiretap revelation could aid Italian seismologists' defence', *Nature News Blog*, 25 January; online at http://blogs.nature.com/news/2012/01/wiretap-revelation-could-aid-italian-seismologists-defence.html

Nowotny, H. (1980) 'Experten in einem Partizipationsversuch. Die Österreichische Kernenergiedebatte', *Soziale Welt*, 31, 4: 442–458.

Nowotny, H., Scott, P. and Gibbons, M. (2001) *Re-Thinking Science: Knowledge and the Public in an Age of Uncertainty*, Cambridge: Polity Press.

Park, H. -J. (2001) 'The creation–evolution debate: carving creationism in the public mind', *Public Understanding of Science*, 10, 2: 173–186.

Peters, H. P. (1994) 'Wissenschaftliche Experten in der öffentlichen Kommunikation über Technik, Umwelt und Risiken', in F. Neidhardt (ed.) *Öffentlichkeit, öffentliche Meinung, soziale Bewegungen*, Opladen: Westdeutscher Verlag, 163–190.

Peters, H. P. (1995) 'The interaction of journalists and scientific experts: co-operation and conflict between two professional cultures', *Media, Culture & Society*, 17, 1: 31–48.

Peters, H. P. (1999) 'Das Bedürfnis nach Kontrolle der Gentechnik und das Vertrauen in wissenschaftliche Experten', in J. Hampel and O. Renn (eds) *Gentechnik in der Öffentlichkeit*, Frankfurt am Main: Campus, 225–245.

Peters, H. P. (2000) 'The committed are hard to persuade: recipients' thoughts during exposure to newspaper and TV stories on genetic engineering and their effect on attitudes', *New Genetics & Society*, 19, 3: 367–383.

Peters, H. P. (2013) 'Gap between science and media revisited: scientists as public communicators', *Proceedings of the National Academy of Sciences*, 110, Supplement 3: 14102–14109.

Peters, H. P. and Heinrichs, H. (2005) *Öffentliche Kommunikation über Klimawandel und Sturmflutrisiken. Bedeutungskonstruktion durch Experten, Journalisten und Bürger*, Jülich: Forschungszentrum.

Peters, H. P., Brossard, D., de Cheveigné, S., Dunwoody, S., Kallfass, M., Miller S. and Tsuchida, S. (2008) 'Science communication: interactions with the mass media', *Science*, 321, 5886: 204–205.

Petersen, I., Heinrichs, H. and Peters, H. P. (2010) 'Mass-mediated expertise as informal policy advice', *Science, Technology & Human Values*, 35, 6: 865–887.

Ren, J., Peters, H. P., Allgaier, J. and Lo, Y. -Y. (2012) 'Similar challenges but different responses: media coverage of measles vaccination in the UK and China', *Public Understanding of Science*, published online before print 10 May; doi:10.1177/0963662512445012.

Rothman, S. (1990) 'Journalists, broadcasters, scientific experts and public opinion', *Minerva*, 28, 2: 117–133.

Rowan, K. E. (1999) 'Effective explanation of uncertain and complex science', in S. M. Friedman, S. Dunwoody and C. L. Rogers (eds.) *Communicating Uncertainty*, Mahwah, NJ: Lawrence Erlbaum, 201–223.

Salomone, K. L., Greenberg, M. R., Sandman, P. M. and Sachsman, D. B. (1990) 'A question of quality: how journalists and news sources evaluate coverage of environmental risk', *Journal of Communication*, 40, 4: 117–131.

Sarkki, S., Niemela, J., Tinch, R., van den Hove, S., Watt, A. and Young, J. (2013) 'Balancing credibility, relevance and legitimacy: a critical assessment of trade-offs in science-policy interfaces', *Science and Public Policy*, published online 28 August; doi: 10.1093/scipol/sct046.

Shepherd, R. G. (1981) 'Selectivity of sources: reporting the marijuana controversy', *Journal of Communication*, 31, 2: 129–137.

Singer, E. (1990) 'A question of accuracy: how journalists and scientists report research on hazards', *Journal of Communication* 40, 4: 102–116.

Singer, E. and Endreny, P. M. (1993) *Reporting on Risk: How the Mass Media Portray Accidents, Diseases, Disasters, and Other Hazards*, New York: Russell Sage Foundation.

Somerville, M. A. and Rapport, D. J. (2003) *Transdisciplinarity: Recreating Integrated Knowledge*, Oxford: EOLSS.

Spinner, H. F. (1988) 'Wissensorientierter Journalismus. Der Journalist als Agent der Gelegenheitsvernunft', in L. Erbring (ed.) *Medien ohne Moral?*, Berlin: Argon, 238–266.

Stehr, N. (2005) *Knowledge Politics: Governing the Consequences of Science and Technology*, Boulder, CO: Paradigm.

Turner, S. (2001) 'What is the problem with experts?', *Social Studies of Science*, 31, 1: 123–149.

Weber, M. (1919) *Politik als Beruf*, München: Duncker & Humblot.

Wilkins, L. (1993) 'Between facts and values: print media coverage of the greenhouse effect, 1987–1990', *Public Understanding of Science*, 2, 1: 71–84.

Wynne, B. (1996) 'May the sheep safely graze? A reflexive view of the expert-lay knowledge divide', in S. Lash, B. Szerszynski and B. Wynne (eds) *Risk, Environment and Modernity: Towards a New Ecology*, London: Sage, 44–83.

7
Scientists in popular culture
The making of celebrities

Declan Fahy and Bruce V. Lewenstein

Introduction

Four days before Stephen Hawking's 70th birthday on 8 January 2012, *New Scientist* published an exclusive interview with the man it called 'one of the world's greatest physicists'. Asked about the most exciting development in physics during his lifetime, Hawking said it was the confirmation of the Big Bang. Asked to name his biggest scientific blunder, he said it was his mistaken view that black holes destroyed the information they swallowed. Asked what he would do if he were a young physicist again, he said he would formulate a new idea that would open a novel field. Asked, finally, what he thought about most during the day, he said: 'Women. They are a complete mystery' (*New Scientist* 2012).

News outlets worldwide – including CBS news, *The Guardian*, *The Telegraph*, News Corp's Australian website, India's *The Hindu,* and *The Huffington Post* – angled their reports around this final answer, catalysing days of media coverage up to a special symposium held on Hawking's birthday at the University of Cambridge to celebrate his life and work. London's Science Museum (2012) marked the occasion with a display about Hawking, which featured previously unseen photographs of him, handwritten research notes, a little-seen 1978 portrait by artist David Hockney, the annotated script of his 1999 appearance on *The Simpsons* and the blue flight suit he wore for a 2007 zero gravity flight. A new biography, *Stephen Hawking: An Unfettered Mind* was published in the same month. The events and publications illustrated the prominent cultural status of Hawking, whom scholars have labeled 'one of the few scientists ever to have become a media celebrity' (Coles 2000: 3) and a scientist who became 'part of popular culture' (Leane 2007: 35).

Yet scientists have an enduring presence in popular culture – not just fictional scientists such as Dr Victor Frankenstein, but increasingly real scientists who become public celebrities: Marie Curie, Albert Einstein, Margaret Mead, Carl Sagan, M. S. Swaminathan, Rita Levi-Montalcini, C. N. R. Rao and many others. Such scientists are frequent contributors to public discourse, are respected advisors to national governments and have instant recognisability to broad public audiences. Although they usually receive their first public notice because of their scientific achievements, some of them actively cultivate a public presence. This chapter explores how the presence of scientists in popular culture has increased over the last century (in part through those active efforts to participate in public life) as well as the effect of popular culture on the scientists themselves and their own image as scientists. A key aspect of understanding scientists in contemporary popular culture will be to understand the functions of celebrity in science.

Historical perspectives

It is not new that scientists are present in public culture. By the mid-1700s, a children's book about Isaac Newton and his work was available (Secord 1985). Thomas Edison was an inventor, engineer and industrialist but his public image emphasised his scientific achievements (Kline 1995; Pretzer 1989). In Britain, Edison's competitor Sebastian Ferranti was portrayed as a colossus standing astride London (Hughes 1983). Tesla, Marconi, and other inventors at the beginning of the twentieth century were widely known public figures (Kline 1992; Douglas 1987). In the 1920s and 1930s, Paul de Kruif's *Microbe Hunters* (1926) became a worldwide bestseller, portraying scientists as heroes creating the modern world (Henig 2002).

But in the second half of the twentieth century, the growing relevance of science in public life led to new relationships between scientists and popular culture. That growth has been best documented in the United States and Britain, on which much of this chapter is based; but, where possible, evidence from other countries will highlight similar phenomena and the potential for new research on the relationship of scientists and popular culture. Our goal is to identify the features that move scientists past being publicly *visible* – the primary analytical category previously identified (Goodell 1977; Bucchi 2014) – and into a broader role as full participants in modern celebrity culture.

Since the early 1800s, science's presence in the public sphere has fluctuated but, beginning in the 1970s, the strength of media and other cultural presence has grown substantially, reaching, by the mid-2000s, a level statistically two standard deviations above the mean for the previous 200 years (Bauer 2012). In the United States, in the 1970s and 1980s, science sections were created in several newspapers, multiple glossy popular science magazines were launched and new weekly television series devoted to science were inaugurated (Lewenstein 1987). The number of science books on the *New York Times* bestseller lists moved through an inflection point in the mid-1970s, going from rarely having *more* than ten new titles per year to rarely having *fewer* than ten new titles per year (Lewenstein 2009; see also Bell and Turney in this volume). Similar trends appear to hold in Britain (Bauer *et al.* 1995), Italy (Bucchi and Mazzolini 2003) and in Bulgaria, while it was under the influence of the Soviet Union from 1945 to 1989 (Bauer *et al.* 2006). These trends, along with public opinion data documented elsewhere in this volume (see Bauer and Falade), suggest that science has had a growing role as a social authority in popular culture throughout the twentieth century, with an especially evident increase near the end of the century.

Along with that growth in the presence of science came a growing presence of *scientists* in general culture – the humans who create reliable knowledge about the natural world, the people who constitute the institutions of science. Media reporters and producers had long understood that focusing on a particular scientist was a powerful way to humanise science, to provide the dramatic core for good stories that reading, listening or watching audiences would want to follow (LaFollette 1990, 2008 and 2013). Although Einstein's theory of relativity had been well covered by newspapers and magazines after its confirmation in 1919, the process of making him a celebrity involved a complex interplay that highlighted his characteristics as a person – including his political views – even more than his science (Missner 1985). Similarly, de Kruif's *Microbe Hunters* focused on scientists' personal characteristics, describing them 'in a style that was breathless in its admiration and at the same time pitiless in its unmasking of these heroes as neither more nor less than complicated and flawed human beings' (Henig 2002).

In the years after World War II, an increasing number of *visible scientists* became figures in the public presence of science, such as Glenn Seaborg and Margaret Mead. Seaborg had discovered many of the first transuranium elements and helped create the modern periodic table; he received the Nobel Prize in Chemistry in 1951. As a member of the scientific establishment, he became known to the general public through his work on a committee dealing with university athletics, then as Chancellor of the University of California Berkeley, and finally as Chairman of the Atomic Energy Commission. Mead had come to public notice with her 1928 book *Coming of Age in Samoa,* and quickly became a prominent anthropologist. Unlike Seaborg, she was not a government insider; she used her prominence to comment on public issues of the day, especially issues of sex and sex roles, drawing on her credibility as an anthropologist whose work spoke directly to these topics. They and other visible scientists rose to popular culture prominence because they had what Goodell calls five salient *media-orientated* characteristics, personal and professional features that conformed to the values and requirements of popular media: these scientists were articulate, controversial, had a credible reputation, a colourful image and worked on hot topics. For Goodell (1977), the visible scientists she examined, including Mead and entomologist Paul Ehrlich, became significant cultural figures also because they used popular media to advocate for political issues, illustrating the increased importance of the mass media as a venue in which scientists could raise and discuss scientific issues, potentially influencing the formation of public opinion around science and the development of science policy.

Not all of the newly prominent scientists were formally trained as scientists or held traditional academic or research positions. For example, Jacques Cousteau was an inventor and explorer. A former French naval pilot, he had helped develop the Aqua-Lung breathing apparatus that was the first practical implementation of modern scuba diving. His goal was to spend more time underwater so that he could take pictures and create films that would reveal the wonders of undersea life. He wrote magazine feature stories and had his first cover story in the *National Geographic* in 1953, when he also released his first book, *Silent World*, which was on the *New York Times* bestseller list for 30 weeks. A film of the same name, co-directed by Louis Malle, was released to worldwide acclaim in 1956, winning the Palme d'Or and Academy Award for Best Documentary. Combining technological development with underwater exploration, underwater archaeology, and popular writing and film-making, Cousteau (called by journalists a 'man-fish') was among the first impresarios of science (Dugan 1948: 30).

Another variety was the scientist who became a prominent populariser, but not necessarily about his or her own work. Among the most prominent of these was Jacob Bronowski, who began his career as a mathematician. During World War II and for some years afterwards, he worked on statistical issues and in operations research. But he also began publishing in literary magazines and by the 1950s was appearing regularly on the British television programme *Brains Trust*, where he became known for the depth and breadth of his knowledge. He wrote philosophical works about science, including the widely read *Science and Human Values* (1956). Bronowski's *Ascent of Man* series was broadcast by the BBC in 1972 and 1973, and rebroadcast in the United States in 1974. The scripts of the shows were published as a book and spent more than 40 weeks on the *New York Times* bestseller list (Moss 1997).

The success of *Ascent of Man* led the US Public Broadcasting System to commission an even broader exploration of the universe from astronomer Carl Sagan, who had already won a Pulitzer Prize for his bestselling *Dragons of Eden* (1977). Sagan had 'the good fortune to be an astronomer at a time when space exploration began' (Bucchi and Trench 2008: 1) and *Ascent of Man* producer Adrian Malone, hired for the new series, deliberately set out to create a 'star' out of the astronomer (Davidson 1999: 321). He succeeded, perhaps too well: much of the criticism of the 1980 series, *Cosmos*, focused on Sagan's immense ego. The show was a big success and the accompanying book (this time written separately, not simply reprinting the scripts) was on the *New York Times* bestseller list for more than 70 weeks (Davidson 1999; Poundstone 1999). Sagan demonstrated an increasingly important phenomenon: the photogenic, media-savvy scientist as object of public worship.

Joy Adamson's *Born Free* (1960), describing her life with a lion cub, was on the *New York Times* bestseller list for 34 weeks before becoming a 1966 feature film that contributed directly to the wildlife conservation movement (though Adamson was, like Cousteau, not formally a scientist). These authors show the complex relationship between the content of their books and their celebrity. Cousteau was as admired for his physique as his science; Adamson's maternal reflections on life in the jungle were as important as her ethological observations. Even Sagan himself was probably famous as much for his appearances on the *Tonight Show with Johnny Carson* as he was for his printed words. Sagan marked the shift from *visible* scientist to *celebrity* scientist; he was a celebrity within a general culture that increasingly valued celebrity for its own sake (Maddox 1988; Rodgers 1992). His biographers say that, for Sagan, even a family outing to a restaurant required special efforts to avoid his fans (Davidson 1999).

LaFollette (2013), in her history of science on American television, argues that becoming famous in the television era marked a significant change in how scientists attracted public attention. For her, television has tended to feature scientists who conformed to its demands for the revelation of personal details and its emphasis on personal appearance. She praises the memorable work of accomplished researchers such as Sagan, Cousteau and Bronowski who exploited the medium's potential to communicate science. But she argues that, in general, television trivialised scientists by portraying them as celebrities akin to other figures in popular entertainment – and this was one reason why the medium did not fulfil its potential to educate broader society about science. This argument echoes a long-standing theme in research on fame, which views celebrity as corrosive to the quality of public culture (Evans and Hesmondhalgh 2005).

Certain scientific events and controversies also brought particular scientists – or those identified as scientists – into public consciousness. The most prominent were the astronauts of the Mercury, Gemini and Apollo space missions, lauded publicly for their contributions to new knowledge despite their largely technical role as operators of powerful vehicles and experimental apparatus (Shayler and Burgess 2007). Controversies over issues of sociobiology, computers and artificial intelligence, and the linkages of race and intelligence, also made scientific stars of researchers such as E. O. Wilson, Stephen Jay Gould and Richard Dawkins (Wilson 1975; Gould 1981; Dawkins 1976; Lewontin and Levins 2002; Weizenbaum 1976). Scientists such as James Watson, co-discoverer of the structure of DNA, turned to popular writing to create their own personae in part to recruit new scientists for the fields they championed (Watson 1968; Yoxen 1985).

Some scientists used their public presence as a way to argue for scientific theories that they were having trouble having accepted by the scientific community. Fred Hoyle, who had coined the term *big bang* as a derisory reference to an astronomical theory that he opposed, eventually found many of his own theories blocked from publication by the critiques of other scientists. He chose, instead, to write popular articles and science fiction novels, and to appear frequently on BBC radio, as a means of articulating his ideas for broader audiences. These activities gave him a public prominence not directly connected to the value attributed to his theories by his scientific colleagues (Gregory 2005). Stephen Jay Gould used the public platform provided by his regular columns in *Natural History* magazine to become a bestselling author and from that publishing success drew more readers to his particular interpretation of the fossils of the Burgess Shale. The approbation attached to his celebrity made his scientific opponents uneasy, such that one of his key antagonists in the Burgess Shale story wrote a competing popular book to challenge Gould's interpretation (Gould 1989; Conway Morris 1998; Conway Morris and Gould 1998).

Scientists in celebrity culture

Although celebrity is not solely a phenomenon of the twentieth century, the dramatic increase in the variety of media forms – including popular newspapers, cinema, television and, most recently, new digital media – has meant that celebrities are now 'a ubiquitous aspect of contemporary Western culture' (Drake and Miah 2010: 49). In this cultural environment, by the final decade of the twentieth century, the potential for scientific celebrity was clear. The increased popularity of popular science books in the 1980s and 1990s meant several scientist authors, such as Sagan and Hawking, had become celebrities (Turney 2001a). *Vogue* wrote that scientists who wrote popular books attracted 'hero worship' at a time when 'serious science became sexy' (Turner 1997: 41). Britain's *The Independent* said it was an age when science was 'dominated by its media superstars' (Connor 2001: 11).

Journalists and writers labelled individual scientists in this milieu as celebrities. As well as Hawking, Richard Dawkins, who in 1995 became Oxford University's first Professor for the Public Understanding of Science, was called 'an authentic celebrity' (Kohn 2005: 319). *The New York Times* described American astrophysicist Neil deGrasse Tyson, who directs New York's Hayden Planetarium and is the author of several popular astrophysics books, as 'a space-savvy celebrity' (Martel 2004: E5). *Nature* called Susan Greenfield, Oxford neuroscience professor, former head of the Royal Institution of Great Britain and author of many popular neuroscience books, a 'celebrity neuroscientist' (Nature 2004: 9). *Science* called her a science 'rock star' who came 'alive in the spotlight' (Bohannon 2005: 962).

Scientists prominent in non-anglophone countries also exhibit these characteristics. In France, for example, the Canadian-born astrophysicist Hubert Reeves has, for decades, popularised science in books and on television and is also known for his public role as environmental advocate. *Le Monde* (2010: 16) called Reeves the 'storyteller of the stars' and described his love of classical music. Mathematician Cédric Villani, who won the prestigious Fields Medal and communicates about his field to non-specialists, has cultivated a distinctive, dandyish public image. *Le Monde* described meeting him in an interview where he wore his two 'legendary' trademark accessories: a large spider-shaped brooch and an ascot necktie (Clarini 2012: 3).

A notable aspect of these celebrities is the over-representation of males among them. A recurring theme in recent research has been the cultural portrayal of scientists in gendered and sexualised terms. Examining the portrayals of female scientists on television, LaFollette (2013) says that women were not prominently featured in dramas, documentaries, news reports and talk shows, suffering as a result a form of 'cultural invisibility'. The female scientists who did appear, she argues, were portrayed largely as stereotypical superwomen consumed by their careers or as 'romantic, adventurous celebrities like Margaret Mead and Jane Goodall or impossibly adroit fictional superheroines' (LaFollette 2013: 186). But this representation, she argues, began to change by the 1990s, in part because of an increased number of female television producers. Any change was not dramatic, however. For example, Chimba and Kitzinger (2010) demonstrated that half of the profiles of women scientists in British newspapers referred to their appearance, clothing, physique or hairstyle – compared to just over one in five profiles of men. The issue of gender and sexuality may also have affected the portrayal of men: a discourse analysis found that physicist and television host Brian Cox, one of the most high-profile scientists in Britain, was portrayed in sexualised terms (Attenborough 2013). *The Daily Mail* newspaper reported that 'fan clubs have been cropping up all over the internet drooling over "Prof Cox the Fox"' (Fryer 2010: 15).

How can we explain the increasing celebrity of scientists, with its complexities of historical context, topical focus and gendered characteristics? In what follows, we suggest some key issues.

The transformation of scientists into celebrities reflects an intensification in the cultural role and reach of the media. The media-orientated characteristics of the visible scientists made them early examples of what has been called the mediatisation of science: the ongoing historical process whereby the relationship between science and the media has, in the last several decades, become closer and more coupled (Franzen et al. 2012). This process is embedded within a wider cultural shift in which the media have become not only increasingly important for the formation of public opinion in public affairs generally, but also increasingly powerful agents in their own right, shaping the public meanings of science and other areas of public life (Hansen 2009; Collini 2006). As a consequence, the dynamics of popular media govern increasingly how civic affairs – including those involving science – are framed and articulated in public life. This process leads to a public culture where 'human faces are plastered on every idea and event', argues cultural critic Leo Braudy (1997: 601), who added that '[c]omplex phenomena wear the reduced features of emblematic individuals'. As a result, celebrities have become a key place for the creation and examination of meanings about culture and society (Turner 2004). As Bucchi (2014) notes, the media's intensified focus on scientists and other public figures with popular influence and appeal leads to a situation where 'discussion on the priorities and implications of science in society is reduced to comparison among voices and figures of public importance'. The idea of celebrity, then, builds on the idea of visibility.[1] This conceptual shift mirrors a historical shift from *visible* scientists to *celebrity* scientists.

Cultural studies scholar Graeme Turner (2004: 9) defines celebrity as 'a genre of representation and a discursive effect; it is a commodity traded by the promotions, publicity and media industries that produce these representations and their effects; and it is a cultural formation that has a social function' (see also Rojek 2001; Evans and Hesmondhalgh 2005). Celebrity as a genre of representation means that there is an intense personalisation in an individual's media portrayal. He or she is represented as a distinctive individual whose public and private selves merge in the portrayals. Celebrities are commodities in that their fame involves the commoditisation of reputation (Hurst 2005). They not only become commodities to sell their books, broadcasts and other cultural products, but they can also promote other commodities (including policies), their names serving as a promotional booster for those other products (Wernick 1991). For celebrity scientists, the product can be their books and broadcasts, and also science itself.

The idea that famous figures have a social function is crucial to understanding celebrity. It means that celebrities help audiences make sense of the social world, as famous figures articulate and represent values and beliefs that are often implicit, supplying 'a human dimension to the public world, personifying or personalizing things that may otherwise be quite abstract' (Evans 2005: 6). This symbolic nature of celebrity is crucial; as Holmes (2005: 12) argues, celebrities with lasting popularity have a 'deep, structural relationship with the ideological contexts of their times', meaning that their image becomes a means of working through useful questions of their eras. The process through which a public figure becomes a celebrity – how media attention focuses on their private life, how they become commodities, how they symbolise wider cultural issues – has been labelled with the unwieldy neologisms celebrification or celebritisation (Evans 2005). The portrayal of scientists as celebrities is important because, as the public representation of elite figures focuses on their individual, idiosyncratic distinctiveness and private lives, celebrity drives out other forms of elite status from the public sphere

(Turner *et al.* 2000). Without celebrity, scientists would no longer have a prominent place in popular culture.

Framework for analysing scientists as celebrities

By synthesising the constituent features of celebrity generally with the features of post-World War II scientists who became prominent in popular culture, six salient characteristics of celebrity scientists can be identified. These characteristics form a framework that can be used to structure in-depth analysis of individuals or collectives of culturally prominent scientists as celebrities. We use several English-language scientists described as celebrities – Hawking, Dawkins, Greenfield and Tyson – to illustrate and illuminate the features of scientific fame.

1. The scientist's represented image features a blurring of his or her public and private lives

Celebrification occurs when media coverage moves from reporting a scientist's public life to reporting his or her private life, a feature that was part of the coverage of Sagan and other visible scientists but which stretches back to coverage of Darwin (Turner 2004; Browne 2003). *The Guardian* called Dawkins charismatic and messianic with the 'fierce, hawkish good looks of a forties film star' (Radford 1996: T2). Kohn (2005: 319) described Dawkins's manner as 'assiduously courteous' and highlighted how the scientist's 'widely noted good looks [are] set off to best effect by grooming and trimmings chosen with a stylish eye'. Kohn notes, furthermore, that Dawkins enjoys 'the luxury that comes with celebrity, in the splendid house... which he shares with his third wife, Lalla Ward'.

Tyson's private life was reported in a range of profiles published in 2000, including in *People* (2000: 92), the magazine devoted to celebrity news, which named him its 'sexiest astrophysicist', adding that 'the 6'2" Tyson indulges his love of wine and gourmet cooking while succumbing to the gravitational pull of his wife of 12 years, mathematical physics Ph.D. Alice Young, 44, who is expecting their second child next month'. *Wine Spectator* magazine described his 700-bottle wine collection (Meltzer 2000). *Ebony* noted that with 'his thick, helmet-like Afro, his quirky star-strewn ties and vests, and his infectious enthusiasm for all-things cosmic, Tyson... cuts quite a telegenic figure' (Whitaker 2000: 58). Tyson's Twitter feed has merged his public and private lives: the majority of his tweets consist of cosmological facts and trivia, but they also allow 'backstage access' (Marwick and Boyd 2011: 144), as they provide personal details that enhance intimacy between him and his followers. For example, on 25 July 2012, he wrote: '@adinasauce: My uncle passed tonite. Your thoughts on our atoms & the universe comforted me. Thank you.//No, thank the cosmos.'

Susan Greenfield's private and public lives also merged in coverage, notably in 1999 when she appeared in *Hello!* magazine to promote the Royal Institution of which she was then Director. She was pictured with her then husband, chemist Peter Atkins, and she described their relationship in an interview: '[w]e're close without being sentimental – best friends and soulmates' (Kingsley 1999: 109). Her private life was opened to further public scrutiny after her marriage to Atkins ended: their divorce 'scandalised not only him but the entire scientific and political establishments' (Churcher 2003: 22). Stephen Hawking's former wife Jane wrote a book about their life together, *Music to Move the Stars* (2000), featuring intimate details of their emotional and romantic lives. *Vanity Fair* reported details of Hawking's second marriage, including points about the police investigation into a series of apparently suspicious injuries that Hawking suffered during the relationship (Bachrach 2004).

2. The scientist is a tradable cultural commodity

Communicating with broader audiences through publishing of popular books or commercial television involves the branding and marketing of the scientist as author or presenter, often involving synergies where variations of the same product are sold across different media formats. For example, the release of the film, *A Brief History of Time,* in 1992 was accompanied by the release of *Stephen Hawking's A Brief History of Time: A Reader's Companion,* which *Vanity Fair* called 'a book of the movie of the book' (Lubow 1992: 76). Turney (2001b: 8), in a review of Hawking's *The Universe in a Nutshell* (2001), argued that because physics had not advanced significantly since 1988, there was little need for this new book 'apart from [publisher] Bantam's urge to keep the franchise going'. A review of the television programme, *Stephen Hawking's Universe,* in Australia's *The Age* commented that Hawking's name was used as a promotional booster. It said Hawking 'tops and tails each episode and chimes in with the occasional comment, but he is mainly used here for name value' (Schembri 1998: 2). In another example, Richard Dawkins's *The God Delusion* was published the same year, 2006, as the television documentary on his work *Root of All Evil?* was aired on Britain's Channel 4. Similarly, Tyson's co-authored book *Origins* (2004) about the evolution of the cosmos was tied to a two-part Public Broadcasting Service show on the same topic broadcast, also called *Origins.*

3. The scientist's public image is constructed around discourses of truth, reason and rationality

The fame of various professions is tied to different discourses. For example, the film star is constructed around discourses of individuality and freedom, the television star around discourses of familiarity and widespread acceptability, and the popular music celebrity around discourses of authenticity (Marshall 1997). The celebrity scientist is associated with discourses of truth, reason and rationality. The image is also associated with progress (Lewis 2001). Scientists' public images have epistemological dimensions, being aligned with truth and the view that the methods of science make it uniquely positioned to uncover truths about the natural world (Fahy 2012). For example, a recurring pattern of the representation of Hawking – a pattern that has recurred in the depiction of iconic historical scientists such as Isaac Newton (Lawrence and Shapin 1998) – presents his mind as existing outside his body, in another realm where knowledge is found. 'Even as he sits helpless in his wheelchair', wrote *Time* magazine (1978a), 'his mind seems to soar ever more brilliantly across the vastness of space and time to unlock the secrets of the universe'. In 1983, *The New York Times* found significant the fact that Hawking's ideas changed conceptions of the universe while his body failed, a contrast that shaped him 'increasingly into a cerebral being' (Harwood 1983: 16). This has become his dominant public image, despite the fact that, as Hélène Mialet demonstrated in *Hawking Incorporated* (2012), he actually depended on a vast network of support services – including nurses and graduate students – to allow him to live and work. The media coverage systematically created an image that supported *reason* as the base of his celebrity, effacing the larger social system that sustained him. Dawkins's public image is similarly tied to ideas of truth. He wrote (2004: 43), 'If I am asked for a single phrase to characterise my role as Professor of Public Understanding of Science, I think I would choose Advocate for Disinterested Truth'. After early writings that focused chiefly on evolution, Dawkins's public career, over time, moved towards a rationalist 'crusade' that has not been linked to efforts by organised programmes created by scientific communities, but rather has been tied to the advocacy work of atheists, rationalists and sceptics (Trench 2008: 122). He established the Richard Dawkins Foundation for Reason and Science and he positioned its mission as supporting 'scientific education, critical thinking and evidence-based understanding of the natural world in the quest to overcome religious fundamentalism, superstition, intolerance and suffering' (Richard Dawkins Foundation for Reason and Science 2013). An earlier version of the mission statement for the Foundation read, in part: 'The enlightenment is under threat. So is reason. So is truth' (cited in Trench 2008: 122). For Hawking and Dawkins, celebrity was based on a commitment to the reliable knowledge about the natural world that goes by the label of truth or science.

4. The scientist has a structural relationship with the ideological tensions of their times

Famous figures who have lasting popularity are emblematic of the social, cultural and political tensions of their times. Tyson, an African American, in his autobiography *The Sky is Not the Limit* (2004), framed his life and career partly in terms of how he came of age in post-1960s America. He describes how he became an astrophysicist in spite of the space agency NASA, which in its early years was sending white astronauts into space while poverty grew in inner-city black communities. He writes about how he dealt with racial stereotypes because he was a black man who did not occupy a public role as an athlete or entertainer and about how he became a public expert in a topic that had nothing to do with race. *The New York Times*, in its review, said the book is strongest when Tyson connects his personal life history to his career development (Knowles 2004). Likewise, journalists profiled Hawking at length for publications across the cultural spectrum including *New Scientist, Time, Reader's Digest* and *Vanity Fair* from the late 1970s, a period when he undertook his pioneering work on black holes and relativistic cosmology experienced a surge in interest. His specialist topic of black holes resonated with wider cultural concerns. *Time* magazine (1978b) said black holes were linked with the contemporaneous 'faddish craze for the likes of parapsychology, the occult, UFOs, thinking plants... and other pseudoscientific hokum'. Dawkins's public image was formed in and through a series of social controversies around science. As well as being part of the sociobiology debate in the late 1970s, Dawkins has been involved with Gould and other figures in controversies, that were played out in popular books and articles, over different interpretations of particular mechanics of evolution. These writings also had cultural and political dimensions as they were tied to ongoing controversies over the influence of biology on human behaviour (Segerstråle 2000). With the publication of *The God Delusion* (2006), Dawkins became a key figure in the social controversies in the first decade of the twentieth century over the claimed conflict between science and religion. The book formed a cornerstone of what became known as the 'new atheism', and his online presence has become a hub for a wider online community of secularists (Cimino and Smith 2011).

5. The scientists' representations feature the tensions and contradictions inherent in fame

The visible scientists of the 1960s and 1970s, Goodell (1977: 202) notes, had an 'overrated credibility with the public', a characteristic similar to one that recurs in the representation of celebrity scientists: a tension between their scientific status and public renown. Physicists argue that Hawking's popular fame far exceeds his reputation within the field (Coles 2000). Neuroscientists, likewise, question the quality of Greenfield's work. One anonymous scientist, for example, told *The Observer*: 'A lot of what she says does not pass muster academically. Britain is very strong on neuroscience and compared to the leaders in the field, she is simply not in the same league. She is never cited in research papers' (O'Hagan 2003: 5). Historian of science Fern Elsdon-Baker (2009: 223) argues that some of Dawkins's peers do not regard him as a scientist and it was 'generally accepted in some research communities that he does not speak for the discipline'. The repetition of this argument indicates that this tension is likely a structural feature in scientific celebrity.

6. The scientists' celebrity status allows them to comment on areas outside their realm of expertise

A recurrent criticism of *The God Delusion* was that Dawkins did not give a fair account of the intellectual and theological positions he challenged (for a representative example, see Eagleton 2006). A similar critical reception greeted the arguments in Hawking's 2012 book *The Grand Design*, co-authored with Caltech physicist and writer Leonard Mlodinow, that philosophy was dead and God was not needed to explain the beginning of the universe; experts in both theology and the philosophy of physics sharply criticised the book (Cornwell 2010; Callender 2010). Similarly, Greenfield – an expert in neurodegenerative conditions – made a series of arguments in the early 2010s, in popular books and media, that children's immersion in screen technologies damages their developing brains. Other experts criticised these claims. For example, professor of neuropsychology Dorothy Bishop (2011) wrote on her blog: 'In recent years your speculations have wandered onto my turf and it's starting to get irritating.... I wish you would focus on communicating about your areas of expertise – there's plenty of public interest in neurodegenerative diseases.'

Concluding remarks

Scientists have played an increasing role in popular culture over the last century – exemplified by their growing celebrity – as public culture, intensively shaped by the media, has become a celebrity culture. As individuals, celebrity scientists are agents linking science and broader culture. They are emblematic of science itself, embodying complex ideas and concepts through their distinctive public personae. They are, in the classic sociological sense, *boundary objects*, able to operate in multiple cultures with the same actions being interpreted in different ways. To the science community, they are ambassadors and preachers, carrying the ideas of science to communities beyond science. In those communities, however, celebrity scientists are independent actors, partaking of and shaping the culture just as much as celebrity actors, politicians and business people do. In this culture, the scientists who successfully communicate over a long period of time with broad audiences will inevitably become celebrities because it is through celebrity that ideas are largely portrayed in public.

Identifying the characteristics of celebrity scientists helps us understand the ways that science participates in public affairs. The shift from visible scientists of the mid-twentieth century to celebrity scientists of the twenty-first century marks the increasing integration of science into popular culture. In an era of celebrity, this integration means that scientists must be celebrities. Celebrity scientists have the prominence to circulate scientific ideas and concepts through culture, shaping public discussion around issues and controversies of public relevance such as evolution and intelligent design – although their actual impact in terms of public attitudes and increased awareness remain largely unproven.

Several issues for future research arise from this framing of scientists as celebrities. This chapter has focused largely on scientists from Britain and the US, so future studies could examine if prominent scientists in other countries or regions conform to the characteristics of celebrity scientists outlined here. Do different national and cultural contexts, for instance, produce other, particular features of scientific fame? Further studies could examine also the different paths through which a scientist becomes prominent and then becomes a celebrity; this is an important process to illuminate, because the way a Nobel laureate becomes famous (Bucchi 2012; Baram-Tsabari and

Segev 2013) is likely to be different from the way an entrepreneurial scientist such as Craig Venter achieves public renown.

The examination of scientists as celebrities could also assess the effect their fame might have on scientific practice itself, contributing to debates on the mediatisation of science. There are indications that celebrity scientists' cultural force may, in turn, affect the science that is embedded in culture. Celebrity scientists have a status (and thus power) *within* science that sometimes outweighs the details of their technical achievements. When Tyson and the Hayden Planetarium left Pluto out of their solar system, they catalysed the scientific discussion about how to define planets (Messeri 2010). When Hawking became a scientific star, physicist Jeremy Dunning-Davies argued that colleagues of his had papers rejected by journals 'simply because the end result disagrees with Hawking'; he wrote (1993: 85): 'Papers which challenge Hawking on purely scientific grounds are not successful because his reputation has in some sense gone beyond the purely scientific'. Science is increasingly embedded in popular culture and, by looking at celebrity, we can see how popular culture affects the production of knowledge itself.

Key questions

- Identify a scientist prominent in popular culture in your country who could be considered a celebrity scientist. How do they conform to the characteristics described in this chapter? Are there other characteristics?
- How are contemporary female scientists portrayed in popular culture? Are they portrayed differently from male scientists?
- How have science books, magazines, and television fed a *science culture*?

Note

1 Our attention to the differences between *visibility* and *celebrity* is key to understanding the development of the scholarly literature in the last 40 years. Goodell (1977) used celebrity and visibility as synonyms, but ideas that have emerged since the late 1970s from the field of celebrity studies provide a new set of conceptual tools to examine scientific stardom. Nevertheless, the characteristics, that Goodell identified, of visible scientists – the fact that they are articulate, controversial, have a colourful image, a credible reputation and work on hot topics – remain useful in broad terms for describing a set of personal and professional attributes necessary for a scientist to earn wider cultural visibility. But as a scientist becomes visible, he or she may or may not become a celebrity. Celebrity as a concept therefore builds on visibility to describe particular features of scientists' portrayal as public personalities, the way their reputation can be commoditised and acquire a symbolic value in a popular culture.

References

Attenborough, F.T. (2013) 'Discourse analysis and sexualization: a study of scientists in the media', *Critical Discourse Studies*, 10, 2: 223–236.

Bachrach, J. (2004) 'A beautiful mind, an ugly possibility', *Vanity Fair*; online at www.vanityfair.com/culture/features/2004/06/hawking200406; accessed 2 June 2013.

Baram-Tsabari, A. and Segev, E. (2013) 'The half-life of a "teachable moment": the case of Nobel laureates', *Public Understanding of Science*, published online first 21 June; doi: 10.1177/0963662513491369.

Bauer, M. W. (2012) 'Public attention to science 1820–2010: a "longue durée" picture', in S. Rödder, M. Franzen and P. Weingart (eds) *The Sciences's Media Connection: Public Communication and Its Repercussions*, Dordrecht: Springer, 35–58.

Bauer, M. W., Durant, J., Ragnarsdottir, A. and Rudolfsdottir, A. (1995) *Science and Technology in the British Press, 1946–1992. The Media Monitor Project, Vols 1–4*, London: The Science Museum and Wellcome Trust for the History of Medicine.

Bauer, M. W., Petkova, K., Boyadjieva, P. and Gornev, G. (2006) 'Long-term trends in the public representation of science across the "Iron Curtain": 1946–1995', *Social Studies of Science*, 36, 1: 99–131.

Bishop, D. (2011) 'An open letter to Baroness Susan Greenfield', *BishopBlog*, 4 August; online at http://deevybee.blogspot.com/2011/08/open-letter-to-baroness-susan.html; accessed 2 June 2013.

Bohannon, J. (2005) 'The baroness and the brain', *Science*, 310, 5750: 962–963.

Braudy, L. (1997) *The Frenzy of Renown: Fame and its History*, second edition, Oxford: Oxford University Press.

Browne, J. (2003) 'Charles Darwin as a celebrity', *Science in Context*, 16, 1/2: 175–194.

Bucchi, M. (2012) 'Visible scientists, media coverage and national identity: Nobel Laureates in the Italian daily press', in B. Schiele, M. Claessens and S. Shi (eds) *Science Communication in the World: Practices, Theories and Trends*, New York: Springer, 259–268.

Bucchi, M. (2014) 'Norms, competition and visibility in contemporary science: the legacy of Robert K. Merton', *Journal of Classical Sociology* (in press).

Bucchi, M. and Mazzolini, R. G. (2003) 'Big science, little news: science coverage in the Italian daily press, 1964–1997', *Public Understanding of Science*, 12, 1: 7–24.

Bucchi, M. and Trench, B. (2008) 'Introduction', in M. Bucchi and B. Trench (eds) *Handbook of Public Communication of Science and Technology*, London and New York: Routledge, 1–3.

Callender, C. (2010) 'Stephen Hawking says there's no theory of everything', *New Scientist*, 2 September; online at www.newscientist.com/blogs/culturelab/2010/09/stephen-hawking-says-theres-no-theory-of-everything.html; accessed 2 June 2013.

Chimba, M. and Kitzinger, J. (2010) 'Bimbo or boffin? Women in science: an analysis of media representations and how female scientists negotiate cultural contradictions', *Public Understanding of Science*, 19, 5: 609–624.

Churcher, S. (2003) 'Peter's brain was a real aphrodisiac, but now I can wear short skirts without being nagged', *Daily Mail*, 4 May: 22–23.

Cimino, R. and Smith, C. (2011) 'The new atheism and the formation of the imagined secularist community', *Journal of Media and Religion*, 10, 1: 24–38.

Clarini, J. (2012) 'Cédric Villani: Se guider à travers l'océan des possibles', *Le Monde*, 14 September, Livres: 3.

Coles, P. (2000) *Hawking and the Mind of God*, New York: Totem Books.

Collini, S. (2006) *Absent Minds: Intellectuals in Britain*, Oxford: Oxford University Press.

Connor, S. (2001) 'Boy from Bingley "lobbed intellectual grenades" at science', *The Independent*, 23 August: 11.

Conway Morris, S. (1998) *The Crucible of Creation: The Burgess Shale and the Rise of Animal*, Oxford and New York: Oxford University Press.

Conway Morris, S. and Gould, S. J. (1998) 'Showdown on the Burgess Shale', *Natural History*, 107, 10: 48–55.

Cornwell, J. (2010) 'What's God got to do with it?', *The Daily Telegraph*, 18 September: 27.

Davidson, K. (1999) *Carl Sagan: A Life*, New York: J. Wiley.

Dawkins, R. (1976) *The Selfish Gene*, Oxford and New York: Oxford University Press.

Dawkins, R. (2004) *A Devil's Chaplain: Selected Essays*, second edition, London: Phoenix.

De Kruif, P. (1926) *Microbe Hunters*, New York: Harcourt Brace and Company.

Douglas, S. (1987) *Inventing American Broadcasting, 1899–1922*, Baltimore: Johns Hopkins University Press.

Drake, P. and Miah, A. (2010) 'The cultural politics of celebrity', *Cultural Politics*, 6, 1: 49–64.

Dugan, J. (1948) 'The first of the menfish', *Science Illustrated*, 3, 12: 30–32; 68–69.

Dunning-Davies, J. (1993) 'Popular status and scientific influence: another angle on "the Hawking phenomenon"', *Public Understanding of Science*, 2, 1: 85–86.

Eagleton, T. (2006) 'Lunging, flailing, mispunching', *London Review of Books*, online at www.lrb.co.uk/v28/n20/terry-eagleton/lunging-flailing-mispunching; accessed 2 June 2013.

Elsdon-Baker, F. (2009) *The Selfish Genius: How Richard Dawkins Rewrote Darwin's Legacy*, London: Icon Books.

Evans, J. (2005) 'Celebrity: what's the media got to do with it?', in J. Evans and D. Hesmondhalgh (eds) *Understanding Media: Inside Celebrity*, Maidenhead: Open University Press, 1–10.

Evans, J. and Hesmondhalgh, D. (eds) (2005) *Understanding Media: Inside Celebrity*, Maidenhead: Open University Press.

Fahy, D. (2012) 'Science and celebrity studies: towards a framework for analysing scientists in public', in M. Bucchi and B. Trench (eds) *Quality, Honesty and Beauty in Science Communication, Book of Papers of the 12th International Public Communication of Science and Technology Conference*, Vicenza: Observa Science in Society: 295–299.

Franzen, M., Weingart, P. and Rödder, S. (2012) 'Exploring the impact of science communication on scientific knowledge production: an introduction', in S. Rödder, M. Franzen and P. Weingart (eds) *The*

Sciences' Media Connection: Public Communication and its Repercussions. Sociology of the Sciences Yearbook, Vol. xxviii, Dordrecht and New York: Springer, 3–14.
Fryer, J. (2010) 'The man who's making space sexy', *Daily Mail*, 31 March: 15.
Goodell, R. (1977) *The Visible Scientists*, Boston, MA: Little, Brown.
Gould, S. J. (1981) *The Mismeasure of Man*, New York: W. W. Norton.
Gould, S. J. (1989) *Wonderful Life: The Burgess Shale and the Nature of History*, New York: W. W. Norton.
Gregory, J. (2005) *Fred Hoyle's Universe*, Oxford and New York: Oxford University Press.
Hansen, A. (2009) 'Science, communication and media', in R. Holliman, E. Whitelegg, E. Scanlon, S. Smidt, and J. Thomas (eds) *Investigating Science Communication in the Information Age*, Oxford: Oxford University Press, 105–127.
Harwood, M. (1983) 'The universe and Dr. Hawking', *The New York Times Magazine*, 23 January: 16–64.
Hawking, J. (2000) *Music to Move the Stars: A Life with Stephen*, second edition, London: Pan Books.
Henig, R. M. (2002) 'The life and legacy of Paul de Kruif', *APF Reporter*, 20; online at http://aliciapatterson.org/stories/life-and-legacy-paul-de-kruif; accessed 23 August 2013.
Holmes, S. (2005) '"Starring... Dyer?": re-visiting star studies and contemporary celebrity culture', *Westminster Papers in Communication and Culture*, 2, 2: 6–21.
Hughes, T. P. (1983) *Networks of Power: Electrification in Western Society, 1880–1930*, Baltimore and London: The Johns Hopkins University Press.
Hurst, C. E (2005) *Living Theory: The Application of Classical Social Theory to Contemporary Life*, second edition, Boston, MA: Allyn and Bacon.
Kingsley, M. (1999) 'Making her mark in the male-dominated world of science', *Hello!*, 583, 26 October: 106–109.
Kline, R. R. (1992) *Steinmetz: Engineer and Socialist*, Baltimore and London: The Johns Hopkins University Press.
Kline, R. (1995) 'Construing "technology" as "applied science": public rhetoric of scientists and engineers in the United States, 1880–1945', *ISIS*, 86, 2: 194–221. Knowles, S. G. (2004) 'Books in brief', *New York Times*, 1 August: 12.
Kohn, M. (2005) *A Reason for Everything: Natural Selection and the English Imagination*, London: Faber & Faber.
LaFollette, M. C. (1990) *Making Science Our Own: Public Images of Science, 1910–1955*, Chicago: University of Chicago Press.
LaFollette, M. C. (2008) *Science on the Air: Popularizers and Personalities on Radio and Early Television*, Chicago: University of Chicago Press.
LaFollette, M. C. (2013) *Science on American Television: A History*, Chicago: University of Chicago Press.
Lawrence, C. and Shapin, S. (1998) *Science Incarnate: Historical Embodiments of Natural Knowledge*, Chicago: University of Chicago Press.
Leane, E. (2007) *Reading Popular Physics: Disciplinary Skirmishes and Textual Strategies*, Aldershot: Ashgate.
Le Monde (2010) 'Au concert avec: Hubert Reeves', *Le Monde*, Culture, 25 July: 16.
Lewenstein, B. V. (1987) 'Was there really a popular science "boom"?' *Science, Technology & Human Values*, 12, 2: 29–41.
Lewenstein, B. V. (2009) 'Science books since 1945', in D. P. Nord, J. S. Rubin and M. Schudson (eds) *The Enduring Book: Print Culture in Postwar America*, Chapel Hill: University of North Carolina Press, 347–360.
Lewis, T. (2001) 'Embodied experts: Robert Hughes, cultural studies and the celebrity intellectual', *Continuum*, 15, 2: 233–247.
Lewontin, R. C and Levins, R. (2002) 'Stephen Jay Gould: What does it mean to be a radical?', *Monthly Review*, 54, 6; online at http://monthlyreview.org/2002/11/01/stephen-jay-gould; accessed 2 December 2013.
Lubow, A. (1992) 'Heart and mind', *Vanity Fair*, June: 72–86.
Maddox, J. (1988) 'The big Big Bang book', *Nature*, 336, 6196: 267. Marshall, D. P. (1997) *Celebrity and Power: Fame in Contemporary Culture*, Minneapolis: University of Minnesota Press.
Martel, N. (2004) 'Mysteries of life, time and space (and green slime)', *The New York Times*, 28 September: E5.
Marwick, A. and Boyd, D. (2011) 'To see and be seen: celebrity practice on Twitter', *Convergence*, 17, 2: 139–158.
Meltzer, P. D. (2000) 'Star collector', *Wine Spectator*, 31 May: 19–20.
Messeri, L. R. (2010) 'The problem with Pluto: conflicting cosmologies and the classification of the planets', *Social Studies of Science*, 40, 2: 187–214.
Mialet, H. (2012) *Hawking Incorporated: Stephen Hawking and the Anthropology of the Knowing Subject*, Chicago: University of Chicago Press.

Missner, M. (1985) 'Why Einstein became famous in America', *Social Studies of Science*, 15, 2: 267–291.
Moss, S. (1997) 'The ascent of Jacob Bronowski', *Science and Public Affairs* (summer): 27–29.
Nature (2004) 'Popularizer Greenfield is blackballed by peers', *Nature*, 429, 699: 9.
New Scientist (2012) 'Stephen Hawking at 70: exclusive interview', *New Scientist*; online at www.newscientist.com/article/mg21328460.500-stephen-hawking-at-70-exclusive-interview.html#.Uv7CFShwkyE; accessed 2 June 2013.
O'Hagan, S. (2003) 'Desperately psyching Susan: sexy or serious?' *The Observer*, Review, 7 September: 5.
People (2000) 'Neil DeGrasse Tyson: sexiest astrophysicist', *People*, 13 November: 92; online at www.people.com/people/archive/article/0,20132902,00.html; accessed 2 June 2013.
Poundstone, W. (1999) *Carl Sagan: A Life in the Cosmos*, New York: Henry Holt.
Pretzer, W. S. (ed.) (1989) *Working at Inventing: Thomas A. Edison and the Menlo Park Experience*, Dearborn, MI: Henry Ford Museum & Greenfield Village.
Radford, T. (1996) 'Astounding stories', *The Guardian*, 17 July: T2.
Richard Dawkins Foundation for Reason and Science (2013) The Richard Dawkins Foundation (US); online at www.richarddawkins.net/home/about; accessed 29 July 2013.
Rodgers, M. (1992) 'The Hawking phenomenon', *Public Understanding of Science*, 1, 2: 231–234.
Rojek, C. (2001) *Celebrity*, London: Reaktion Books.
Schembri, J. (1998) 'Stephen Hawking's universe', *The Age*, Green Guide, 19 February: 2.
Science Museum (2012) *Stephen Hawking: A 70th birthday celebration*, The Science Museum; online at www.sciencemuseum.org.uk/hawking; accessed 29 July 2013.
Secord, J. A. (1985) 'Newton in the nursery: Tom Telescope and the philosophy of tops and balls, 1761–1838', *History of Science*, 23: 127–151.
Segerstråle, U. (2000) *Defenders of the Truth: The Sociobiology Debate*, Oxford: Oxford University Press.
Shayler, D. J and Burgess, C. (2007) *NASA's Scientist-Astronauts*, Berlin: Springer.
Time (1978a) 'Science: soaring across space and time', *Time*, 4 September; online at www.time.com/time/magazine/article/0,9171,912132,00.html; accessed 2 June 2013.
Time (1978b) 'Science: those baffling black holes', *Time*, 4 September; online at www.time.com/time/magazine/article/0,9171,912131,00.html; accessed 2 June 2013.
Trench, B. (2008) 'Towards an analytic framework of science communication models', in D. Cheng, M. Claessens, T. Gascoigne, J. Metcalfe, B. Schiele and S. Shi (eds) *Communicating Science in Social Contexts: New Models, New Practices*, Dordrecht: Springer, 119–135.
Turner, G. (2004) *Understanding Celebrity*, London: Sage.
Turner, G., Bonner, F. and Marshall, D. P. (2000) *Fame Games: The Production of Celebrity in Australia*, Cambridge: Cambridge University Press.
Turner, J. (1997) 'Scientific sex appeal', *Vogue*, 2385, 163, April: 40–43.
Turney, J. (2001a) 'Telling the facts of life: cosmology and the epic of evolution', *Science as Culture*, 10, 2: 225–247.
Turney, J. (2001b) 'Strung out', *The Guardian*, Saturday Pages, 10 November: 8.
Tyson, Neil deGrasse (2004) *The Sky is Not the Limit: Adventures of an Urban Astrophysicist*, Amherst: Prometheus Books.
Watson, J. D. (1968) *The Double Helix: A Personal Account of the Discovery of the Structure of DNA*, New York: Atheneum.
Weizenbaum, J. (1976) *Computer Power and Human Reason: From Judgment to Calculation*, San Francisco: Freeman.
Wernick, A. (1991) *Promotional Culture: Advertising, Ideology and Symbolic Expression*, London and New York: Sage.
Whitaker, C. (2000) 'Super stargazer: Neil deGrasse Tyson is the nation's astronomical authority', *Ebony*, August: 58–62.
Wilson, E. O. (1975) *Sociobiology: The New Synthesis*, Cambridge, MA: Belknap Press of Harvard University Press.
Yoxen, E. (1985) 'Speaking out about competition: an essay on The Double Helix as popularisation', in T. Shinn and R. Whitley (eds) *Expository Science*, Dordrecht, Boston and Lancaster: Reidel.

8
Science and technology in film
Themes and representations

David A. Kirby

Introduction[1]

Science and movies have been intertwined from the beginning of cinema. The cinematic apparatus actually emerged out of the scientific research of Eadweard Muybridge and Etienne-Jules Marey, who were looking for technological means for studying animal movement in the late nineteenth century (Tosi 2005). Yet, until relatively recently, there was a dearth of studies examining science in fictional cinema. The current upsurge in the number of academic studies explicitly addressing the use of science in cinema can be attributed to a number of factors. For one, we are living in what could be described as a *golden age* for science in movies and on television. Many of the most financially successful films of all time have science at their core and were made in the last decade, including *Spider-Man* (2002), *Finding Nemo* (2003), and the all-time box office champion *Avatar* (2009). Similarly, a significant number of the most popular television shows of the last decade are immersed in science and technology, including *CSI* (2000–), *House* (2005–2012) and *The Big Bang Theory* (2007–). The increased popularity of science in entertainment media coincided with the rise of *geek culture* on the Internet. There are now a vast number of websites, such as io9 and Boing Boing,[2] celebrating the use of science in fiction.

Most significantly, the growth in scholarship on science in cinema can be attributed to a new understanding within the science communication community that the *meanings* of science, not scientific knowledge, may be the most significant element contributing to public attitudes towards science (Nisbet and Scheufele 2009). According to Alan Irwin (1995), the public makes sense of science – constructs their *science citizenship* – in the context of their everyday lives, pre-existing knowledge, experience and belief structures. Popular films significantly influence people's belief structures by shaping, cultivating or reinforcing the cultural meanings of science. One result of this intellectual shift away from a concern about science literacy towards a focus on the cultural meanings of science has been that several high-profile scientific organisations, including the US National Academy of Sciences and the Wellcome Trust in Britain, have embraced movies and television as legitimate vehicles for science communication by facilitating scientific involvement in the production of films and television programmes.

Science in cinema is not defined solely as factual information; it encompasses what I term the 'systems of science'; these include the methods of science, the social interactions among scientists, laboratory equipment, science education, industrial and state links, along with aspects of science that exist, in part, outside the scientific community, such as, science policy, science communication and cultural meanings. In the end, scholarship on science and cinema should be aimed at understanding how the systems of science are depicted in cinema, how these depictions have developed over time, how contemporary film-making practices contribute to these depictions and how these depictions impact the real-world systems of science.

Although there is a need for more work to be done, there now exists an established literature on science in film.[3] These works draw upon a wide variety of approaches and methodologies from numerous disciplines including communication, sociology, history, film studies, cultural studies, literature and science fiction studies. As with studies of science and news media, exploration of science communication in popular films revolves around four basic research questions: 1) How is science representation constructed in the production of cinematic texts? (*Production*); 2) How much science, and what kind of science, appears in popular films? (*Content analysis*); 3) What are the historic and contemporary cultural interpretations of science and technology in popular films? (*Cultural meanings*); and 4) What effect, if any, does the fictional portrayal of science have on science literacy, public awareness of, and attitudes towards, science? (*Media effects*). In this chapter, I summarise what scholars have uncovered about science and cinema in these four areas as well as pointing to areas that still require academic attention.

Production

Scholarship on the production of cinema has primarily focused on the role that scientists and scientific organisations have played as science consultants during film productions. Film-makers' use of science consultants goes all the way back to the earliest days of cinema in films such as *A Blind Bargain* (1922) and *The Lost World* (1925) (Kirby 2011).[4] Several studies on scientists' involvement in film production have looked at specific movies. For example, several studies on the production of the *The Beginning or the End* (1947) show how Manhattan Project scientists exhibited substantial control over the final version of the film, having the power to veto any portion of the script with which they disagreed (Reingold 1985). Public health officials, physicians and medical researchers frequently co-operated with film-makers in making issue-based dramatic films. The US Surgeon General was involved in the making of the Paul Ehrlich biopic (biographical picture) *Dr. Ehrlich's Magic Bullet* (1940) (Lederer and Parascandola 1998). The film proved useful to the US Public Health Service who convinced Warner Brothers to make a revised version of the film three years after its release for educational purposes. Joseph Turow (2010) also shows how the American Medical Association wielded significant power in determining the images of doctors in fictional movies and television shows. Likewise, Martin Pernick (1996) explores the use of dramatic films as both pro- and anti-eugenic propaganda in early cinema with physicians and public health officials often serving as consultants.

Science consulting has become prevalent for entertainment media. In fact, it would be surprising to have a contemporary film production or television show with scientific content that did not employ a science consultant (Kirby 2011). Several scientists have written personal recollections about their consulting experiences such as Frederick Ordway for *2001: A Space Odyssey* (1968), Ian Lipkin (2011) for *Contagion* (2011), Donna Nelson (Science and Entertainment Exchange 2011) for *Breaking Bad* (2008–2013) and David Saltzberg (2011) for *The Big Bang Theory*. For my book, *Lab Coats in Hollywood* (2011), I interviewed a number of scientists and film-makers about their experiences working together in the production of Hollywood films. I found that scientists assist film-makers in a number of ways including fact-checking, shaping visual iconography, advising actors, enhancing plausibility, creating dramatic situations and placing science in its cultural contexts. I also found that scientists and scientific organisations benefit from this arrangement as popular films can promote research agendas, stimulate technological development, contribute to scientific controversies and even stir citizens into political action.

Scientific organisations concerned about entertainment media's perceived impact on science literacy and public attitudes towards science have recently developed programmes to facilitate scientific involvement in the production of films and television programmes. The most prominent of these programmes include the US National Academy of Sciences' Science and Entertainment Exchange, the US National Science Foundation's collaboration with the Entertainment Industries Council, USC's Hollywood Health and Society programme, the American Film Institute's Sloan Science Advisor programme, NASA's partnership with the Hollywood Black Film Festival, and the German Federal Ministry of Education and Research's MINTiFF initiative. The development of these programmes has dramatically increased the level of scientists' collaboration in the production of entertainment media. Many of these organisations, and individual scientists, have even begun creating their own music videos for outreach purposes (Allgaier 2013). There are also new funding sources for screenwriters who wish to incorporate science into their movie projects. The Alfred P. Sloan Foundation's Science-in-Film Initiative supports the development of films exploring scientific themes, while the Wellcome Trust and British Film Institute have developed a screenwriting prize for the best movies inspired by biology and medicine. Several international film festivals have also emerged that focus on science-based films; these include the Imagine Science Film Festival, the Pariscience International Science Film Festival and the European Science TV and New Media Festival. Even the mainstream Sundance film festival has begun having special screenings of science-based films sponsored by the Sloan Foundation (Valenti 2012).

Academic works studying the phenomenon of science consultants argue that entertainment media presentations of science reveal a tension not only between the narrative forms of media and those of science, but also between the needs of the entertainment industry and those of the scientific community (Kirby 2011; Kirby 2003a; Frank 2003). Scientists and scientific organisations who work on popular films would like film-makers to maintain the accuracy of scientific depictions. Film-makers, on the other hand, only need to claim accuracy for their films and ask scientists to help them maintain an acceptable level of verisimilitude. This discrepancy in goals clearly leads to multiple interpretations of the term *accuracy*. For scientists, accuracy requires an adherence to scientific verisimilitude over an entire film. Film-makers consider a film to be scientifically accurate if it has any scientific verisimilitude within the constraints of budget, time and narrative. This is why I argue that *authenticity*, rather than accuracy, serves as a better lens through which to see science in cinema (Kirby 2011). Authentic science does not have to be accurate science, and a focus on authenticity allows scientists and film-makers to achieve a shared goal that is mutually satisfactory.

Content analysis

Science communication researchers have primarily relied on content analysis of newspapers to determine what science, and how much of it, appears in news media. Very few studies of fiction, however, have used this methodology outside of television (e.g. Dudo *et al.* 2011). In terms of fictional cinema, there have been two wide-ranging quantitative studies of science in cinema. Film scholar Andrew Tudor (1989) undertook a comprehensive content analysis of 990 horror films 1931 to 1984. Horror films elicit fear by introducing a *monstrous* threat into a stable situation. Tudor found that *science* is historically the most frequent type of monstrous threat in horror films (251 out of 990, or 25 per cent). There has been, however, a broad decline in the proportion of science-based horror films after 1960. This decline does not necessarily indicate a change in public attitudes towards science, but it does represent a change in the production of horror films where psychological horror took over as the dominant threat in the 1970s.

Peter Weingart and colleagues' (2003) quantitative study of 222 films of all genres created over 80 years looks at both recurring themes and changing patterns in the depiction of science in cinema. Unsurprisingly, given its dominance in news media (see Pellechia 1997), medical science is the most common research field depicted in films followed by the physical sciences (chemistry and physics). These fields are also the most likely to be shown as *ethically problematic* as in *Coma* (1978) and to have scientist characters working in secret laboratories as in *Hollow Man* (2000). In addition, Weingart *et al.* (2003) found that depictions of scientists are predominantly white, male and American. The overwhelming picture that both of these studies paint is a cinematic history expressing deep-rooted fears of science and scientific research in the twentieth century.

Both of these studies concerned mainstream fiction film primarily emanating out of Hollywood. There have been some quantitative content analyses of science and technology in non-US cinema, but these studies have focused exclusively on non-fiction films. Francesco Paolo de Ceglia (2012), for example, analyses science and technology in Italian documentary films from the first half of the twentieth century. He found that, as in Weingart *et al*.'s (2003) study of fiction films, these non-fiction films focus on the biological sciences as well as micro-naturalistic phenomena. Rosa Medina-Doménech and Alfredo Menéndez-Navarro (2005) found that depictions of medical technologies in Spanish newsreels that were produced during Francisco Franco's dictatorship were constructed in such a way as to break with the past and legitimate Franco's regime.

Cultural meanings

The studies discussed above fall under the category of *traditional* or *quantitative* content analysis. However, many researchers utilise a broader definition of content analysis which encompasses qualitative methods, including framing analysis. In fact, the most active area of research into the representation of science in cinema has been in what Jon Turney (1998) refers to as the *cultural history of images* where textual analysis of fictional films provides researchers with a gauge of social concerns, social attitudes and social change regarding science and technology. Popular cultural products, like fictional films, not only reflect ideas about science and technology, but they also construct perceptions for both the public and scientists in a mutual shaping of science and culture.

Movie scientists

The phrase *movie scientist* does not often have positive connotations. For most people the phrase conjures up images of Colin Clive as Dr. Frankenstein maniacally repeating "He's alive!" as his creature is brought to life in the classic 1931 Universal horror film *Frankenstein*. The mad scientist may be the most recognisable movie scientist, but it is not the only image of scientists on the screen. In her comprehensive study of scientist representations in literature and film, Roslynn Haynes (1994) identifies six recurrent scientist stereotypes: the alchemist/ mad scientist, the absent-minded professor, the inhuman rationalist, the heroic adventurer, the helpless scientist and the social idealist. Depictions of scientists are particularly important as they represent the public face of science (Pansegrau 2008). Schibeci and Lee (2003) argue that cinematic images of scientists play a significant role in constructing students' *science citizenship* by putting science in its sociocultural context. These stereotypes recur in cinema because they possess narrative utility. Stereotypes are cinematic shorthand. Audiences easily recognise these scientist caricatures, so film-makers do not need to take up valuable screen time establishing character backgrounds (Merzagora 2010). While these six basic stereotypes recur in cinema, they do not appear equally across genres. Horror films feature the mad scientist, comedies are the realm of the absent-minded professor and dramas predominantly feature social idealists. Likewise, action films incorporate heroic scientists, while science fiction films embrace inhuman rationalists and helpless scientists. The prevalence of these cinematic stereotypes varies over time. The helpless scientist who loses control of his experiments was a common subject in films of the early twentieth century such as *Reversing Darwin's Theory* (1908). Although the stereotype of the scientist losing control of his experiments continues throughout cinema history, it takes on more ominous overtones as experiments have more dire consequences. The 1920s and 1930s, on the other hand, were the heyday of the mad scientist character of which Clive's Dr. Frankenstein is representative (Tudor 1989; Skal 1998; Frayling 2005). Unlike the helpless scientist, the character of the mad scientist has become so recognisable as a stereotype that the character now exists mainly in self-referential parodies (e.g. *Young Frankenstein* (1974)) and satires (e.g. *Dr. Strangelove* (1964)).

The 1930s and 1940s were also the peak of the scientist biopic in Hollywood. Hollywood's approach to scientist biopics of this time can be summed up by two words: miracle and tragedy (Elena 1997). *The Story of Louis Pasteur* (1936) exemplifies the standard biopic formula and the film's success started the Hollywood scientist biopic trend that lasted until the mid-1940s. In the film, Pasteur has to overcome dogmatic scientific thinking and personal tragedies in order to bring the *truth* of bacteriology to the public. In addition, the narratives of scientist biopics, especially those of inventors like Thomas Edison, link the work of science into the capitalist system by depicting science as the underlying source of mass production (Böhnke and Machura 2003).

The perceived motivations of scientists in the Manhattan Project fuelled a host of films in the 1950s featuring amoral rationalist scientists who deny any responsibility for the consequences of their research (Jones 2001; Vieth 2001; Frayling 2005; Weingart 2008; Wiesenfeldt 2010). *The Thing From Another World* (1951) exemplifies the depiction of the inhuman rationalist and the danger this represents to humanity. It is the scientists' insistence on studying the frozen alien body which creates the film's crisis situation. As one character claims, 'Knowledge is more important than life!'. Absent-minded professors, as in *The Nutty Professor* (1963) and *The Absent Minded Professor* (1961), joined inhuman scientists as scientist stereotypes who appeared regularly in films of this time period (Terzian and Grunzke 2007).

The 1990s and 2000s saw the ascendance of the heroic scientist stereotype in film. The popularity of the disaster film genre provided numerous opportunities to depict heroic scientists, as in *Dante's Peak* (1997) and *The Core* (2003) (King 2000). What is unique to this period is that many of the heroic scientist characters are women (Flicker 2003). Several studies have questioned historic gender representations in cinema particularly in regards to primatology (Kanner 2006), the environment (Jackson 2011) and the *Jurassic Park* (1993–) film series (Franklin 2000). Joceyln Steinke (2005) surveys 74 science-based Hollywood films of the 1990s and found that 33 per cent (25 films) featured female scientists and engineers. Contrary to previous depictions, female scientist characters in the 1990s were more realistic and did not always conform to traditional gender stereotypes. However, female scientists still corresponded to traditional notions of femininity in appearance and dress, and romance was a dominant theme in these films. In addition, female characters reinforced social and cultural assumptions about the role of women in science and engineering.

Table 8.1 Dominant stereotypes, scientific themes, and representative films across time

Time Period	Scientist Stereotypes	Scientific Fields	Representative Films
1900–1910	Helpless Scientists	Electricity X-Rays Evolution	*X-Rays* (1897) *Reversing Darwin's Theory* (1908)
1911–1920	Helpless Scientists	Eugenics	*Damaged Goods* (1914) *The Regeneration of Margaret* (1916)
1921–1930	Mad Scientists	Glands Engineering	*A Blind Bargain* (1920) *Metropolis* (1926)
1931–1940	Mad Scientists Biopics	Medicine	*Frankenstein* (1931) *The Story of Louis Pasteur* (1936)
1941–1950	Biopics	Medicine Psychology	*Shining Victory* (1941) *Madame Curie* (1944)
1951–1960	Amoral Scientists	Space Science Nuclear Science	*Destination Moon* (1950) *Them* (1954)
1961–1970	Absent-Minded Professors	Space Science	*The Nutty Professor* (1961) *2001: A Space Odyssey* (1969)
1971–1980	Amoral Scientists	Ecology	*Silent Running* (1971) *Soylent Green* (1973)
1981–1990	Helpless Scientists	Computer Science	*War Games* (1983) *Robocop* (1987)
1991–2000	Heroic Scientists	Genetic Engineering Astronomy	*Jurassic Park* (1993) *Deep Impact* (1998)
2000–present		Biomedical Sciences Nanotechnology	*Splice* (2009) *I, Robot* (2004)

Scientific research fields in popular films

As with scientific stereotypes, the prevalence of specific scientific disciplines in cinema varies over time. Many scientific themes in cinema between 1900 and 1930 emerged out of scientific discoveries made at the end of the nineteenth century. Louis Lumière patented his cinematograph in the same year, 1895, that William Roentgen discovered X-rays. It did not take long for film-makers to exploit X-rays in films such as *X-Rays* (1897). Electricity also captured the minds of film-makers and numerous films of the 1900s and 1910s incorporated electricity as a *miracle* substance, as in *The Wonderful Electric Belt* (1907). The endocrinologist Dr Serge Voronoff became an international celebrity in 1919 for implanting monkey glands to *rejuvenate* wealthy older men, which led to endocrinology becoming a staple in horror films of the 1920s, such as *A Blind Bargain* (1922). In addition, a number of films in the 1920s and 1930s highlighted chemistry's *dark side* after the use of chemical warfare in World War I (Griep and Mikasen 2009). There were also dramatic films in this period dealing with scientific topics but these were propagandistic films about controversial scientifically based social issues including eugenics (Pernick 1996).

Science was seeing unprecedented growth in its activities and prestige in the 1950s with the rise of the military-industrial complex. Society was looking at scientific progress as the means for leading post-war society towards a utopian existence. Despite science's overall increased visibility, it was a single event, the dropping of atomic bombs on Japan, which shaped the predominant portrayal of science in 1950s cinema (Weart 1988; Shapiro 2002). Films of this time period are about science as power for good or evil. The opposition between science's destructive power and its progressive possibilities play out in various science fiction films of the time period including the trendsetting *Them!* (1954).

Nuclear science was not the only field heavily featured in films of the 1950s and 1960s. Starting with the groundbreaking *Destination Moon* (1950), space science became a major theme in cinema (Kirby 2011). Space films significantly shaped American space policy through their impact on American public opinion by showing space as an exciting and, most of all, technologically achievable adventure (McCurdy 1997). While radiation and space science dominated the 1950s and 1960s, popular cinema did address other aspects of scientific research including the discovery of the double helical nature of DNA (Kirby 2003b) and advances in the human sciences (Vieth 2001).

By the end of the 1960s, radiation was no longer the top scientific concern, at least according to the movies. Sparked by Rachel Carson's *Silent Spring* (1962), films of the 1970s show an overriding concern for ecological disaster (Lambourne *et al.* 1990; Ingram 2000; Brereton 2005). There were a multitude of environment-based science fiction, eco-horror, and revenge-of-nature films in the early 1970s including *Frogs* (1972) and *Soylent Green* (1973). Many of these films focus on issues of human overpopulation and resource use, and convey an impression that governmental ineptness or inaction is to blame for these problems. By the 1980s and into the 1990s the trend shifted to more serious dramas that moved away from an emphasis on government action towards corporate responsibility and individual responsibility in films such as *Silkwood* (1983) and *Erin Brockovich* (2000). In the 2000s the focus remained on personal actions and corporate greed but the main film genres shifted to animated children's films, including *Happy Feet* (2006) and *Wall-E* (2008), and big-budget action blockbusters such as *The Day After Tomorrow* (2004) and *Avatar* (2009).

Computer science emerges as a strong theme in fictional films of the 1980s as cinema grappled with two distinct aspects of our relationship to digital technologies. In the first instance, these films question the notion that humanity is really in control of our cybernetic creations as in *War Games* (1983) and *The Terminator* (1984) (Dinello 2006). Other films feature human-like robot/android/cyborg characters including *Blade Runner* (1982) and *Robocop* (1987). Such artificially created humans in cinema represent the most effective way to gauge the range of definitions of humanness as audiences must decide if these characters are actually *human* (Telotte 1995; Wood 2002). As Donna Haraway (1991) contends, *cyborg bodies* show how the boundary between organisms and machines has eroded to the point of being invisible.

The biomedical sciences have been crucial to the plots of a high number of films in the 1990s and 2000s including Alzheimer's research in *Deep Blue Sea* (1999), cancer research in *The Fountain* (2006) and pharmaceutical research in *Splice* (2009). Even the plots of many superhero films, which was the dominant genre of the 2000s, involved biomedical research in the creation of the superheroes as in *Captain America* (2011) or of the villains as in *Spiderman 2* (2004). This focus on the biomedical sciences also coincided with the emergence of nanotechnology as a central cinematic science in the 2000s (Thurs 2007). As happened with nuclear science in the 1950s, nanotechnology has become the go-to science for creating cinematic monsters as in *Hulk* (2003), *I, Robot* (2004) and *The Day the Earth Stood Still* (2008).

The 1990s and 2000s are a well-studied period with regards to cinema and its impact on the cultural meanings of genomics and genetic engineering. Although cinema is not the focus of their study, Nelkin and Lindee's (1995) groundbreaking work on genetics in popular culture demonstrates fiction's considerable input into shaping the *cultural meaning* of DNA. Several films over the last 25 years have influenced the cultural meanings of genomics and genetic engineering including *Boys From Brazil* (1978), *Twins* (1988), and *The Island of Dr Moreau* (1996) (Van Dijck 1998; Jörg 2003; Kirby 2007; Stacey 2010).

Cloning films represent a distinct subgenre of genetic-engineering-based films (Haran et al. 2008; Eberl 2010). Cloning films are overwhelmingly negative in their depictions despite the fact that most news coverage of the technology presents it as a positive development (Jensen 2008). Human clones are either depicted as monstrous beings as in *Godsend* (2004) or as people unaware of their status who undergo identity crises when they learn about their origins as in *The Island* (2005). Although horror and revulsion have been the dominant cinematic portrayal of cloning, its framing in movies made since the cloning of Dolly the sheep in 1996 has shifted to include the themes of hope and medical cures as well (O'Riordan 2008).

Two films, in particular, stand out for the quantity of scholarly attention they have received over the last ten years: *Jurassic Park* (1993) and *GATTACA* (1997). *Jurassic Park*'s perceived impact on public perceptions of biotechnology explains a good deal of the scholarly commentary on the film (Franklin 2000; Stern 2004). *Jurassic Park* is about the power of genetic engineering to unwittingly unleash monsters, while *GATTACA* is about our power to shape humanity itself (Kirby 2000; Wood 2002; Stacey 2010). *GATTACA*, in fact, is a rarity among film in its serious exploration of the bioethical issues surrounding human genetic manipulation. Most films support the idea that humanity's fundamental nature lies within its genome and could be improved by technological means (Kirby 2007). *GATTACA*, on the other hand, contains the messages that we are more than the sum of our genes and that being human means that we are able to *transcend* our genetic obstacles. Ultimately, genetic engineering films of the 1990s and 2000s are about genetics as a science of information, control, transformation and identity.

Audience research and media effects

Although the difficulties and limitations of media effects research are well documented, several empirical studies of science in the media suggest that fictional representations can have an influence on public attitudes towards science (Greenbaum 2009; Nisbet and Dudo 2010). This influence is why scientific organisations have become proactive in assisting film-makers during movie productions, as discussed above. Scientists have also been reacting to what they perceive as *bad movie science* by releasing 'real science of …' reviews. A 'real science of …' analysis consists of a scientist critiquing a movie in terms of what they see as inaccurate scientific content, as physicist Sidney Perkowitz did in *Hollywood Science* (2007). The US National Institutes of Health also maintains a long-running public film series involving scientists' critiques of science in films. Despite preliminary evidence showing that entertainment media can negatively impact science literacy and public perceptions of science, the effects of science in cinema on the public still remains a relatively sparse area of research. There has been a growing number of traditional audience reception studies examining entertainment media's impact on science literacy, attitudes and behaviours, as well as some recent work on science education; but most studies on this question are sociologically based, and a growing area of research examines the impact fictional films have on awareness of scientific issues.

Audience reception studies

Audience reception studies on science and films are limited so it is difficult to say exactly what impact fictional movies have on public opinion. It has been demonstrated that audience readings of films are always plural and that audience interpretations change within different social contexts, making it difficult to determine how a film has influenced attitudes or behaviour. Massarani and Moreira (2005), for example, found that movies could negatively impact attitudes towards human genetic engineering, but their results were complicated by pre-existing attitudes towards the issue. Despite this difficulty, several studies have shown that movies can strongly influence the public's attitudes towards science by shaping, cultivating or reinforcing the *cultural meanings* of science and scientists. Losh (2010) showed how the shift in entertainment texts to more heroic representations of scientists in the 1990s and 2000s has influenced public perceptions of scientists. Steinke and colleagues (2009) also found that teaching media literacy skills helped people recognise media stereotypes of scientists. The difficulty of separating out fictional influences is not as much of a factor if a researcher's only concern is with identifying how audiences take on board factual information. One recent study along this line (Barriga et al. 2010) examined the ability of audiences to identify the accuracy of scientific facts in a film depending on science's centrality to the storyline. They identified gender differences with men being more likely to identify factual errors when science was central to the plot while women detected factual errors more often when science was important for character relationships but peripheral to the plot.

Another difficulty with audience reception work on fictional film is separating out a movie's influence from all other media coverage as well as from other cultural contexts. Researchers need to be able to study audience responses before and after immediate exposure. This requires identification of an appropriate television programme or movie with sufficient time to arrange for surveys, focus groups and interviews. One film for which this was accomplished was *The Day After Tomorrow* with survey- and focus-group-based studies of public attitudes about global warming before and after the release of the film in Germany (Reusswig et al. 2004), Britain (Balmford et al. 2004; Lowe et al. 2006) and the US (Leiserowitz 2004). These studies showed mixed, culturally specific impacts of the film on public opinion towards climate change (Schiermeier 2004; Nisbet 2004). There was little change in the opinions of US audiences towards climate change, positive impact on British audiences with stronger motivations to act on climate change, and a negative impact on belief in climate change in Germany. All of these studies found, however, that the film raised awareness of climate change as an issue.

Although audience reception work on cinema is limited, there has been a significant amount of work done on forensic-based television and the supposed '*CSI* effect'. One of the concerns about the *CSI* effect is that people watching forensic-science-based television shows have developed unrealistic expectations regarding the prosecution's access to complex forensic evidence. Although the news media version of the *CSI* effect has no support, there is substantial evidence from survey, focus group and mock trial studies that television programmes are a major source of public knowledge about forensic science and that these programmes influence jurors in meaningful ways such as overestimating the legitimacy of any class of scientific evidence and – a potentially contrasting effect – lowering juror standards for conviction (e.g. Schweitzer and Saks 2007; Shelton et al. 2007). Nonetheless, *CSI* viewers were more likely to critique weak forensic evidence in mock trial studies than non-*CSI* viewers (Schweitzer and Saks 2007). However, there is no evidence that there have been higher acquittal rates in the US since the development of *CSI* in 2000 (Cole and Dioso-Villa 2007).

Science education

Although it seems counter-intuitive, the public's difficulty in distinguishing fact from fiction has rendered cinema a useful tool within conventional pedagogical situations and for informal science education (ISE). The idea is that movies could help attract students to science courses because of their popularity, that cinema's visual nature can be used to grab their attention while they are in the classroom and that the ability of students to point out factual errors in the fiction raises their confidence (Dubeck et al. 2004; Barnett and Kafka 2007). The National Science Teachers Association (NSTA 2012) recognises the prominent role that entertainment media play in informal science learning, and the organisation includes television and movies in their position statement on ISE.

Studies exploring the efficacy of movies as educational tools fall into two main focus areas. One area concerns educators and scientists discussing, at an anecdotal level, their best practices for using entertainment media in science classrooms (e.g. Rose 2003; Efthimiou and Llewellyn 2007). The second area of research involves experimental ISE studies empirically demonstrating how fictional films improve students' understanding of scientific concepts. Educational researchers have found that films can create lasting mental images of concepts that are correlated to the underlying scientific theory (Knippels *et al.* 2009). Movies can help students better understand, and retain, many of the abstract concepts that are covered in the sciences, particularly chemistry, physics and geology concepts (Barnett *et al.* 2006). Another advantage is that movies provide a connection between concepts and applications that can help avoid the disconnect that often occurs when students learn a concept and are then expected to apply that concept in real-world situations (Dubeck *et al.* 2004).

Entertainment Education

Work in the area of *entertainment education* has also shown that science in entertainment media can significantly impact the public's behaviour, especially regarding health issues. Entertainment education involves the intentional use of fiction to raise awareness of social issues and change individual behaviour (see Singhal *et al.* 2004). These works are almost exclusively related to using television as a means for changing individual behaviour regarding public health issues. It may well be that television is a more effective medium for entertainment education than cinema because audiences feel they *know* characters they encounter on a weekly basis. There are several organisations involved in entertainment education including the University of Southern California's Hollywood Health and Society, the Kaiser Family Foundation and the Environmental Media Association. In cinema's early days, movies were frequently used by public health advocates in attempts to alter behaviours regarding matters of health. Cinematic cautionary tales about public health issues were numerous in the early twentieth century, and many were made with the co-operation of public health officials, physicians and medical researchers (Pernick 1996; Lederer and Parascondola 1998).

Raising awareness, agenda setting, and framing

Fictional films can influence science by enhancing funding opportunities, promoting research agendas, influencing public controversies and playing a role in intra-specialist communication (Kirby 2011). Kay (2000) offers the useful notion of the 'technoscientific imaginary' to account for shared representational practices both within science and in the broader culture. Technoscientific imaginary encompasses all the narratives, both scientific and public, that frame an issue and give it its cultural value. Thus, the assumed impact of a film on public opinion can give it utility within the political arena, as was the case for *The Day After Tomorrow* and *Contagion*, whether or not these films had any real impact on the public because they became part of the technoscientific imaginary (Nisbet 2004).

One of the biggest impacts of movies on public perceptions of science and technology has come through their ability to raise public awareness of an issue or scientific field (Kirby 2011). Entertainment media serve an agenda-setting function for news outlets that use the release of a new movie or the success of a new television programme to provide news content. The ability of entertainment media to raise the visibility of a scientific issue has led to an inordinate amount of influence for movies on national science policies (Greenbaum 2009) including debates over cloning and stem cells (Haran *et al.* 2008), near-Earth objects (Kirby 2011; Mellor 2007), nuclear power (Sjöberg and Engelberg 2010) and emerging viruses (Tomes 2000). Movies have also proved to be an effective means for creating public excitement about undeveloped technologies which then move from the fictional into the real world (Bleecker 2009; Kirby 2010).

Science in cinema, however, rarely exists as a solitary entity. One need only look at *Jurassic Park* in its incarnations as a novel, film, comic book, computer game, television documentaries and news articles to see the high degree of intertextuality in science-based media. The interplay between popular texts and formal scientific discourse is what Schell (1997) calls *genre interpenetration*. Genre interpenetration was clearly evident in the case of *Outbreak* (1995), as popular science texts, documentaries, political treatises and scientific works all borrowed imagery and narratives from the film. News media incorporated the film's images and narratives in their coverage of a real-life outbreak of Ebola in Zaire that occurred while the film was in cinemas in the US and Europe (Vasterman 1995; Ostherr 2005). Clearly, cinema fits into Bruce Lewenstein's (1995) web model of science communication in that cinema, other mass media and technical media interact in complex ways, informing and referring to each other.

Concluding remarks

The work outlined in this chapter shows that the presence of science and technology in film is a powerful cultural force that can have a major impact on our concept of science communication and public attitudes towards science. Cinematic depictions of science involve the production and presentation of an *image* of science whether or not the image has anything to do with *real science*. Scholarship on scientists' role in movie production reveals that film-makers take a flexible approach in determining what *authenticity* means in the context of fiction; scientific accuracy always takes a back seat to storytelling. The point of movies is not to devise accurate or educational communications about science, but to produce images of science that are entertaining. Despite the increased number of studies into scientists' role in movie production, there is still a need for scholars to uncover exactly how, and why, film-makers produce filmic images of science. How do scriptwriters approach science? What role does science play in storytelling? How important is science for special effects technicians? What role does science play for production designers or the art department? What does *scientific accuracy* mean in the context of fictional films?

In terms of media effects, scholars have begun considering science and cinema outside of this medium's impact, if any, on science literacy. Even those who write 'real science of ...' reviews of cinematic science understand that these exercises are meant to be entertainment or to be used for pedagogical purposes in ISE, rather than as serious critiques of how film-makers approach science in movies. Filmic images can have an impact on the public's conceptions of science by provoking reactions, from encouraging excitement to instilling fear about science and technology, and sometimes both. The question still remains, however: exactly what, and how strong, is this impact? This question requires innovative studies on media effects of fictional cinema and science that move beyond traditional quantitative approaches to audience reception. These approaches should include explorations into how specific cultural groups, such as scientists, religious communities and policymakers, respond to, and appropriate, science-based movies.

The landscape for science and cinema has changed dramatically since 2008 when the US National Academy of Sciences introduced their Science and Entertainment Exchange initiative. The involvement of major scientific organisations like National Academy of Science and Wellcome Trust in film and television production is an exciting development in the field of science communication. All the recent programmes emerging from scientific organisations, including science film festivals, provide ample opportunities to study cinematic science through production, text or reception. For a start, science communication scholars can analyse whether new programmes like the Science and Entertainment Exchange are effective. Have these initiatives significantly altered the ways in which science is presented in film? Have they increased Hollywood's capacity to make better films? Have these programmes changed public attitudes towards science or modified behaviour? Science communication researchers are well placed to provide these scientific organisations with the evidence they need to improve or modify their attempts to communicate science through movies and television.

Key questions

- What kinds of advice, beyond factual information, would you expect scientists to provide filmmakers?
- How straightforward is the notion of scientific *accuracy* in fictional movies?
- Are scientist stereotypes problematic for science?
- In what ways, outside of science literacy, can we measure the impact of movies on the public or on science itself?

Online resources

- National Academy of Sciences' Science and Entertainment Exchange www.scienceandentertainmentexchange.org/
- Entertainment Industries Council www.eiconline.org/
- American Film Institute's Sloan Science Advisor program www.afi.com/conservatory/admissions/sloanadvisors.aspx
- USC's Hollywood Health and Society hollywoodhealthandsociety.org/
- Imagine Science Film Festival www.imaginesciencefilms.org/

Notes

1 Since the publication of the first version of this chapter (Kirby 2008) there have been some dramatic changes in both the study of science in cinema and the involvement of scientific organisations in movie and television productions. This revised version takes account of new academic work in audience reception, including work on *informal science education*, science literacy, the *CSI effect*, and entertainment education. It takes account of initiatives by high-profile scientific organisations such as the National Academy of Sciences whose Science and Entertainment Exchange facilitates scientific advice for filmmakers and television producers, and of the rise of the science film festival and the growing use of short fictional films for outreach in the age of YouTube.
2 http://io9.com; http://boingboing.net
3 My discussion in this chapter relates predominantly to popular fictional films. While there is a good deal of similarity between cinema and television as visual media, television has its own production practices, marketing, dissemination routes, sites of reception and cultural contexts. As such, science on television has its own body of scholarly literature. Likewise, the same issues and lessons can be applied to other fictional media including literature. In addition, it must be noted that this study focuses primarily on mainstream Hollywood cinema.
4 It is not possible to give complete descriptions of films within this chapter. For more information about a film please visit the Internet Movie Database (www.imdb.com).

References

Allgaier, J. (2013) 'On the shoulders of YouTube: science in music videos', *Science Communication*, 35, 2: 266–275.
Balmford, A., Manica, A., Airey, L., Birkin, L., Oliver, A. and Schleicher, J. (2004) 'Hollywood, climate change, and the public', *Nature*, 305, 5691: 1713.

Barnett, M., Wagner, H., Gatling, A., Anderson, J., Houle M. and Kafka, A. (2006) 'The impact of science fiction film on student understanding of science', *Journal of Science Education and Technology*, 15, 2: 179–191.

Barnett, M. and Kafka, A. (2007) 'Using science fiction movie scenes to support critical analysis of science', *Journal of College Science Teaching*, 36: 31–35.

Barriga, C., Shapiro, M. and Fernandez, M. (2010) 'Science information in fictional movies: effects of context and gender', *Science Communication*, 32, 1: 3–24.

Bleecker, J. (2009) 'Design fiction: a short essay on design, science, fact and fiction', Near Future Laboratory; online at http://nearfuturelaboratory.com/2009/03/17/design-fiction-a-short-essay-on-design-science-fact-and-fiction/; accessed 28 April 2013.

Böhnke, M. and Machura, S. (2003) 'Young Tom Edison – Edison, the man: biopic of the dynamic entrepreneur', *Public Understanding of Science*, 12, 3: 319–33.

Brereton, P. (2005) *Hollywood Utopia*, Bristol: Intellect Books.

Cole, S. and Dioso-Villa, R. (2007) 'CSI and its effects: media, juries, and the burden of proof', *New England Law Review*, 41, 3: 435–470.

de Ceglia, F. (2012) 'From the laboratory to the factory, by way of the countryside: fifty years of Italian scientific cinema (1908–1958)', *Public Understanding of Science*, 21, 8: 949–967.

Dinello, D. (2006) *Technophobia!*, Austin, TX: University of Texas Press.

Dubeck, L., Moshier, S. and Boss, J. (2004) *Fantastic Voyages: Learning Science Through Science Fiction Films*, New York: Springer.

Dudo, A., Brossard, D., Shanahan, J., Scheufele, D., Morgan, M. and Signorielli, N. (2011) 'Science on television in the 21st century: recent trends in portrayals and their contributions to public attitudes toward science', *Communication Research*, 38, 4: 754–777.

Eberl, J. (2010) 'I, clone: how cloning is (mis)portrayed in contemporary cinema', *Film and History*, 40, 2: 27–44.

Efthimiou, C. and Llewellyn, R. (2007) 'Cinema, Fermi problems and general education', *Physics Education*, 42, 3: 253–261.

Elena, A. (1997) 'Skirts in the lab: Madame Curie and the image of the woman scientist in the feature film', *Public Understanding of Science*, 6, 3: 269–278.

Flicker, E. (2003) 'Between brains and breasts – women scientists in fiction film: on the marginalization and sexualization of scientific competence', *Public Understanding of Science*, 12, 3: 307–318.

Frank, S. (2003) 'Reel reality: science consultants in Hollywood', *Science as Culture*, 12, 4: 427–469.

Franklin, S. (2000) 'Life itself: global nature and genetic imaginary', in S. Franklin, C. Lury and J. Stacey (eds) *Global Nature, Global Culture,* London: Sage, 188–227.

Frayling, C. (2005) *Mad, Bad and Dangerous*, London: Reaktion.

Greenbaum, D. (2009) 'Is it really possible to do the Kessel Run in less than twelve parsecs and should it matter? Science and film and its policy implications', *Vanderbilt Journal of Entertainment and Technology Law*, 11, 2: 249–333.

Griep, M. and Mikasen, M. (2009) *ReAction!: Chemistry in the Movies*, Oxford: Oxford University Press.

Haraway, D. (1991) *Simians, Cyborgs and Women*, London: Routledge.

Haran, J., Kitzinger, J., McNeil, M. and O'Riordan, K. (2008) *Human Cloning in the Media: From Science Fiction to Science Practice*, New York: Routledge.

Haynes, R. (1994) *From Faust to Strangelove*, Baltimore: Johns Hopkins University Press.

Ingram, D. (2000) *Green Screen*, Exeter: University of Exeter Press.

Irwin. A. (1995) *Citizen Science*, London: Routledge.

Jackson, J. (2011) 'Doomsday ecology and empathy for nature: women scientists in "B" horror movies', *Science Communication*, 33, 4: 533–555.

Jensen, E. (2008) 'The Dao of human cloning: utopian/dystopian hype in the British press and popular films', *Public Understanding of Science*, 17, 2: 123–143.

Jones, R. (2001) 'Why can't you scientists leave things alone?: Science questioned in British films of the post-war period (1945–1970)', *Public Understanding of Science*, 10, 4: 1–18.

Jörg, D. (2003) 'The good, the bad, the ugly: Dr. Moreau goes to Hollywood', *Public Understanding of Science*, 12, 3: 297–305.

Kanner, M. (2006) 'Going on instinct: gendering primatology in film', *Journal of Popular Film and Television*, 33, 4: 206–212.

Kay, L. (2000) *Who Wrote the Book of Life?*, Stanford, CA: Stanford University Press.

King, G. (2000) *Spectacular Narratives*, London: I.B. Tauris.

Kirby, D. (2000) 'The new eugenics in cinema: genetic determinism and gene therapy in *GATTACA*', *Science Fiction Studies*, 27, 2: 193–215.
Kirby, D. (2003a) 'Scientists on the set: science consultants and communication of science in visual fiction', *Public Understanding of Science*, 12, 3: 261–278.
Kirby, D. (2003b) 'The threat of materialism in the age of genetics: DNA at the drive-in', in G. Rhodes (ed.) *Horror at the Drive-In: Essays in Popular Americana*, Jefferson, NC: McFarland, 241–258.
Kirby, D. (2007) 'The devil in our DNA: a brief history of eugenic themes in science fiction films', *Literature and Medicine*, 26, 1: 83–108.
Kirby, D. (2008) 'Cinematic science', in M. Bucchi and B. Trench (eds) *Handbook of Public Communication of Science and Technology*, London and New York: Routledge, 41–56.
Kirby, D. (2010) 'The future is now: diegetic prototypes and the role of popular films in generating real-world technological development', *Social Studies of Science*, 40, 1: 41–70.
Kirby, D. (2011) *Lab Coats in Hollywood: Science, Scientists, and Cinema*, Cambridge, MA: MIT Press.
Knippels, M. -C., Severiens, S. and Klop, T. (2009) 'Education through fiction: acquiring opinion-forming skills in the context of genomics', *International Journal of Science Education*, 31, 15: 2057–2083.
Lambourne, R., Shallis, M. and Shortland, M. (1990) *Close Encounters?*, New York: Adam Hilger.
Lederer, S. and Parascandola, J. (1998) 'Screening syphilis: Dr. Ehrlich's Magic Bullet meets the Public Health Service', *Journal of the History Of Medicine*, 53, 4: 345–370.
Leiserowitz, A. (2004) 'Before and after The Day After Tomorrow: a US study of climate change risk perception', *Environment*, 46, 9: 22–37.
Lewenstein, B. (1995) 'From fax to facts: communication in the cold fusion saga', *Social Studies of Science*, 25, 3: 403–436.
Lipkin, I. (2011) 'Professor Ian Lipkin brings science to Hollywood's "Contagion"', Columbia University press release; online at www.mailman.columbia.edu/news/media/professor-ian-lipkin-brings-science-hollywoods-contagion; accessed 28 April 2013.
Losh, S. (2010) 'Stereotypes about scientists over time among US adults: 1983 and 2001', *Public Understanding of Science*, 19, 3: 372–382.
Lowe, T., Brown, K., Dessai, S., de França Doria, M., Haynes, K. and Vincent, K. (2006) 'Does tomorrow ever come? Disaster narrative and public perceptions of climate change', *Public Understanding of Science*, 15, 4: 435–457.
Massarani, L. and de Castro Moreira, I. (2005) 'Attitudes towards genetics: a case study among Brazilian high school students', *Public Understanding of Science*, 14, 2: 201–212.
McCurdy, H. (1997) *Space and the American Imagination*, Washington, DC: Smithsonian.
Medina-Doménech, R. and Menéndez-Navarro, A. (2005) 'Cinematic representations of medical technologies in the Spanish official newsreel, 1943–1970', *Public Understanding of Science*, 14, 4: 393–408.
Mellor, F. (2007) 'Colliding worlds: asteroid research and the legitimisation of war in space', *Social Studies of Science*, 37, 4: 499–531.
Merzagora, M. (2010) 'Reflecting imaginaries: science and society in the movies' in A. Smelik (ed.) *The Scientific Imaginary in Visual Culture*, Göttingen, Germany: V&R Unipress, 39–52.
National Science Teacher Association (2012) *NSTA position statement: Learning Science in informal Environments*; online at www.nsta.org/about/positions/informal.aspx; accessed 28 April 2013.
Nelkin, D. and Lindee, S. M. (1995) *The DNA Mystique*, New York: W.H. Freeman.
Science and Entertainment Exchange (2011) 'Scientist spotlight: Donna Nelson', Science and Entertainment Exchange; online at www.scienceandentertainmentexchange.org/article/scientist-spotlight-donna-nelson; accessed 28 April 2013.
Nisbet, M. (2004) 'Evaluating the impact of The Day After Tomorrow: can a blockbuster film shape the public's understanding of a science controversy?', *Skeptical Inquirer*, 16 June; online at www.csicop.org/specialarticles/show/evaluating_the_impact_of_the_day_after_tomorrow; accessed 28 April 2013.
Nisbet, M. and Dudo, A. (2010) 'Science, entertainment, and education: a review of the literature', Report for the National Academy of Sciences, Washington, DC: National Academy of Sciences.
Nisbet, M. and Scheufele, D. (2009) 'What's next for science communication? Promising directions and lingering distractions', *American Journal of Botany*, 96, 10: 1767–1778.
Ordway, F. I. (undated) '2001: A Space Odyssey in retrospect'; online at www.visual-memory.co.uk/amk/doc/0075.html; accessed 28 April 2013.
O'Riordan, K. (2008) 'Human cloning in film: horror, ambivalence, hope', *Science as Culture*, 17, 2: 145–162.
Ostherr, K. (2005) *Cinematic Prophylaxis*, Durham, NC: Duke University Press.

Pansegreau, P. (2008) 'Stereotypes and images of scientists in fiction films', in P. Weingart and B. Huppauf (eds) *Science Images and Popular Images of the Sciences*, New York: Routledge, 257–266.

Pellechia, M. (1997) 'Trends in science coverage: a content analysis of three US newspapers', *Public Understanding of Science*, 6, 1: 49–68.

Perkowitz, S. (2007) *Hollywood Science*, New York: Columbia University Press.

Pernick, M. (1996) *The Black Stork*, Oxford: Oxford University Press.

Reingold, N. (1985) 'Metro-Goldwyn-Mayer meets the atom bomb', in T. Shinn and R. Whitley (eds) *Expository Science*, Dordecht: D. Reidel, 229–245.

Reusswig, F., Schwarzkopf, J. and Pohlenz, P. (2004) Double impact: the climate blockbuster '*The Day After Tomorrow*' and its impact on the German cinema public, *PIK Report, 92*, Potsdam: Potsdam Institute for Climate Impact Research: online at www.pik-potsdam.de/research/publications/pikreports/.files/pr92.pdf; accessed 28 April 2013.

Rose, C. (2003) 'How to teach biology using the movie science of cloning people, resurrecting the dead, and combining flies and humans', *Public Understanding of Science*, 12, 3: 289–296.

Saltzberg, D. (2011) The Big Blog Theory; online at www.thebigblogtheory.wordpress.com/; accessed 28 April 2013.

Schell, H. (1997) 'Outburst! A chilling true story about emerging-virus narratives and pandemic social change', *Configurations*, 5, 1: 93–133.

Schibeci, R. and Lee, L. (2003) 'Portrayals of science and scientists, and "science for citizenship"', *Research in Science & Technological Education*, 21, 2: 177–192.

Schiermeier, Q. (2004) 'Disaster movie highlights transatlantic divide', *Nature*, 431, 7004: 4.

Schweitzer, N. and Saks, M. (2007) 'CSI effect: popular fiction about forensic science affects the public's expectations about real forensic science', *Jurimetrics Journal*, 47, 3: 357–364.

Shapiro, J. (2002) *Atomic Bomb Cinema*, New York: Routledge.

Shelton, A., Kim, D., Young, S. and Barak, G. (2007) 'Study of juror expectations and demands concerning scientific evidence: does the CSI effect exist?', *Vanderbilt Journal of Entertainment & Technology Law*, 9, 2: 331–368.

Singhal, A., Cody, M., Rogers, E. and Sabido, M. (eds) (2004) *Entertainment-education and Social Change: History, Research, and Practice*, Mawwah, NJ: Lawrence Erlbaum.

Sjöberg, L. and Engelberg, E. (2010) 'Risk perception and movies: a study of availability as a factor in risk perception', *Risk Analysis*, 30, 1: 95–106.

Skal, D. (1998) *Screams of Reason*, New York: Norton.

Stacey, J. (2010) *The Cinematic Life of the Gene*, Durham, NC: Duke University Press.

Steinke, J. (2005) 'Cultural representations of gender and science: portrayals of female scientists and engineers in popular films', *Science Communication*, 27, 1: 27–63.

Steinke, J., Lapinski, M., Long, M., Van Der Maas, C., Ryan, L. and Applegate, B. (2009) 'Seeing oneself as scientist: media influences and adolescent girls' science career-possible selves', *Journal of Women and Minorities in Science and Engineering*, 15, 4: 279–301.

Stern, M. (2004) 'Jurassic Park and the moveable feast of science', *Science as Culture*, 13, 3: 347–372.

Telotte, J. (1995) *Replications*, Chicago: University of Illinois Press.

Terzian, S. and Grunzke, A. (2007) 'Scrambled eggheads: ambivalent representations of scientists in six Hollywood film comedies from 1961 to 1965', *Public Understanding of Science*, 16, 4: 407–419.

Thurs, D. (2007) 'Tiny tech, transcendent tech: nanotechnology, science fiction, and the limits of modern science talk', *Science Communication*, 29, 1: 65–95.

Tomes, N. (2000) 'The making of a germ panic, then and now', *American Journal of Public Health*, 90, 2: 191–198.

Tosi, V. (2005) *Cinema before Cinema: The Origins of Scientific Cinematography*, London: British Universities Film and Video Council.

Tudor, A. (1989) *Monsters and Mad Scientists*, Oxford: Basil Blackwell.

Turney, J. (1998) *Frankenstein's Footsteps*, New Haven, CT: Yale University Press.

Turow, J. (2010) *Playing Doctor: Television, Storytelling and Medical Power*, Ann Arbor: University of Michigan Press.

Valenti, J. (2012) 'Sundance 2012: robots, dying lakes, and sci-fi disasters', *Science Communication*, 34, 2: 292–295.

Van Dijck, J. (1998) *Imagenation: Popular Images of Genetics*, London: Macmillan.

Vasterman, P. (1995) 'The Hollywood plague', *Albion Monitor*, 19 August; online at www.monitor.net/monitor/8-19-95/virus.html; accessed 28 April 2013.

Vieth, E. (2001) *Screening Science*, Lanham, MD: Scarecrow.
Weart, S. (1988) *Nuclear Fear*, Cambridge, MA: Harvard University Press.
Weingart, P. (2008) 'The ambivalence towards new knowledge: science in fiction film', in P. Weingart and B. Huppauf, B. (eds) *Science Images and Popular Images of the Sciences*, New York: Routledge, 267–282.
Weingart, P. with C. Muhl and P. Pansegrau (2003) 'Of power maniacs and unethical geniuses: science and scientists in fiction film', *Public Understanding of Science*, 12, 3: 279–287.
Wiesenfeldt, G. (2010) 'Dystopian genesis: the scientist's role in society, according to Jack Arnold', *Film & History*, 40, 1: 58–74.
Wood, A. (2002) *Technoscience in Contemporary American Film*, Vancouver, BC: University of British Columbia Press.

9

Environmentalists as communicators of science

Advocates and critics

Steven Yearley

Introduction

In 2003 British newspaper readers were faced with full-page advertisements showing a figure closely resembling Michelangelo's David, complete with its tiny phallic endowment. The small text beneath the image revealed that the images were being displayed by Greenpeace in conjunction with its report on the human and environmental impacts of man-made chemicals (Greenpeace 2003). The accompanying information suggested that people should begin to worry about threats to men's reproductive capacity owing to the environmental release of hormone-mimicking substances (for one of the leading early accounts of this topic, see Cadbury 1998). Chemicals used in plasticisers and other applications could be feminising the environment and leading to declining male fertility, in humans and in wild animals too, it was claimed.

This advertisement can be seen as symptomatic of Greenpeace's strategy. It expressed, in an arresting way, the supposed facts of the case: here was a new form of harm arising from a novel and unanticipated type of environmental pollution. At the same time, it was also somewhat misleading since the chemicals were unlikely, in anyone's view, to lead to a threat to the size of male members – the likely harm was to the health of sperm. This advertisement encapsulated a key challenge in the public communication strategy of environmental NGOs: the need to balance powerful, evocative images with the perceived demands of accuracy.

Their opponents have frequently criticised Greenpeace and other environmental organisations for favouring the slick image over the accurate message but this criticism – though interesting and important – implicitly acknowledges something even more fundamental. The key point is that environmental pressure groups can be called to account on this issue precisely because the persuasiveness of their message depends on the notion that their claims have a basis in factual accuracy, that they are not matters of opinion. Environmentalists, more than any other type of campaigner, need to persuade the public that things are *in fact* the way they say things are (see Yearley 1992) even when some of the claims they are making seem – at first glance at least – to be counter-intuitive or implausible: that plastics can make you (and fish) infertile or that burning coal, gas and oil can unsettle the entire global climate.

Thus, in what is clearly today's pre-eminent environmental debate, environmentalists are keen to assert that climate change is in fact taking place and that it has been human-caused alterations in the make-up of the atmosphere that are responsible. Indeed, so pivotal is the dependence on scientific evidence in relation to climate change that the environmentalist and author Mark Lynas announced a high-profile about-turn in his views on genetically modified crops. In his 2013 statement he made clear that deference to science had been key for him. He states that the reason for his change of heart was:

> fairly simple: I discovered science, and in the process I hope I became a better environmentalist. Having written two books about the science of global warming, I came to understand that defending climate science was incompatible with attacking the science of biotechnology.
>
> (Lynas 2013, n.p.)

The decisive point here is that Lynas felt it was so vital to support the scientific community's claims about climate change that he was driven to re-evaluate his scepticism about scientists' views on agricultural biotechnology.

In short, my suggestion underlying this chapter is that there is an elective affinity between environmental campaign organisations and scientific claims that is to a large degree distinctive among pressure groups. This gives environmentalists and green campaigning bodies an urgent interest in science communication issues and makes them significant science communication actors.[1]

Climate change as a science communication challenge

Environmental campaign organisations have been important in supplying arguments about and publicising problems in relation to a very large number of environmental issues. Rather than trying to conduct a review of all these, this chapter focuses first on one leading example, derives some points of principle from it and then assesses their generalisability by applying them to a contrasting case. The case I shall examine first is that of climate change.

Scientists have been aware, for over a century, that climate undergoes significant variation and there has long been a concern that human society could not count on a stable climate forever. As such climate research was refined, in part, thanks to the growth in computer power in the 1970s and 1980s, the majority opinion endorsed the earlier suggestion that enhanced warming driven by the build-up of atmospheric carbon dioxide was presenting a problem for humankind in the short to medium term. Environmental groups are reported to have been initially wary of campaigning around this issue (Pearce 1991) since it seemed such a long shot and with such high stakes. With acid rain on the agenda and many governments active in denying scientific claims about even this comparatively straightforward effect, the risks seemed too high in the 1980s to declare publicly that emissions might be sending the whole climate out of control. Worse still from a campaigner's point of view, at a time when environmentalists were looking for concrete successes, the issue seemed almost designed to provoke and sustain controversy. The records of past temperatures across the globe as a whole were not good and there was the danger that rising trends in urban air-temperature measurements in the West were simply an artefact: perhaps cities had simply become warmer as they grew in size. Others doubted that additional carbon dioxide releases would lead to a build-up of the gas in the atmosphere since the great majority of carbon is in soils, trees and the oceans, so sea creatures and plants might simply sequester more carbon. And even if the scientific community was correct about the build-up of carbon dioxide in the atmosphere, it was fiendishly difficult to work out what the implications of this would be in order to build campaigns with local resonance.

Hart and Victor (1993) tracked the interaction between climate science and US climate policy from the 1950s up to the mid-1970s by which time greenhouse emissions had begun to be 'positioned as an issue of pollution' (668); the climate, 'scientific leaders discovered, could be portrayed as a natural resource that needed to be defended from the onslaught of industrialism' (667). Subsequently, according to Bodansky (1994), the topic's rise to policy prominence was assisted by other considerations. There was, for example, the announcement of the discovery of the *ozone hole* in 1987; this lent credibility to the idea that the atmosphere was vulnerable to environmental degradation and that humans could unwittingly cause harm at a global level. Also important was the coincidence in 1988 between Senate hearings into the issue and a very hot and dry summer in the USA. Nonetheless, most politicians responded to the warnings in the 1980s with a call for more research.

One significant outcome of this support for research was the setting up, in 1988, of a new form of scientific organisation, the Intergovernmental Panel on Climate Change (IPCC), under the aegis of the World Meteorological Organisation and the United Nations Environment Programme. The aim of the IPCC was to collect together leading figures in all aspects of climate change with a view to establishing, in an authoritative way, the nature and scale of the problem and to identify possible policy responses. This initiative was accorded significant political authority and was novel in important ways. Among its innovations were the explicit inclusion of social and economic analyses, alongside the atmospheric science, and the involvement of governmental representatives in agreeing and authoring report summaries; '[w]hile by no means the first to involve scientists in an advisory role at the international level, the IPCC process has been the most extensive and influential effort so far' (Boehmer-Christiansen 1994: 195).

As is widely known, the IPCC and mainstream climate research have met with determined criticism. There have been scholars and moderate critics who have concerns that the IPCC procedure tends to marginalise dissenting voices and that particular policy proposals (such as the Kyoto Protocol) are maybe not as wise or as cost-effective as proponents suggest (see, for example, Prins and Rayner 2007). There are also very many consultants backed by the fossil fuel industry who throw doubt on claims about climate change (Freudenburg 2000 offers a discussion of the social construction of *non-problems*); these claims-makers have entered into alliance with right-leaning politicians and commentators to combat particular regulatory moves as detailed by McCright and Dunlap (2000; 2003; see also Nisbet in this volume). Informal networks, often web-based,[2] have been set up to allow *climate change sceptics* to publicise their views, and they have welcomed all manner of contributors, whether direct enemies of the Kyoto Protocol or more distant allies such as opponents of wind farms or conspiracy theorists who see climate change warnings as the machinations of the nuclear industry.

Novelist Michael Crichton waded into this controversy with *State of Fear* (2004) having a technical appendix and author's message on the errors in climate science. In his book, Crichton even offers his own estimate of the level of global warming (0.812436 degrees) over the next century (2004: 677). Crichton and others have concentrated not only on the scientific conclusions (and their disagreements with them) but also on putative explanations for the persistence of error in *establishment* science. Meanwhile, mainstream environmental NGOs have argued simply that one should take scientists' word for the reality of climate change. Indeed, at the 2007 camp for climate action at London's Heathrow Airport, environmentalists protesting at plans for further airport development famously carried a huge banner declaring 'we are armed only with peer-reviewed science' (Bowman 2010: 177).

The rhetorical difficulties of speaking up for mainstream science had already been foreshadowed in the strategy of Friends of the Earth in London about 20 years before; campaign staff working on climate change issues were disturbed by a programme aired on the UK's Channel 4 in the *Equinox* series in 1990 that sought to question the scientific evidence for global warming.

The programme even implied that scientists might be attracted to make extreme and sensational claims about the urgency of the problem in order to maximise their chances of receiving research funding. The programme was criticised in the Campaign News section of the Friends of the Earth magazine, *Earth Matters*. An unfavourable comparison was drawn between the sceptical views expressed in the programme and the conclusions of the IPCC, whose scientific analysis Friends of the Earth was generally in agreement with. Friends of the Earth's article invoked the weight of 'over 300 scientists [who] prepared the IPCC's Science Report compared to about a dozen who were interviewed for *Equinox*' (1990: 4).[3] When apparently well-qualified scientists are seen to disagree, it seems like a reasonable alternative to invoke the power of the majority. But, of course, in many areas environmentalists believe themselves to be factually correct when they have been in the scientific minority, at least initially. In March 2007, Channel 4 repeated its attention-seeking strategy, broadcasting a programme unambiguously entitled *The Great Global Warming Swindle*. The argumentational response of NGOs and green commentators was essentially the same: we should trust the advice of the great majority of well-qualified scientists who accept the evidence of climate change. Environmental groups looked to invoke the possible vested interests of the critics in order to make sense of the programme-makers' and contributors' continued scepticism.

In the relationship between the IPCC – indeed the whole climate change regulation community – and its critics, not only the science but the various ways in which the science is legitimated have come under attack (see Lahsen 2005). Critics have been quick to point to the supposed vested interests of this community. Its access to money depends on the severity of the potential harms that it warns about; hence – or so it is argued – it inevitably has a structural temptation to exaggerate those harms. As it was working in such a multidisciplinary area and with high stakes attached to its policy proposals, the IPCC attempted to extend its network widely enough so as to include all the relevant scientific authorities; it was clearly important that the IPCC should not be dominated by meteorologists or atmospheric chemists. But this meant that the IPCC ran into problems with peer-reviewing and perceived impartiality; there were virtually no *peers* who were not already within the IPCC (see Edwards and Schneider 2001). Conventional peer reviewing relies on there being few authors and many (more or less disinterested) peers; in many ways the IPCC reversed this situation. This development also created problems for environmentalists' claims to be 'armed only with peer-reviewed science' (in Bowman's words, 2010: 177) since it pointed to potential limitations with peer review itself.

If challenged, the IPCC tended to fall back in line with the classic script of 'science for policy' (Yearley 2005); the IPCC legitimated itself in terms of the scientific objectivity and impartiality of its members. But critics were able to point out that the IPCC itself selects who is in the club of the qualified experts and, thus, threatens to be a self-perpetuating, elite community. This was exactly the point that Crichton picked up. His principal argument was that the key requirements are a form of independent verification for claims about climate change and the guarantee of access to unbiased information. However well meant, this is clearly an unrealistic demand since there is no one with scientific skills in this area who could plausibly claim to be entirely disinterested. There is no Archimedean point to which to retreat, and environmentalists will very reasonably claim that such demands for a review are primarily ways to put off taking action. Crichton further muddied the water by proposing to offer his own estimate of future climate change to six decimal places; though the ridiculous precision clearly signalled some jocular intent, the idea that even he (a medic turned author) could offer a temperature-change forecast implies that there are lots of people able to make independent judgements. By contrast, there are relatively few people placed to make such judgements, and a central science communication challenge for environmentalists is to distinguish between those who can credibly comment and those who cannot.

Though they have found it hard to participate in the central scientific debate and have been obliged to take up the (for them) unusual position of defending the correctness of mainstream science, environmentalists have found other activities that they have been able to pursue. For example, in the USA they have been active in trying to identify novel ways to press the government to change its position on climate change aside from simply bolstering the persuasiveness of climate science and trying to rebut the claims of critics. In 2006 the Center for Biological Diversity (CBD), the Natural Resources Defense Council and Greenpeace learned that their inventive use of the Endangered Species Act to sue the US government for protection of polar bears and their habitat in Alaska had won concessions from the government. In its campaigning, the CBD had argued that oil exploration in the far north would harm polar bears and their hunting grounds; but they also suggested that ice melting caused by global warming was responsible for additional habitat loss and harm to bears who need large expanses of solid ice in spring for successful hunting.[4] Potentially, the Endangered Species legislation could force the government to examine the impact on polar bears of all actions in the US (such as energy policy), not just activities local to polar bear habitat.

Environmental NGOs' dilemma starts with this: what they see as the world's leading environmental problem is fully endorsed by the mainstream scientific community. Indeed, in January 2004 the UK government's Chief Scientific Adviser Sir David King gave his judgement that climate change posed a greater threat than terrorism.[5] Their principal efforts have accordingly been directed at restating and emphasising official findings, finding novel ways to publicise the message and countering the claims of greenhouse sceptics. What causes the dilemma is that such statements in favour of the objectivity of the scientific establishment's views mean that it is harder for NGOs to distance themselves from scientists' conclusions on other occasions without appearing arbitrary or tendentious.

For environmentalists, the science communication issues around climate change have been further complicated by societal and technical developments over the last decade. In a sense the debate around climate change and the policy options appeared reasonably clear at the very start of the twenty-first century. Apart from countries that did not wish to play along with international agreements, everyone else assumed that the goal was to move away from the standard fossil fuels. The question was simply how quickly and by what means. A key strategy in addressing climate change is to find ways to reduce emissions by switching to other fuels, and complications for environmentalists have arisen around both nuclear and wind energy. Nuclear power has made a strong comeback as a low-carbon, large-scale energy source, though the disaster at the Fukushima plant on Japan's east coast in 2011, when the plant was overwhelmed by a tsunami provoked by a large earthquake, has reawakened enduring concerns over nuclear safety. The Fukushima disaster had far-reaching effects with, for example, the German government opting to withdraw from nuclear energy, even while Sweden and Britain press ahead. Environmentalists have not been able to agree on a response to this issue – some prioritise decarbonisation ahead of worries about nuclear safety while others take the opposing view.

The other major source of energy to be mentioned in this context is wind power. Wind has played a significant role in Denmark for 30 years, and in the last decade Germany has enormously increased its wind energy use, raising the total to just below 10 per cent of electricity production, all this as part of the well-known 'Energiewende' (energy turn). Chinese and US engineering companies are also strongly represented in this sector. In this case there is little discussion of how safe the technology is but the focus is on its acceptability in relation to landscape, amenity and wildlife. In Britain there has been extensive and well-organised opposition to wind farms. The arguments advanced have ranged from debates over the principles for determining where to site turbines to claims that landscape has been undervalued or that damage may be done – and even carbon dioxide released – by the footings required to anchor turbines. Community responses have often focused on the construction phase of wind farms as much as on their eventual operation. Finally, claims have been made about the effects of the use of rotor blades on birdlife and even on bats (Aitken 2010). The move to offshore wind would seem to remove some of the amenity arguments, though the value of seascapes has again been raised and work continues on the impact on coastal birds. In many cases alliances have developed between opposition to wind energy and climate scepticism, with a pastoral conservatism interpreting both things as the imposition of an untrustworthy modern discourse on the rural environment.

Wind and nuclear energies are complicated for environmentalists to evaluate and green advocates have arrived at conflicting positions. But the issues at stake are at least relatively predictable. However, there are three other issues that have come to prominence since the Kyoto Protocol was devised and which create further significant science communication problems for environmentalists. The first arises from the success of developing economies, with China most notably in the lead. These countries were not limited by the Kyoto targets but in the ensuing years their rapid industrial growth has led them to become major emitters. China has already overtaken the US as the world's primary producer of carbon dioxide. It is, of course, difficult to champion a policy that precisely omits the country that makes the largest contribution to the problem. But the deeper complicating factor here is that China has in large part developed so rapidly because it has taken over bits of industrial production that were formerly located in Europe, North America or Japan and it is this production that generates so much of the carbon dioxide. In effect, Europe has allowed China to produce its goods and also generate its greenhouse emissions. To put this another way, at least part of the good emissions performance in Britain, Germany and the Netherlands this century has come about because China has taken over their emissions even while European citizens import Chinese manufactured products (Helm 2012). Helm argues that environmentalists have been much less good at communicating the economics of climate change; their priority has been on the atmospheric science and on certain human rights issues. Tacitly, he argues, environmentalists have signed up to an interpretation of the carbon problem that does not match with economic realities.

Current energy policies have also been decisively affected by innovations in the extraction of fossil fuels. In North America vast new reserves of fossil energy have been identified in oil sands and shales. These non-standard sources have been developed rapidly and in a way that lessens dependence on the global market, reduces domestic energy prices and, at the same time, leads to some carbon benefits if the gas displaces coal, a much heavier carbon emitter. The US has seen its energy output transformed by hydraulic fracturing (given the unlovely sobriquet of *fracking*), the latest development in non-traditional fossil fuel extraction. While several senior policy actors have spoken up in favour of a renewed switch to *fracked* gas, environmentalists have typically resisted these moves and been opposed to oil-sands-derived oil and especially to fracking. Extracting oil from the sands uses significantly more energy than conventional oil wells so the carbon *footprint* is accordingly greater. Because hydraulic fracturing involves using oil industry techniques to split soft rock with pumped water at high pressure so that gas flows out, there has been concern over the fate of the now-polluted water. Anxieties have also been expressed over the possibility that the fracturing of the rock strata could give rise to subsidence and even minor earthquakes. Environmentalists have faced the difficult situation of wishing to leave the gas in the ground, even if the gas could potentially take the place of coal, a much worse contributor to claimed climate change.

Another emerging communication issue around climate is characterised by an even more complicated relationship between environmentalists on the one hand and scientists and engineers on the other: geoengineering. This refers to proposals to address climate change by intervening to deal directly with the consequences – for example, by finding ways to remove carbon from the atmosphere or by offsetting global warming through measures to lessen the amount of solar heat reaching the earth. There are some feasible strategies here that are relatively straightforward, such as colouring roofs and buildings white, though the simple ideas would tend to bring only modest gains. But at the other end of the range there are dramatic interventions including the idea of space-based reflectors designed to cut down the amount of the sun's heat arriving at the earth's surface and the idea of spraying materials high into the atmosphere to mimic the effects of vast volcanic emissions which, in the past, have caused cooling (Hamilton 2013).

The scientists' and engineers' ambition is to find out if they are able to counteract climate change directly. Russia and the US are both keen to have these opportunities explored. However, environmentalists find themselves with a difficult message to communicate. Like advocates of geoengineering, they want to stress that action on climate change is urgent. But they generally wish to disassociate themselves from geoengineering approaches, in part because they are not confident they would succeed but primarily because they fear that if geoengineering came to appear plausible it would take away all the pressure to cut emissions and to decarbonise the economy. For that reason, environmentalists tend to oppose research into the viability of geoengineering and in Britain campaigned against the SPICE (Stratospheric Particle Injection for Climate Engineering) project, part of which was designed to test a way of pumping material into the atmosphere. The researchers were unable to organise any stakeholder dialogue about the device and the research was somewhat disrupted, though the purely engineering aspects continued in the laboratory.

It is clear that, even in the case of climate change, environmentalists are committed to supporting scientists' claims about the reality of the climate problem, but they are keen to take a more independent line on fracking and on geoengineering. They do not want to endorse every possible practical or policy response. In that sense – as Mike Hulme nicely pointed out in *The Guardian* in 2009[6] – the activists at the camp for climate action were equipped with more than peer review alone; they had political and ethical arguments of their own which, at times, set them apart from the views of mainstream scientists.

Communicating safety and risks in relation to GMOs

The case of GMOs – genetically modified (or genetically engineered) organisms – was just the opposite of climate change in the sense that environmental groups were, initially at least, out of line with the views of the scientific establishment; the science communication issues were accordingly very different. In this case the principal issues addressed safety and safety testing. Here was a new product, whether GM crop, animal or bacterium, that needed to be assessed for its implications for consumers and the natural environment. All major industrialised countries had some sort of procedures for testing new foodstuffs but the leading question was how novel were GM products taken to be and, thus, what sorts of tests they should be exposed to. For some, the potential for the GM entity to reproduce itself or to cross with living relatives in unpredictable ways suggested that this was an unprecedented form of innovation that needed unparalleled forms of caution and regulatory care. On the other hand, industry representatives and many scientists and commentators claimed that it was far from unprecedented. People had been introducing agricultural innovations for millennia by crossing animals. Modern (though conventional) plant breeding already used extraordinary chemical and physical procedures to stimulate mutations that might turn out to be beneficial. On this view, regulatory agencies were well prepared for handling innovations in living, reproductive entities (see Jasanoff 2005).

In this case environmental action groups argued that the regulatory system was insufficiently demanding and that the consequences of new technologies were not being examined closely enough. They suggested that governments, keen to promote economic success and to support agribusiness and the farm sector, were not taking enough care of consumers and the environment. Indeed, the protest over GM differed from preceding environmental controversies in just two principal ways: the GM debate combined worries over environmental impacts and over the health consequences of the new technology; the GM controversy in the late 1990s came after a period in which there had been growing co-operation between environmental groups and official bodies, as they often worked together on projects aimed at so-called sustainable development.

Work on genetic engineering has a three-decade-long history and environmental action groups were preparing their arguments before the main range of products came to market in the 1990s. In the US, these products passed relevant food safety and environmental tests relatively quickly. The question for environmental action groups was how to express doubts about the advisability of this new technology. Opponents were worried about specific impacts – the possibility of adverse environmental impacts and conceivable food safety issues – but they also had serious concerns over the potential direct intervention in nature that the technology (at least in principle) offered.

Among the issues that came to be the focus of campaigns were the impact of GM crops on beneficial insects, the likely difficulty of organic growers in keeping their crops free of GM contamination, the possible effects of GM foods on people with allergies (since allergy-promoting aspects of crops might accidentally be crossed into formerly innocuous foodstuffs) and contentious evidence that GM crops might be less nutritious than existing crops in unexpected ways. At the same time, campaigners were aware that there was a danger in offering very specific objections to the new technology since, if these objections were successfully countered, then opposition might begin to crumble. Campaigners feared that any accommodation to the new techniques would open the world to GM. Moreover, even if GM agriculture might arguably be less bad for the environment than present-day intensive farming, there was still a worry that the GM route was a one-way journey to a new relationship with nature since one could not readily imagine how GM 'contamination' could be undone (see Stirling and Mayer 1999).

To maintain the line against GM in Europe, environmental action groups fused their scientific communication with other strategies in largely opportunistic ways (see Priest 2001). A broad anti-GM coalition emerged that spanned groups engaged in direct destruction of trial GM crops to those doing detailed research work (examining, for example, just how far pollen from GM crops could travel). Those with distinctively environmental concerns were joined by antiglobalisation protesters and – particularly in France – by groups devoted to protecting the livelihoods and way of life of smallholders against large seed and agrichemical companies.

The more professionalised campaigning organisations concentrated on publicising and analysing environmental harms. The scientific evidence for adverse health effects was contentious and not easy to campaign around, but there were more readily agreed mechanisms by which GM planting might be causing environmental harms. In Britain, this emphasis was further promoted as an unintended consequence of government policy. The government organised a series of field-scale trials over several years aimed at investigating what the results of GM agricultural practice would be on wild plants, bird species, insects and so on. This strategy focused the debate on impacts on the rural environment and away from impacts on consumers. Official countryside protection agencies and more establishment conservation groups were keen to see such tests done too. Given that today's GM food crops work either by killing off pests or through allowing weeds to be controlled more easily, there was a good chance that – even if GM crops behaved exactly as predicted – they would have a negative impact on wildlife. There would be fewer weeds, therefore fewer seeds and insects and, thus, less to sustain wild birds. Naturally, it was hard to believe that the decline in field weeds was the prime concern of individual consumers who worried about whether to buy GM foods, but it provided a reasonably objective basis for claiming that GM agriculture would have a negative impact on the countryside and thus a sound legitimation for an anti-GM stance.

In practice the debate over GMOs in Europe was distributed over a range of issues, but the fundamental issue in EU legal assessments of GM crops and foodstuffs was the question of risk: were these new crops more risky than existing ones? This framing of the issue was the predominant one in North America too. Disturbed by protestors' success in raising public disquiet about GM products, US companies and allied politicians sought to combat European resistance to GM imports by appeals to the World Trade Organisation (WTO). A formal complaint was lodged in 2003, with the United States hoping to use the WTO to force open European markets to US farm imports and seed companies. Activists anticipated (correctly, as it turned out) that the WTO would rule largely in favour of the US; even if this has little effect in Europe owing to developed consumer resistance, it will discourage people in other parts of the world from trying to regulate against GM agriculture. There was also concern over the basis for the ruling (see Winickoff *et al.* 2005). The WTO decision-makers took a narrow view of the basis for the decision: it should be about risk assessment. Environmental action groups and academic authors had little success in opening up a debate around this issue and in convincing the WTO of the shortcomings of such an approach.

A final distinctive opportunity for science communication around GM arose in relation to the consultative exercises that were run in several EU states and elsewhere (such as New Zealand) as a way of trying to win legitimacy for, or even decide on, public policy (see Hansen 2005). The British exercise, called GM Nation? (see Horlick-Jones *et al.* 2007), was undertaken partly to be seen to do something without actually deciding for or against GM agriculture. Environmental action groups put a lot of effort into encouraging people to participate in the public debate even though it was clear that many participants would be frustrated since the debate would inform – but not determine – national policy.

Participation and public engagement as forums for science communication

The scope for public participation in environmental policy has grown greatly in recent years. The exact rationale for such participation has differed from one context to another, ranging from a wish to give citizens a say in democratic decision-making through to the suggestion that citizens may have insights into their local environment that are unavailable to the customary scientific experts (see Kasemir *et al.* 2003; Yearley 1999). For the purposes of this chapter, the key issue is the response of environmental action groups to such initiatives.

One would expect environmental groups to have an affinity with such moves. Social movements tend to thrive in democratic societies and to espouse democratic principles. On many occasions they call for government to respond to the supposed *will* of the people, for example over GM food in Europe. But, at the same time, these organisations are aware that not all environmental objectives are popular; nor are popular polices necessarily environmentally benign. In Britain, the Blair government's decision to allow official online petitions to be created on the Downing Street website led to an enormous response when a record number, well over a million people, expressed their opposition to road pricing in early 2007. Popular action appeared to favour personal consumption against environmental objectives.

Environmental groups are thus reluctant to relinquish control over the policy agenda for these public consultation initiatives, in part because they fear that such exercises might be manipulated by government (or business) but also because they are concerned that people may not favour the best environmental option. In many respects environmental groups are as loath as governments to hand environmental policy over to the public, even though environmental groups are happy to laud the wisdom of the public when the public happens to favour the same objectives as they do. Citizens are thus deemed wise about GM foods but less so about wind turbines and least of all in their devotion to car ownership and use.

Questions of public consultation and participation come to the fore chiefly in open, pluralistic societies. A contrasting situation confronts environmental action groups in China, a country of enormous importance for global environmental politics. It is believed that China overtook the US to become the largest emitter of carbon dioxide in 2006;[7] China also has a large commitment to GM agriculture, and there are currently plans to introduce GM rice which would move the Chinese diet from a low GM profile to one of the world's most intensive. Though the Chinese state has a large and reasonably well-resourced environmental protection agency, environmental campaigning groups are generally not welcomed. Just a few international NGOs operate in China alongside GONGOs (government organised non-governmental organisations; Yang 2005). Though some education- and membership-based environmental organisations operate, many environmental activists are attracted to alternative methods for communicating their message. Several recent analyses in the Chinese and other East and South-East Asian contexts have focused on the role of online communication (see Yang 2003). As Yang (2005: 59) points out:

> [f]or web-based ENGOs [environmental NGOs], the Internet makes up for their lack of resources and helps to overcome some political constraints. While the restrictive regulations create barriers to registering an NGO, web-based groups can stake out an existence on the Internet.

Mol (2006), too, draws attention to the role of the Internet in China and Vietnam in circulating information and allowing people access to information which they would otherwise have had difficulty accessing.

The potential of the Internet has also been explored in the industrialised North, both as a forum for debate and a means for the provision of technical information: for example, Friends of the Earth in Britain beat the official Environment Agency to provide online map-based information about local chemical pollution. Members of the public could search for possible sources of hazard by entering their postcode. Having shamed the Environment Agency into improving its public information, Friends of the Earth withdrew its site. Increasingly *apps* are used to turn mobile devices into networked distributed environmental monitoring systems, and crowdsourcing techniques are used to collect environmental information.

Concluding remarks

The central claim of this chapter has been that environmentalists, more than most other political and reform movements, are increasingly obliged to act as communicators of science and technology because empirical claims about the state of the natural environment are core to their message. Often they have had to do this communication under circumstances where they disagree with the orientation of large parts of the scientific and technological *establishment* and they have developed argumentational tools for tackling this job. Over climate change, they first had to devise a new strategy for bolstering the IPCC and other mainstream science and then subsequently needed to work out ways of distancing themselves from the scientific mainstream (on fracking and geoengineering) without giving encouragement to sceptics. The Internet has proven to be a rich resource for such communication both because it can handle detailed information and because the user can *personalise* it by entering geographical data (zip codes or postcodes). In contexts where environmental action groups face limitations on their activities, the Internet has become a particularly important means of environmental communication.

Key questions

- How different are environmentalists from other social campaigners (for example, activists concerned with gender equality or homelessness or ethnic differences) in relation to the scientific nature of their messages? Does this tend to favour or hinder environmentalists compared to other activists?
- Some environmentalists have claimed that, in relation to the truth and significance of climate change, they are 'armed only with peer-reviewed science'. Is this an accurate and useful slogan for environmentalists to adopt?
- Environmentalists are increasingly interested in using online platforms to encourage public involvement in environmental monitoring and reporting. In which contexts and for which kinds of issue is this most likely to be beneficial?

Notes

1 In relation to the original version of this chapter (Yearley 2008), this revised version has reworked the argument on environmentalists' science communication dilemmas, incorporated references to more recently emerging issues over fracking, wind power and geoengineering, and considered the growth of the online sphere of debate on science-based environmental issues.
2 The scientific community has responded by establishing its own sites (most famously www.RealClimate.org), though, for research scientists, writing for this medium tends to be less important than authoring for formal publication.
3 There was no author given for this report in *Earth Matters*, Autumn/Winter 1990, 4.
4 According to the CBD website: '"Short of sending Dick Cheney to Alaska to personally club polar bear cubs to death, the administration could not have come up with a more environmentally destructive plan for endangered marine mammals", said Brendan Cummings, ocean program director of the Center. "Yet the administration did not even analyze, much less attempt to avoid, the impacts of oil development on endangered wildlife."' See www.biologicaldiversity.org/swcbd/press/off-shore-oil-07-02-2007.html
5 'US Climate Policy Bigger Threat to World than Terrorism', *The Independent* (9 January 2004).
6 See www.theguardian.com/commentisfree/2009/dec/04/laboratories-limits-leaked-emails-climate
7 See the report of the Dutch MNP (Milieu- en Natuurplanbureau) (last consulted on June 22, 2007) at: www.mnp.nl/en/dossiers/Climatechange/moreinfo/Chinanowno1inCO2emissionsUSAinsecondposition.html

References

Aitken, M. (2010) 'Why we still don't understand the social aspects of wind power: a critique of key assumptions within the literature', *Energy Policy*, 38, 4: 1834–1841.
Bodansky, D. (1994) 'Prologue to the Climate Change Convention', in I. M. Minter and J. A. Leonard (eds) *Negotiating Climate Change: The Inside Story of the Rio Convention*, Cambridge: Cambridge University Press, 45–74.
Boehmer-Christiansen, S. (1994) 'Global climate protection policy: the limits of scientific advice. Part 2', *Global Environmental Change*, 4, 2: 185–200.
Bowman, A. (2010) 'Are we armed only with peer-reviewed science? The scientization of politics in the radical environmental movement', in S. Skrimshire (ed.), *Future Ethics: Climate Change and Apocalyptic Imagination*, London: Continuum, 173–196.
Cadbury, D. (1998) *The Feminization of Nature*, Harmondsworth: Penguin.
Crichton, M. (2004) *State of Fear*, London: HarperCollins.
Edwards, P. N. and Schneider, S. H. (2001) 'Self-governance and peer review in science-for-policy: the case of the IPCC Second Assessment Report', in C. A. Miller and P. N. Edwards (eds) *Changing the Atmosphere: Expert Knowledge and Environmental Governance*, Cambridge, MA: MIT Press, 219–246.

Freudenburg, W. R. (2000) 'Social constructions and social constrictions: toward analyzing the social construction of "the naturalized" as well as "the natural"', in G. Spaargaren, A. P. J. Mol and F. H. Buttel (eds) *Environment and Global Modernity*, London: Sage, 103–119.

Greenpeace UK (2003) *Human Impacts of Man–Made Chemicals*, London: Greenpeace.

Hamilton, C. (2013) *Earthmasters: The Dawn of the Age of Climate Engineering*, New Haven, CT: Yale University Press.

Hansen, J. (2005) *Framing the Public: Three Case Studies in Public Participation in the Governance of Agricultural Biotechnology*, PhD thesis, European University Institute, Florence.

Hart, D. M. and Victor, D. G. (1993) 'Scientific elites and the making of US policy for climate change research', *Social Studies of Science*, 23, 4: 643–680.

Helm, D. (2012) *The Carbon Crunch*, New Haven, CT: Yale University Press.

Horlick-Jones, T., Walls, J., Rowe, G., Pidgeon, N. F., Poortinga, W., Murdock, G. and O'Riordan, T. (2007) *The GM Debate: Risk, Politics and Public Engagement*, London: Routledge.

Jasanoff, S. (2005) *Designs on Nature*, Princeton, NJ: Princeton University Press.

Kasemir, B., Jäger, J., Jaeger, C. C. and Gardner, M. T. (eds) (2003) *Public Participation in Sustainability Science: A Handbook*, Cambridge: Cambridge University Press.

Lahsen, M. (2005) 'Technocracy, democracy and US climate politics: the need for demarcations', *Science, Technology and Human Values*, 30, 1: 137–169.

Lynas, M. (2013) Lecture to Oxford Farming Conference, 3 January; online at www.marklynas. org/2013/01/lecture-to-oxford-farming-conference-3-january-2013/ and at (Farmers' Weekly) www. fwi.co.uk/articles/11/01/2013/137081/mark-lynas-why-i-became-pro-gm.htm.

McCright, A. M. and Dunlap, R. E. (2000) 'Challenging global warming as a social problem: an analysis of the conservative movement's counter-claims', *Social Problems*, 47, 4: 499–522.

McCright, A. M. and Dunlap, R. E. (2003) 'Defeating Kyoto: the conservative movement's impact on US climate change policy', *Social Problems* 50, 3: 348–73.

Mol, A. P. J. (2006) 'Environmental governance in the Information Age: the emergence of informational governance', *Environment and Planning*, C, 24, 4: 497–514.

Pearce, F. (1991) *Green Warriors: The People and the Politics Behind the Environmental Revolution*, London: Bodley Head.

Priest, S. H. (2001) *A Grain of Truth: The Media, the Public, and Biotechnology*, Lanham, MD: Rowman and Littlefield.

Prins, G. and Rayner, S. (2007) *The Wrong Trousers: Radically Rethinking Climate Policy*, Oxford and London: James Martin Institute for Science and Civilization, University of Oxford and MacKinder Centre for the Study of Long-Wave Events, London School of Economics and Political Science.

Stirling, A. and Mayer, S. (1999) *Re-Thinking Risk: A Pilot Multi-Criteria Mapping of a Genetically Modified Crop in Agricultural Systems in the UK*, Brighton, UK: Science Policy Research Unit.

Winickoff, D., Jasanoff, S., Busch, L., Grove-White, R. and Wynne, B. (2005) 'Adjudicating the GM food wars: science, risk and democracy in world trade law', *Yale Journal of International Law*, 30, 1: 81–123.

Yang, G. (2003) 'The co-evolution of the Internet and civil society in China', *Asian Survey*, 43, 3: 405–422.

Yang, G. (2005) 'Environmental NGOs and institutional dynamics in China', *The China Quarterly*, 181, (March): 46–66.

Yearley, S. (1992) 'Green ambivalence about science: legal-rational authority and the scientific legitimation of a social movement', *British Journal of Sociology*, 43, 4: 511–532.

Yearley, S. (1999) 'Computer models and the public's understanding of science: a case-study analysis', *Social Studies of Science*, 29, 6: 845–866.

Yearley, S. (2005) *Making Sense of Science: Science Studies and Social Theory*, London: Sage.

Yearley, S. (2008) 'Environmental groups and other NGOs as communicators of science', in M. Bucchi and B. Trench (eds) *Handbook of Public Communication of Science and Technology*, London and New York: Routledge, 159–172.

10
Publics and their participation in science and technology
Changing roles, blurring boundaries

Edna F. Einsiedel

Introduction

Configuring the place and roles of publics and the nature of participation in the context of science-and-society relations has continued to preoccupy academics, policy actors, stakeholder organisations and publics themselves. Examining the place of public participation within the increasingly complex worlds of governance offers one realm of understanding. At the same time, the world of the everyday has seen publics engage with science and technology in different ways, combining a panoply of roles of citizen, consumer, user, or even disinterested bystander.

In this chapter, we give a bird's-eye view of the extensive attention to public participation,[1] its varied meanings and applications, and the different trajectories of our understandings of its associated practices and influences. We suggest that the beginnings of public participation have been examined with spectacles tinted with romanticism and idealism; but more clear-eyed and critical studies have also emerged, some growing out of lessons learned from earlier studies and others increasingly sceptical of extant designs, their assumptions and their outcomes. At the same time, public participation has become more fluid, contested, a creature of shifting norms, just as its associated processes also contribute to the (re)shaping of such norms. It is this idea of the changing nature of public participation as one arena for understanding science-and-society relations that is the primary focus of this chapter.

The construction of publics

There are many ways of conceptualising or considering publics: the debate between the American philosophers Walter Lippmann and John Dewey epitomises the ongoing struggles between the unattainable ideal of an omnicompetent and sovereign citizen – one likened to 'a fat man who tries to be a ballet dancer' (Lippmann 1922), and that of the more hopeful possibilities of an enlarged public sphere imbued with the promise and hopes of a democratic polity that valued the *practical wisdom* of the citizen (Dewey 1927). Publics are also deployed as analytical categories and *performed* (Michael 2009) as part of the reconfiguration of science and society (Irwin and Michael 2003).

Publics have also been assigned roles or emerge as social categories – as citizens, as consumers, as users or non-users, as constructions of the self, whether on the basis of affectedness or proclaimed identities. These 'Publics-in-Particular', as distinguished from 'Publics-in-General' (Michael 2009: 617), emerge through some common condition or shared fate, or identify themselves as having a specific stake in a particular science and technology issue. While these publics can be identified spatially (e.g. those affected by some local environmental problem), they can also be dispersed (those with a common condition or who share some global concern) but are increasingly connected through the Internet. They can be mobilised by expert interests for reasons that include 'publicity, financial resources, provision of volunteers for studies, and the making of a market' (Michael 2009: 623). They can also, in turn, mobilise others, including science and technology experts, for their own ends and interests. They are increasingly bodies of expertise themselves with their own knowledge production and dissemination mechanisms (Epstein 2007; Einsiedel 2013).

The roles taken by different publics assume a variety of enactments and meanings – the consumption of (scientific and technological) knowledge, the construction and display of identities, the conduct of a particular form of citizenship such as the *scientific citizen* (Irwin 2001), the *biocitizen* (Rose and Novas 2004) or the *environmental citizen* (Dobson and Bell 2005), to name a few. The boundaries between different role identities are often blurred. For example, it has become increasingly difficult to keep distinct the practices of citizenship and consumption (Michael 1998). Just as citizenship has been brought to bear on consumption practices, the practices of consumption have been increasingly deployed for political ends (see e.g. Barnett *et al.* 2011).

While we use *publics* as noun, we recognise that the adjectival use of the term (the public sphere, public interest) is also a means of demarcation, pointing to the shifting boundaries between the private and the public, the collective and the individual, the exclusive and inclusive within professional or policy communities.

Understanding participation

This discussion on publics provides an entry point to understanding the many different forms of participation entailed in public participation. It is not possible to be exhaustive in presenting these varied forms but we present one way of categorising participation to elucidate the shifting boundaries we have identified as an overall theme. We can consider participation on science and technology according to three purposes (see Table 10.1): for policymaking, for public dialogue (an inclusive category that incorporates education, entertainment, persuasion in various dialogic forms) and for knowledge production. These categories focus on primary purposes for public participation but are not mutually exclusive categories. This means public participation for policymaking may include dialogue events and knowledge production may have unplanned

Table 10.1 Categorising public participation

Purposes	Primary actors/sponsors	Examples
Policymaking	Government, research institutions, international organisations, stakeholder organisations, citizen panels	Consensus conferences, citizen juries, deliberative polls, negotiated rule-making, crowdsourcing
Dialogue	Government, research institutions, scientist/research networks, stakeholder organisations, public collectives	Science cafés, festivals, art/science exhibits, online discussion boards
Knowledge production	Scientists/research networks, community groups, citizens	Citizen science, traditional knowledge, crowdsourcing

or unanticipated consequences for policymaking. Other categorisation systems have employed a spectrum of knowledge production mapped against categories of sponsorship and spontaneity (Bucchi and Neresini 2008).

The reference to negotiated rule-making in Table 10.1 is to a specific US regulatory process referring to an advisory committee established by a federal agency 'to consider and discuss issues for the purpose of reaching a consensus in the development of a proposed rule'; such committees typically consist of government agency representatives and stakeholder community representatives with 'interests likely to be significantly affected by the rule'.[2] There remains a debate regarding the impacts of *reg-neg*, as it has been referred to informally, and interest groups in the US still remain reliant on the courts for mobilising regulatory changes.

Rowe and Frewer (2000) discriminate between participation and communication by describing the former as soliciting public views as well as engaging in active dialogue while the latter is described as 'information transfer':

> Public participation may be loosely defined as the practice of consulting and involving members of the public in the agenda-setting, decision-making, and *policy-forming activities* of the organizations or institutions responsible for such functions.
>
> (Rowe *et al.* 2004)

The International Association for Public Participation (IAP2) has this definition which attempts to incorporate a range of activities under its term:

> Public participation means to involve those who are affected by a decision in the decision-making process. It promotes sustainable decisions by providing participants with the information they need to be involved in a meaningful way, and it communicates to participants how their input affects the decision. The practice of public participation might involve public meetings, surveys, open houses, workshops, polling, citizens' advisory committees and other forms of direct involvement with the public.[3]

While such descriptions focus primarily on the top-down approaches that have typified the policymaking process, we recognise that policy decisions and institutional change can also be driven from the ground, as seen in the roles of patient groups, environmental organisations or even individuals.[4]

Such differentiation by degrees (and quality) of participation has underpinned a preferred value system where what counts as *true participation* involves power-sharing in decision-making, the classic metaphor being the *ladder of participation* (Arnstein 1969). We prefer a more flexible description, reflected in Table 10.1, where degrees of participation are recognised in different contexts, for different goals or purposes, without the requirement of *true participation* that occurs only when there is defined impact on policy decisions. Indeed, while the deficit model outlined by Wynne (1993) has helpfully interrogated the assumptions behind the construction of *public understanding* and *expertise*, we also recognise that participation can be as much a creature of agency and choice or preferences (see Mejlgaard and Stares 2013) as it is an outcome of structural opportunities and constraints.

The interest in policymaking and governance is understandable and this interest has spawned a considerable literature on public participation in such contexts. At the same time, events that have been considered to be information exchange or simply one-way communications (Rowe and Frewer 2000; OECD 2001) have proponents who suggest these are also dialogic events (Davies *et al.* 2009) and have their place in the realm of public participation.

The engagement of publics in policymaking arenas, also including publics as respondents on public opinion surveys (still commonly used by policymakers), has been elaborated through an increasingly popular approach: the use of a variety of participation designs that rest on deliberation. These have typically been focused on technology questions – from specific policy questions such as how to dispose of nuclear waste in Britain (Chilvers 2007) to broader questions around how to deal with new or emerging controversial technologies. These have also been referred to as *participatory technology assessment*, most frequently practiced in Europe under the aegis of technology assessment institutes. They are manifestations of the recognition of *wicked problems* posed by science and technology which demand broader, socially distributed forms of expertise (Funtowicz and Ravetz 1993).

Deliberative participatory practices and policymaking

What has been described as *the participatory turn* in policymaking has been driven by several theoretical imperatives. One push came from theorists of deliberative democracy who were motivated, in part, by the limitations of representative democracy and liberal individualist or economistic understandings of democracy (Chambers 2003). Writers in this vein emphasised deliberative public participation as foundational to democracy, as providing a critical base for ensuring legitimacy, transparency and decision-making accountability, and as a means to produce policies considered to be just and fair (e.g. Habermas 1989; Guttman and Thompson 1996). Theorists in this camp have maintained that 'deliberation under the right conditions will have a tendency to broaden perspectives, promote toleration and understanding between groups, and generally encourage public-spirited attitudes' (Chambers 2003: 318). At the same time, empirical studies have tested these assumptions against the limitations of deliberative contexts (e.g. Mutz 2006, 2008). The contingencies for deliberative participation have offered important lessons that go towards developing a clearer sense of such activities within the normative expectations of democratic theorists (Thompson 2008).

A second area of theorising was in the field of policy and governance where growing dissatisfaction with 'dominant technocratic empiricist models' considered inadequate for 'policy decisions that combined technical knowledge with intricate and often subtle social and political realities' created a push for the discursive turn in policies and politics (Fischer 2003: 17–18). Many of the debates were around the political questions surrounding technologies, and the emergence of participatory technology assessment combined the engagement of a broader range of stakeholders and publics in their socio-technical assessments.

Observations about the changing contexts of science and technology and the associated changes in their knowledge production processes – including the broader social distribution of these processes and practices (Gibbons *et al.* 1994; Nowotny *et al.* 2001) and the challenges posed by the "wicked problems" of many science-connected political issues – led to postulations about the need for an expertise base that was more extended and differentiated, and for approaches more suited to the challenges normal science was unable to cope with (Funtowicz and Ravetz 1993).

The literature on public participation in general and more particularly that of deliberative public participation has exploded in the last three decades. Institutional initiatives to engage publics in policy processes on environmental issues were notable in the 1980s and 1990s through such mechanisms as public hearings and negotiated rule-making, practices most prominent in the US (see Fiorino 1990). Internationally, the biotechnology problem became the issue around which a host of deliberative public participation initiatives were conducted. As an illustrative case, focusing only on the trajectory of both this issue as well as the application of deliberative public participation approaches, our review showed 18 countries conducting over 40 public participation initiatives, most of which were on biotechnology's most controversial application, GM food, and a minority on biomedical applications (Einsiedel 2012). Many of these initiatives also involved the use of the consensus conference model. This model and that of the citizen jury or study circles were considered innovations for their focus on education, the opportunity for discussions and deliberation, and interactions with a range of representative stakeholder expertise (Konisky and Beierle 2001).

While most deliberative consultation attempts were carried out by European countries, with a few in Canada and the US, similar initiatives were also undertaken in Japan (Hirakawa 2001), South Korea (Korean National Commission for UNESCO 1998), Taiwan (Chen and Lin 2006) and India (Wakeford et al. 2008). In quite a few instances, the process involved some adaptation to local circumstances; for example, the goal of reaching consensus was replaced by that of representing the spectrum of opinions on the issue (van Est et al. 2002; Skorupinski et al. 2007). In others, research questions around design furthered understanding of effectiveness (Pellegrini 2009; Hamlett 2002).

The first decade and a half can be considered a period of social experimentation and learning as well as moves toward institutionalisation, which has been most prominent in Europe with the rise of technology assessment institutes or reformulation of the mandates of existing ones. While there were three institutes of technology assessment in Europe in the 1980s, there are currently 18 (Sclove 2010). France and Germany, which had traditional technology assessment institutes, expanded their remits to include publics in assessment initiatives. Beyond structural changes, attempts at institutionalising public participation may also incorporate efforts at social learning – from specific policy initiatives (see e.g. Jones and Einsiedel 2011) to broader reflexive assessments such as have been undertaken in Britain – which best exemplifies long-term processes of reflection, experimentation and reflexive reformulations on publics and participation (see Chilvers 2012).

The second phase can be seen as taking stock and applying lessons learned to emerging technologies such as nanotechnology (see e.g. Godman and Hansson 2009; Rogers-Hayden and Pidgeon 2008) and synthetic biology (Royal Academy of Engineering 2009). These initiatives reflected efforts to be more anticipatory by moving public engagement upstream (Wilsdon and Willis 2004). Further reflections were being carried out on the varied meanings of impacts on policy communities and institutions (Hennen 2012; Jones and Einsiedel 2011) and understanding the different contexts and cultures of policymaking that foster different approaches to, and degrees of, public participation[5] (see e.g. Griessler 2012; Degelsegger and Torgersen 2011; Dryzek and Tucker 2008). Increasing recognition of the significant time lags involved in the development of novel innovations and their shifting governance contexts and challenges encouraged longer-range perspectives and emphasised the importance of continuous reflexivity and adaptation. Such initiatives have been clustered under the label of responsible innovation (Owen et al. 2012) or anticipatory governance (Karinen and Guston 2010).

Governance cultures have also helped to explain modes of public participation. A study of EU countries identified six forms of governance among member states, including *discretionary* (with few interactions with the public); *corporatist* (negotiations based on stakeholder interests); *educational* (based on the deficit model); *market* (based on principles of demand and supply); *agonistic* (governance in the context of confrontation and conflict); and *deliberative*, which recognised the importance of open debate and discussion (Hagendijk and Irwin 2006). Each of these models highlights different approaches to public participation and deliberation (see also Howlett and Migone 2010). Importantly, different models may also be found within the same country at different times.

The increasing prominence of global issues has encouraged the experimentation with international collaborative public participation initiatives and allowed exploration of governance challenges beyond the perimeters of the state. A unique initiative managed by the Danish Board of Technology saw 100 citizens from each of 44 countries participate in a one-day consultation on the issue of climate change as a way to provide citizen input into the international policy discussions around the UN Framework on Climate Change Conference in Copenhagen (see a collection of articles reflecting on this initiative in Worthington et al. 2011). More recently, over 30 groups in 25 countries participated in a similar initiative relevant to the UN discussions on the Convention on Biological Diversity.[6] While multi-country deliberative public participation initiatives have been held in the European region, these were the first international initiatives involving northern and southern hemisphere countries.

Public dialogue as an end

In this section we refer to processes of communication exchange which can vary from simple information transmission to information exchange or critical dialogue. Publics can enact themselves as receivers of information for purposes of learning, entertainment or some instrumental goal such as learning about a disease, or as more active participants in such exchanges. These activities can occur in institutional settings such as art and science museums or in places where particular publics may be found – from truck stops in Kenya to coffee houses in Buenos Aires. Some view these processes as opportunities to critically reflect on science and technology, a way of talking about, with, and back to, science (Dallas 2006). We use the example of science cafés as an illustrative case where discussions about science and technology can take place in informal settings, where learning can take place about a topic and where scientists might also learn of potential public values, preferences and concerns and acquire different skill sets for knowledge translation. The long tradition of science cafés in Europe has been extended to North America, Latin America, Africa and Asia.

In Africa, the adoption of the science café has been an opportunity to consider important social problems from the HIV virus or cervical cancer vaccines in Uganda (Nakkazi 2012) to discussing parasites in Kenya (Mutheu and Wanjala 2009). In Argentina, there have been café initiatives to inspire interest in science and technology careers and promote their roles in the country's industrialisation project.[7] A Brazilian science café helped bring about collaboration between scientists and a samba group that resulted in a science-themed activity during Carnival (Dallas 2006).

These events in informal places and spaces have been more recently represented in terms of correcting information deficits – not only of publics but also of scientists in terms of their understandings of publics and their approaches to practicing science communication. Japanese scientists interviewed after their participation in such cafés suggested these events could be viewed initially as troublesome or time-consuming because they were outside the scope of their work or responsibilities, because they imposed (unwanted) pressures to represent science well or because the scientists had to deal with apprehensions about engaging effectively with publics; but enjoyment of the experience and positive public feedback typically helped to allay these concerns (Mizumachi et al. 2011). Science cafés can also be sites for research policy agenda setting, an interesting move to the front end of the policy cycle (Higashijima et al. 2012). The variety of their purposes and formats are fitting reminders of the cultural moorings of science and society.

Artistic media have afforded other avenues for scientists, artists and publics to engage in dialogue. Images produced during the course of scientific research have been a channel for scientists to look at their work through different lenses or 'to get in touch with their inner artist' (Gewin 2013: 537) and to share with various publics the beautiful images that emerge – be they cellular scaffolds, molecular movements or robotic systems. Scientific tools have also been appropriated by artists through the practice of 'tactical media' as a way to critique science and technology (Rogers 2011: 102), 'a qualified form of humanism (that is) an antidote to the emerging forms of technocratic scientism' (Garcia and Lovink 1997, quoted in Rogers 2011: 101). A recent compendium of an array of artistic works from around the globe incorporating music, dance and computer-controlled video performances that draw on scientific and technological developments illustrates the attempts to bridge two cultures, in part with the help of digital systems and an open-source environment that has encouraged such border crossings (Wilson 2010). Sometimes, the artistic presentation can trump the dialogue mode when it comes to learning (Lafreniere and Cox 2012).

Activities and arenas in this category can include other informal practices such as science festivals, an increasingly prevalent forum within the broader spectrum of public engagement on and with science. These events try to embrace learning, entertainment, outreach, recruitment and innovation extensions (see e.g. Bultitude et al. 2011; Jormanainen and Korhonen 2010).

Public participation and knowledge production

Central to the production of knowledge in modernity has been the enterprise of scientific knowledge production. While such an enterprise and its associated practices have been typically closed off in the interest of compliance with *sound scientific practice* – what Gibbons and his colleagues (1994) have called Mode 1 science – Mode 2 knowledge is less bounded (transdisciplinary), more socially accountable and more reflexive. It has, at the same time, become more complex with multiple sites of knowledge production, a wider range and greater sophistication of tools, and problems that require transdisciplinary skills and multiple actors. Such complexity has encouraged the opening up of spaces and practices in science and technology at the same time that the questions of what counts as knowledge and whose knowledge counts have challenged the pre-eminence of scientific knowledge.

As indicators of such changes, we point to examples of citizen science that illustrate the opening up of science knowledge production through one form of public participation. Citizen science generally refers to 'a form of research collaboration involving members of the public in scientific research projects to address real-world problems' (Wiggins and Crowston 2011: 1). Such citizen participation has also incorporated a spectrum, from providing assistance in data collection – called a contributory project which is primarily researcher driven, or where citizen participants have been considered 'mere data drones' (Hemment et al. 2011: 63) – to participating in a broader range of activities. The latter has been called collaborative citizen science where involvement might range from data analysis and interpretation to co-creation, where participants may be involved in all stages of the research including helping to define the research questions.[8] The early forms of citizen science were primarily of the first type, growing from the need to extend data collection and monitoring requirements. The early growth of citizen science was prompted by the increasing realisation among professional scientists that 'the public represent a free source of labour, skills, computational power, and even finance' (Silvertown 2009: 467). The types of projects involving citizen scientists have grown since the early 1990s (Catlin-Groves 2012) and have impressively taken place within a wide variety of fields, from ecology and conservation to astronomy, earth sciences, paleontology, microbiology and molecular biology (Wiggins and Crowston 2011; Catlin-Groves 2012).

Participation in front-end problem-solving is illustrated by the Foldit initiative – a study intended to explore how linear chains of amino acids curl up into three-dimensional shapes that minimise internal stresses and strains (Hand 2010), one of the basic problems in structural biology (Dill and MacCallum 2012).[9] How proteins fold and why they do this so rapidly is one of the key steps to accelerating development of new drugs. By developing a protein-folding game designed to entice video game players to solve the problem, the scientists found top-ranked Foldit players 'could fold proteins better than a computer – they came up with entirely new folding strategies' (Hand 2010: 685).

In addition to participating in knowledge production, citizen scientists have also been lauded for their roles in extending the knowledge translation process along with the scientists, which can increase the profile of issues being researched (Couvet *et al.* 2008). There have also been instances of citizen science-based projects contributing to policy development (see e.g. Crabbe 2012).

Interrogations into the meanings of partnerships and collaborations and their social-political outcomes have been occurring between communities and researchers as demonstrated in such areas as community-based research partnerships (as they are known in North America) or participatory action research in international development contexts (see Fals-Borda and Rahman 1991; Whyte 1991). These are typically collaborations between affected communities and groups of researchers with an explicit emphasis on partnership in the knowledge production process and the goal of social change. Israel and colleagues (1988: 177) describe community-based participatory research (CBPR) in these terms:

> A collaborative process that *equitably* involves *all partners* in the research process and recognises the unique strengths that each brings. CBPR begins with a specific topic of importance to the community with the aim of combining knowledge and action for social change to improve community health and eliminate health disparities.

Other forms of partnership have emerged in the development of science shops where citizen groups in a community approach institutions such as universities to assist in the exploration of community problems – from possible contamination of water supplies to evaluation of care centers for the elderly or the disabled (see Leydesdorff and Ward 2005).

Finally, the emergence of a host of communication technologies, from mobile phones to social media tools, and the rise of data-intensive sciences (data gathered over large spatial and temporal scales or contributed by numerous data input and collection activities) have expanded the opportunities for citizen science (see Young *et al.* 2013; Haklay 2013). Publications about citizen science have also featured increasingly in science journals and studies on citizen science initiatives have investigated the risks and benefits of such collaborations from the perspective of researchers (Whyte and Pryor 2011). The benefits include greater efficiency in the research process, the development of new research capabilities (including capabilities to find new questions and analyse evidence), and extended opportunities for knowledge exchange and impacts (ibid.). However, concerns have also arisen around data reliability and quality (Flanagin and Metzger 2008; Catlin-Groves 2012).

On the part of citizen participants, outcomes have been found that ranged from greater content knowledge, better understanding of scientific processes and greater involvement in related policy questions and initiatives (Bonney *et al.* 2009; Catlin-Groves 2012). Such involvement has also been viewed as 'encouraging doubt rather than promoting blind acceptance of fact' (Paulos 2009). Questions remain about how far to extend open disclosure and more equitably share the ownership of research results (Delfanti 2010). These opportunities and challenges are generally more reflective of the ongoing boundary work, border crossings and renegotiations of expertise occurring in knowledge production processes.

An alternative pathway to knowledge production is further represented by rising interest in and increased recognition of traditional or indigenous knowledge – a departure from the 1950s and 1960s when such knowledges was seen to be inefficient, inferior, and *unscientific*. Such knowledges has been increasingly integrated with a large variety of scientific fields – from ecology, soil science, botany, zoology, agronomy, agricultural economics and veterinary medicine, to forestry, human health, aquatic science, management, rural sociology, mathematics, fisheries, range management, information science, wildlife management and water resource management (Warren *et al*. 1991). Most recently, such knowledges have been institutionally legitimated in the United Nations' efforts through the Convention on Biological Diversity, recognising its validity and the rights to its protection.[10]

Patient organisations have similarly contributed to the reshaping of knowledge production. Starting with Epstein's (2007) seminal work done on AIDS organisations in the US, research on such mutual shaping or co-production has also been carried out with other *disease groups*. The evolution of a partnership model between the French Muscular Dystrophy organisation and health professionals and scientists contributed to restructured power relations between these groups (Rabeharisoa 2003; Callon and Rabeharisoa 2003), occurring through participation in strategic decisions from funding and steering research directions to elaborating on the dynamics of disease progression. In the case of rare diseases, early models of co-production have been demonstrated with experiences such as those of Genetic Alliance whose knowledge production activities have included the creation of tissue repositories, steering of research through funding, establishing intellectual property arrangements that included their own patents on disease genes, redesigning organisational structures through alliance of small groups and extending circuits of knowledge circulation through establishment of journals (Einsiedel 2013).

These illustrations of participation similarly blur into the policymaking world. Such experiences show patients 'do not simply speak to science' (Nowotny *et al*. 2001: 199) but also contribute to complex political and regulatory negotiations and collaborate in knowledge production (Crompton 2007). In a larger context, contributions to the knowledge production process which began with cross-disciplinary work (Wuchty *et al*. 2007) and expanded to stakeholder organisations and publics have forced reconsiderations of how knowledge is generated, evaluated and communicated (Stodden 2010). Such roles as *scientific peers* and the extended peer review projected by advocates of post-normal science (Funtowicz and Ravetz 1993) are being exemplified through these instances of collaboration that have been likened to approaches to open innovation elsewhere (Stodden 2010).

The elaboration of research on knowledge co-production has coalesced around several theoretical questions: 'At what levels of social aggregation (laboratories, communities, cultures, the nation, the state, all of humanity), and in what kinds of institutional spaces or structures does it make sense to look for co-production?' (Jasanoff 2013: 5). We add to this the questions: In what forms and formats, under what conditions, and in what ways do these co-constitution practices take place? When we look at such questions through the lens of citizen scientists, traditional knowledge communities or stakeholder groups as they work alongside scientists, or as they challenge established modes and outcomes of knowledge production, we are in essence attempting to understand the cultural practices of science and technology in terms of the construction of knowledge, its forms of legitimation, its practices of standardisation (or their reconfiguration), the bridging practices between science and society – and the blurring of the various boundaries around these domains.

The spaces and sponsors of participation

While governments have been the primary sponsors of public dialogue and deliberation events and may remain so for some time, sponsorship has become increasingly dispersed among other institutional and non-state actors and agencies. In the face of the growing number of science and technology controversies in Britain, such sponsorship has come from research councils and scientific professional societies interested in funding controversial science but perhaps wary of their questionable reception among the public. Public participation around nanotechnology, synthetic biology and geoengineering has been sponsored by such groups. Similar activities are also increasingly dispersed in post-industrial and industrialising countries.[11] Stakeholder organisations (see Fowler and Allison 2008) and professional ethics councils (e.g. the Nuffield Institute) have also turned to public participation initiatives to engage, consult, inform, enroll, mobilise and collaborate with.

No longer in meeting rooms isolated from where many publics emerge or congregate, conversations and disputes occur in pubs, festival spaces, art museums, parks and roadsides where science and society can mutually offer ideas and interrogations. Publics can also be found in the privacy of their homes with their smart phones, tablets or desktops, bound together by some common interest or question. The Internet has increasingly provided the means to define publics and forms of participation and for publics to help define their identities as participants, whether in processes of policy design, leisure or the creation of new knowledge.

The growth of the online arena has also contributed to the social shaping of public participation; publics have, in turn, used online spaces for reconfiguring participation. As with many emerging technologies, the promises have initially been outpacing their capacities. Speaking about public participation on urban planning, Evans-Cowley and Hollander suggest:

> Today, technology allows for an entirely new generation of forms and practices of public participation *that promise to elevate the* public *discourse in an unprecedented manner while providing an interactive, networked environment for decision-making.* This is occurring with asynchronous communities interacting with one another on a variety of planning subjects, which allows for more democratic planning and more meaningful participation.
>
> (2010: 399; italics added)

The promissory expectations have paralleled those that heralded new technologies and proclaimed the obsolescence of those being replaced:

> The age of the public sphere as face-to-face talk is clearly over: the question of democracy must henceforth take into account new forms of electronically mediated discourse.
>
> (Poster 1997: 209)

Others have not been so sanguine, suggesting the Internet simply reinforces pre-existing social relations (Tyler 2002), whether between state and society or between various groups in society. Still others have also acknowledged the complex combination of both liberation and control.

Despite these differing prognostications, the online environment has replicated, extended and modified other traditional forms of participation, while at the same time providing new experimental forms and outcomes. There clearly has been an expansion of collaborative knowledge production opportunities, from crowdsourcing for science funding and collection of samples to the expansion of data collection and analysis through the various tools of the Internet, social media, and other technologies such as GPS, smart phones, sensor networks and cloud computing. Such tools have incorporated and extended problem analyses and solutions and provided opportunities for focusing on problems ranging from the micro level (genetic) to the macro global level and even to the far reaches of the universe.

Empirical studies on deliberative online participation have been more modest in their assessments, showing modest gains in developing opinions, grasping key arguments for or against, and willingness to engage opponents (Price 2006). Other limitations include the question of a broader accommodation of ways of communication that include the symbolic resources that are part of one's community.

The question of the Internet and its efficacy as a platform and set of technologies for public participation recalls the earlier stages of public participation as a whole. That is, its use marks a period of experimentation and ongoing social learning and an exploration of its possibilities and limitations. Many of its current uses that might be folded under the rubric of public participation are still emerging and under examination.

Concluding remarks

This overview of public participation has focused on the theme of blurring boundaries. Public participation was explored in its broadest sense, incorporating formal and informal sites, formats, sponsorships, intersections with information and communication technologies, and increasing multi-directionality. In taking this approach, we note that these social practices of public participation emerge from their times and places but also reflect assumptions about publics and science.

The chapter has drawn a picture of public participation as reflecting the ongoing reconfiguration of science and society, 'the continuation of scientific debate by other means' (Bucchi 2008: 61), the production of knowledges that has grown out of new and different collaborative enterprises involving scientific communities and different publics and other sites of knowledge production external to, and sometimes collaborating with, these communities. The agora has demonstrated heterogeneous and overlapping actors and activities with often unpredictable outcomes and, as Nowotny and colleagues (2001: 210) observed, 'the grounds of the agora are shifting continuously, as are (its many and varied) interconnections'.

Key Questions

- How and why have participatory methods developed for science/public interactions?
- What social factors are reshaping the forms of public participation?
- What are the possibilities and limits of knowledge co-production between scientists and citizens?
- How effective are Internet platforms for facilitating public participation with emerging science and technology?

Notes

1 The earlier version of this chapter focused primarily on deliberative approaches (Einsiedel 2008).
2 See www.archives.gov/federal-register/laws/negotiated-rulemaking/562.html; accessed 31 March 2013.
3 www.iap2.org; accessed 20 April 2013.
4 For example, the role played by the founder of Mothers Against Drunk Driving (MADD) and the role of that organisation in policy, cultural and institutional change have been detailed by Fell and Voas (2006). The organisation built on the developing scientific foundation of the relationship between alcohol consumption and blood alcohol concentration as well as the relationship of alcohol to highway crashes, while at the same time it mobilised the passions of mothers who had experienced a child's death from drunk driving.

5 For a concerted examination of public participation practices and policy cultures on one topic, xenotransplantation, see *Science and Public Policy* special issue, 38, 8 (2011).
6 See www.wwviews.org/; accessed 21 November 2013.
7 www.acercandonaciones.com/en/cultura/cafe-cultura-vuelve-a-bariloche-para-presentar-cafe-de-las-ciencias.html; accessed 21 November 2013.
8 These categories are from the Center for Advancement of Informal Science Education report (Bonney *et al.* 2009).
9 Proteins are one of the key building blocks in the body's biochemistry: they demonstrate 'a remarkable relationship between structure and function at the molecular level' (Dill and MacCallum 2012: 1042).
10 See The Nagoya Protocol on Access and Benefit Sharing, www.cbd.int/abs/; accessed 21 November 2013.
11 See for example the *Yearbook of Nanotechnology in Society* whose scope includes considerations of equity, equality and development (Cozzens and Wetmore 2011), or convergences of different scientific fields whose social implications are considered upstream (Ramachandran *et al.* 2011).

References

Arnstein, S. R. (1969) 'A ladder of citizen participation', *Journal of the American Institute of Planners*, 35, 4: 216–224.
Barnett, C., Cloke, P., Clarke, N. and Malpass, A. (2011) *Globalizing Responsibility: The Political Rationalities of Ethical Consumption*, New York: John Wiley.
Bucchi, M. (2008) 'Of deficits, deviations and dialogues – Theories of public communication of science', in M. Bucchi and B. Trench (eds) *Handbook of Public Communication of Science and Technology*, London and New York: Routledge, 57–76.
Bucchi, M. and Neresini, F. (2008) 'Science and public participation', in E. Hackett, O. Amsterdamska, M. Lynch and J. Wajcman (eds) *Handbook of Science and Technology Studies*, Cambridge, MA: MIT Press, 449–472.
Bonney, R., Ballard, H., Jordan, R., McCallie, E., Phillips, T., Shirk, J. and Wildermann, C. (2009) *Public Participation in Scientific Research: Defining the Field and Assessing Its Potential for Informal Science Education. A CAISE Inquiry Group Report*, Washington DC: CAISE.
Bultitude, K., McDonald, D. and Custead, S. (2011) 'The rise and rise of science festivals: an international review of organized events to celebrate science', *International Journal of Science Education, Part B*, 1, 2: 165–188.
Callon, M. and Rabeharisoa, V. (2003) 'Research "in the wild" and the shaping of new social identities', *Technology in Society*, 25, 2: 193–204.
Catlin-Groves, C. L. (2012) 'The citizen science landscape: from volunteers to citizen sensors and beyond', *International Journal of Zoology*, 2012; online at ; accessed 21 November 2013.
Chambers, S. (2003) 'Deliberative democratic theory', *Annual Review of Political Science*, 6: 307–326.
Chen, D. S. and Lin, K. (2006) 'The prospects of deliberative democracy in Taiwan', in M. H. Huang (ed.), *Asian New Democracies: The Philippines, South Korea, and Taiwan Compared*, Taipei: Center for Asia-Pacific Area Studies, RCHSS, Academia Sinica, 289–304.
Chilvers, J. (2007) 'Towards analytic-deliberative forms of risk governance in the UK? Reflections on learning in radioactive waste', *Journal of Risk Research*, 10, 2:197–222.
Chilvers, J. (2012) 'Reflexive engagement? Actors, learning and reflexivity in public dialogue on science and technology', *Science Communication*, 35, 3: 283–310.
Cozzens, S.E. and Wetmore, J. (eds) (2011) *Nanotechnology and the challenges of equity, equality and development*, Heidelberg: Springer.
Couvet, D., Jiguet, F., Julliard, R., Levrel, H. and Teyssedre, A. (2008) 'Enhancing citizen contributions to biodiversity science and public policy', *Interdisciplinary Science Reviews*, 33, 1: 95–103.
Crabbe, M. J. C. (2012) 'From citizen science to policy development on the coral reefs of Jamaica', *International Journal of Zoology*, 2012; online at http://dx.doi.org/10.1155/2012/102350; accessed 21 November 2013.
Crompton, H. (2007) 'Mode 2 knowledge production: evidence from orphan drug networks', *Science and Public Policy*, 34, 3: 199–211.
Dallas, D. (2006) 'Café scientifique – déjà vu', *Cell*, 126, 2: 227–229.
Davies, S., McCallie, E., Simonsson, E., Lehr, J. L. and Duensing, S. (2009) 'Discussing dialogue: perspectives on the value of science dialogue events that do not inform policy', *Public Understanding of Science*, 18, 3: 338–353.
Degelsegger, A. and Torgersen, H. (2011) 'Participatory paternalism: citizens' conferences in Austrian technology governance', *Science and Public Policy*, 38, 5: 391–401.

Delfanti, A. (2010) 'Editorial: open science, a complex movement', *JCOM – Journal of Science Communication*, 9, 3: online at http://jcom.sissa.it/archive/09/03/Jcom0903(2010)E/Jcom0903(2010)E.pdf

Dewey, J. (1927) *The Public and Its Problems*, New York: Holt.

Dill, K. A. and MacCallum, J. L. (2012) 'The protein-folding problem, 50 years on', *Science*, 338, 6110: 1042–1046.

Dobson, A. and Bell, D. (eds) (2005) *Environmental Citizenship*, Cambridge, MA: The MIT Press.

Dryzek, J. and Tucker, A. (2008) 'Deliberative innovation to different effect: consensus conferences in Denmark, France and the US', in *Public Administration Review*, 68, 5: 864–876.

Einsiedel, E. F. (2008) 'Public participation and dialogue', in M. Bucchi and B. Trench (eds) *Handbook of Public Communication of Science and Technology*, London and New York: Routledge, 173–184.

Einsiedel, E. F. (2012) 'The landscape of public participation on biotechnology', in M. Weitze, M., Puhler, A., Heckl, W. M., Muller-Rober, B. and Renn, O. (eds) *Biotechnologie-Kommunikation: Kontroversen, analysen, aktivitäten*, Berlin: Springer-Verlag, 379–412.

Einsiedel, E. F. (2013) 'Rethinking "publics" and "participation" in new governance contexts: stakeholder publics and extended forms of participation', in K. O'Doherty and E. F. Einsiedel (eds) *Public Engagement and Emerging Technologies*, Vancouver: UBC Press, 279–292.

Epstein, S. (2007) *Inclusion: The Politics of Difference in Medical Research*, Chicago: University of Chicago Press.

Evans-Cowley, J. and Hollander, J. (2010) 'The new generation of public participation: internet-based participation tools', *Planning Practice and Research*, 25, 3: 397–408.

Fals-Borda, O. and Rahman, M. A. (1991) *Action and Knowledge: Breaking the Monopoly with Participatory Action Research*, New York: Intermed Technology/Apex.

Fell, J. C. and Voas, R. B. (2006) 'Mothers Against Drunk Driving (MADD): the first 25 years', *Traffic Injury Prevention*, 7, 3: 196–212.

Fiorino, D. (1990) 'Citizen participation and environmental risk', *Science, Technology and Human Values*, 15, 2: 226–43.

Fischer, F. (2003) *Reframing Public Policy: Discursive Politics and Deliberative Practices*, Oxford: Oxford University Press.

Flanagin, A. and Metzger, M. (2008) 'The credibility of volunteered geographic information', *GeoJournal*, 72, 3–4: 137–48.

Fowler, G. and Allison, K. (2008) 'Technology and citizenry: a model for public consultation in science policy formation', *Journal of Evolution and Technology*, 18, 1: 56–69.

Funtowicz, S. and Ravetz, J. (1993) 'Science for the post-normal age', *Futures*, 25, 7: 739–55.

Gewin, V. (2013) 'Artistic merit', *Nature*, 496, 7448: 537–539.

Gibbons, M., Limoges, C., Nowotny, H., Schwartzman, S., Scott, P. and Trow, M. (1994) *The New Production of Knowledge: The Dynamics of Science and Research in Contemporary Societies*, London: Sage.

Godman, M. and Hansson, S. O. (2009) 'European public advice on nanotechnology: four convergence seminars', *Nanoethics*, 3, 1: 43–59.

Griessler, E. (2012) 'One size fits all? On the institutionalization of participatory technology assessment and its interconnection with national ways of policy-making: the cases of Switzerland and Austria', *Poiesis and Praxis*, 9, 1–2: 61–80.

Gutmann, A. and Thompson, D. (1996) *Democracy and Disagreement*, Cambridge, MA: Harvard University Press.

Habermas, J. (1989) *The Structural Transformation of the Public Sphere*, Cambridge, MA: The MIT Press.

Hagendijk, R. and Irwin, A. (2006) 'Public deliberation and governance: engaging with science and technology in contemporary Europe', *Minerva*, 44, 2: 167–84.

Haklay, M. (2013) 'Citizen science and volunteered geographic information: overview and typology of information', in D. Sui, S. Elwood and M. Goodchild (eds) *Crowdsourcing Geographic Knowledge*, Heidelberg: Springer, 105–122.

Hamlett, P. (2002) 'Adapting the Internet to citizen deliberations: lessons learned', in *Proceedings: Social Implications of Information and Communication Technology, IEEE International Symposium on Technology and Society*, Raleigh, NC: Institute of Electrical and Electronics Engineers, 213–218.

Hand, E. (2010) 'Citizen science: people power', *Nature*, 466, 7307: 685–87.

Hemment, D., Ellis, R. and Wynne, B. (2011) 'Participatory mass observation and citizen science', *Leonardo*, 44, 1: 62–63.

Hennen, L. (2012) 'Why do we still need participatory technology assessment?' *Poiesis and Praxis*, 9, 1–2: 27–41.

Higashijima, J., Miura, Y., Nakagawa, C., Yamanouchi, Y., Takahishi, K. and Nakamura, M. (2012) 'Public opinions regarding the relationship between autism spectrum disorders and society: social agenda construction via science café and public dialogue using questionnaires', *Journal of Science Communication*, 11, 4; online at http://jcom.sissa.it/archive/11/04/Jcom1104%282012%29A03/; accessed 21 November 2013.

Hirakawa, H. (2001) *Provisional Report on the GM Crops Consensus Conference in Japan*, presented to CNADS – National Council for the Environment and Sustainability Development; online at http://hideyukihirakawa.com/GMO/cc_report_lisbon.html; accessed 18 May 2012.

Howlett, M. and Migone, A. (2010) 'Explaining local variation in agri-food biotechnology policies: "green" genomics regulation in comparative perspective', *Science and Public Policy*, 37, 10: 781–795.

Irwin, A. (2001) 'Constructing the scientific citizen: Science and democracy in the biosciences', *Public Understanding of Science*, 10, 1: 1–18.

Irwin, A. and Michael, M. (2003) *Science, Social Theory and Public Knowledge*, London: Open University Press.

Israel, BA, Schulz, A., Parker, E. and Becker, A. (1998) 'Review of community-based rsearch: assessing partnership approaches to improve public health', *Annual Review of Public Health*, 19: 173–202.

Jasanoff, S. (2013) *States of Knowledge: The Co-Production of Science and the Social Order*, London: Routledge.

Jones, M. and Einsiedel, E. F. (2011) 'Institutional policy learning and public consultation: the Canadian xenotransplantation experience', *Social Science and Medicine*, 73, 5: 655–662.

Jormanainen, I. and Korhonen, P. (2010) *Science festivals on computer science recruitment*, Proceedings of the 10th Koli Calling International Conference on Computing Education Research, 72.

Karinen, R. and Guston, D. (2010) 'Towards anticipatory governance: the experience with nanotechnology', in M. Kaiser and M. Kurath (eds) *Nanotechnology and the Rise of an Assessment Regime*, Dordrecht: Springer, 217–232.

Konisky, D. M. and Beierle, T. (2001) 'Innovations in public participation and environmental decision-making: examples from the Great Lakes region', *Society and Natural Resources*, 14, 9: 815–826.

Korean National Commission for UNESCO (1998) *Korean Consensus Conference on the Safety and Ethics of Genetically Modified Food – Citizens' Panel Report*, Seoul: Korean National Commission for UNESCO.

Lafreniere, D. and Cox, S. M. (2012) 'Means of knowledge dissemination: are the café scientifique and the artistic performance equally effective?', *Sociology Mind*, 2, 2: 191–199.

Leydesdorff, L. and Ward, J. (2005) 'Science shops: a kaleidoscope of science-society collaborations in Europe', *Public Understanding of Science*, 14, 4: 353–332.

Lippman, W. (1922) *Public opinion*, New Jersey: Harcourt Brace and Co.

Mejlgaard, N. and Stares, S. (2013) 'Performed and preferred participation in science and technology across Europe: exploring an alternative idea of "democratic deficit"', *Public Understanding of Science*, 22, 6: 660–673.

Michael, M. (1998) 'Between citizen and consumer: multiplying the meanings of the "public understanding of science"', *Public Understanding of Science*, 7, 4: 313–327.

Michael, M. (2009) 'Publics performing publics: of PiGs, PiPs and politics', *Public Understanding of Science*, 18, 5: 617–631.

Mizumachi, E., Matsuda, K., Kano, K., Kawakami, M. and Kato, K. (2011) 'Scientists' attitudes toward a dialogue with the public: a study using science cafés', *Journal of Science Communication*, 10, 4; online at http://jcom.sissa.it/archive/10/04/Jcom1004(2011)A02; accessed 21 November 2013.

Mutheu, J. and Wanjala, R. (2009) 'The public, parasites and coffee: the Kenyan Science Café concept', *Trends in Parasitology*, 25, 6: 245.

Mutz, D. (2006) *Hearing the Other Side: Deliberative Versus Participatory Democracy*, Cambridge: Cambridge University Press.

Mutz, D. (2008) 'Is deliberative theory a falsifiable theory?', *Annual Review of Political Science*, 11: 521–538.

Nakkazi, E. (2012) 'Drinking up science in African cafés', *Scidev.net*, 3 September; online at: www.scidev.net/global/disease/feature/drinking-up-science-in-african-caf-s-1.html

Nowotny, H., Scott, P. and Gibbons, M. (2001) *Rethinking Science: Knowledge and the Public in an Age of Uncertainty*, Cambridge: Polity Press.

OECD (2001) *Citizens as Partners: Information, Consultation, and Public Participation in Policy Making*, Paris: OECD.

Owen, R., Macnaghten, P. and Stilgoe, J. (2012) 'Responsible research and innovation: from science in society to science for society, with society', *Science and Public Policy*, 39, 6: 751–760.

Paulos, E. (2009) 'Designing for doubt: Citizen science and the challenges of change', in *Engaging Data: First International Forum on the Application and Management of Personal Electronic Information*, Cambridge, MA: MIT; online at www.cs.berkeley.edu//%7EPaulos/papers/2009/Designing_for_doubt.pdf; accessed 24 November 2013.

Pellegrini, G. (2009) 'Biotechnologies and communication: participation for democratic processes', *Comparative Sociology*, 8, 4: 517–540.

Poster, M. (1997) 'Cyberdemocracy: Internet and the public sphere', in D. Poert (ed.) *Internet Culture*, London: Routledge, 202–214.

Price, V. (2006) 'Citizens deliberating online: theory and some evidence', in T. Davies and B. Simone Noveck (eds) *Online Deliberation: Design, Research, and Practice*, Stanford, Calif: CSLI Publications, 37–58.

Rabeharisoa, V. (2003) 'The struggle against neuromuscular diseases in France and the emergence of the "partnership model" of patient organization', *Social Science and Medicine*, 57, 11: 2127–2136.

Ramachandran, G., Wolf, S. M., Paradise, J., Kuzma, J., Hall, R., Kokkoli, E., Fatehi, L. (2011) 'Recommendations for oversight of nanobiotechnology: dynamic oversight for complex and convergent technology', *Journal of Nanoparticle Research*, 13, 4: 1345–1371.

Rogers, H. (2011) 'Amateur knowledge: public art and citizen science', *Configurations*, 19, 1: 101–115.

Rogers-Hayden, T. and Pidgeon, N. (2008) 'Developments in nanotechnology public engagement in the UK: "upstream" towards sustainability?', *Journal of Cleaner Production*, 16, 8–9: 1010–1013.

Rose, N. and Novas, C. (2004) 'Biological citizenship', in A. Ong and S. Collier (eds) *Global Assemblages: Technology, Politics and Ethics as Anthropological Problems*, Oxford: Blackwell, 439–463.

Rowe, G. and Frewer, L. (2000) 'Public participation methods: a framework for evaluation' *Science, Technology and Human Values*, 25, 1: 3–29.

Rowe, G., Marsh, R. and Frewer, L. (2004) 'Evaluation of a deliberative conference', *Science, Technology and Human Values*, 29, 1: 88–121.

Royal Academy of Engineering (2009) *Synthetic Biology: Public Dialogue on Synthetic Biology*, London: Royal Academy of Engineering.

Sclove, R. (2010) *Issues in Science and Technology*, New York: National Academy of Sciences.

Silvertown, J. (2009). 'A new dawn for citizen science', *Trends in Ecology and Evolution*, 24, 9: 467–471.

Skorupinski, B., Baranzke, H., Ingenslep, H. W. and Meinhardt, M. (2007) 'Consensus Conferences – A Case Study: Publiforum in Switzerland with Special Respect to the Role of Lay Persons and Ethics', *Journal of Agricultural and Environmental Ethics*, 20, 1: 37–52.

Stodden, V. (2010) 'Open science: policy implications for the evolving phenomenon of user-led scientific innovation', *Journal of Science Communication*, 9, 1; online at http://jcom.sissa.it/archive/09/01/Jcom0901(2010)A05; accessed 21 November 2013.

Thompson, D. F. (2008) 'Deliberative democratic theory and empirical political science', *Annual Review of Political Science*, 11: 497–520.

Tyler, T. (2002) 'Is the internet changing social life? It seems the more things change, the more they stay the same', *Journal of Social Issues*, 58, 1: 195–205.

Van Est, R., van Eijndhoven, J. C. M., Aarts, W. and Loeber, A. (2002) 'The Netherlands: seeking to involve wider publics in technology assessment', in S. Joss and S. Bellucci (eds) *Participatory Technology Assessment: European Perspectives*, London: Centre for the Study of Democracy, 108–125.

Wakeford, T., Murtuja, B. and Bryant, P. (2008) 'Four brief analyses of citizens' juries and similar participatory processes', *Participatory Learning and Action*, 58, 1: 48–55.

Warren, D. M, Slikkerveer, J. and Brokensha, D. (1991) *Indigenous Knowledge Systems: The Cultural Dimensions of Development*, London: Kegan Paul International.

Whyte, W. F. (1991) *Participatory Action Research*, Newbury Park, CA: Sage.

Whyte, A. and Pryor, G. (2011) 'Open science in practice: researcher perspectives and participation', *International Journal of Digital Curation*, 6, 1: 200–213.

Wiggins, A. and Crowston, K. (2011) 'From conservation to crowdsourcing: a typology of citizen science', in *Proceedings of the 44th Hawaii International Conference on System Sciences*, Institute of Electrical and Electronics Engineers.

Wilsdon, J. and Willis, R. (2004) *See-Through Science: Why Public Engagement Needs to Move Upstream*, London: DEMOS.

Wilson, S. (2010) *Art + Science Now: How Scientific Research and Technological Innovation are Becoming Key to 21st Century Aesthetics*, London: Thames and Hudson.

Worthington, R., Rask, M. and Lammi, M. (2011) *Citizen Participation in Global Environmental Governance*, London: Earthscan.

Wuchty, S., Jones, B. F. and Uzzi, B. (2007) 'The increasing dominance of teams in production of knowledge', *Science*, 316, 5827: 1036–1039.

Wynne, B. (1993) 'Public uptake of science: a case for institutional reflexivity', *Public Understanding of Science*, 2, 4: 321–337.

Young, J. C., Wald, D., Earle, P. and Shanley, L. (2013) *Transforming Earthquake Detection and Science Through Citizen Seismology*, Washington, DC: Woodrow Wilson Center; online at www.wilsoncenter.org/sites/default/files/CitizenSeismology_FINAL.pdf; accessed 21 November 2013.

11
Public understanding of science
Survey research around the world

Martin W. Bauer and Bankole A. Falade

Introduction[1]

The term *public understanding of science* (PUS) has a dual meaning. First, it covers a wide field of activities that aim at bringing science closer to the people and to promote public understanding in the tradition of a public rhetoric of science (see Fuller 2001, European Commission 2012, and Miller *et al.* 2002 for inventories of such initiatives). Second, it refers to empirical social research that investigates the public understanding of science and how this might vary across time and context. We concentrate our review on research that uses large-scale representative national and international sample surveys, asking people standard questions from a questionnaire. We review the changing agenda by typifying three *paradigms* of PUS research through the questions they raised, the interventions they supported and the criticisms they attracted. The chapter ends with an afterthought on *deficit concepts* and survey research, and an outlook on future research. This expands on previous reviews of the field (Etzioni and Nunn 1976; Pion and Lipsey 1981; Wynne 1995; Miller 2004; Allum 2010).

A *big data* pool: decades of survey research

Tables 11.1a, 11.1b and 11.1c list the main surveys of public understanding of science among adult populations since 1957, typically with nationally representative samples of 1,000 interviewees and more. The lists show the best-known surveys of scientific literacy, public interests and attitudes to science, many of which are partially comparable because they have often been modelled on the US National Science Foundation indicator (NSF) indicators series since 1979. The Eurobarometer series covers science since 1978 in initially eight and recently 32 European countries, with Britain covered (MORI, ESRC, OST, Wellcome Trust, BIS) since 1985 while the French series (see Boy 2012) reaches back to 1972; Italy started a regular survey in the early 2000s. The earliest of these PUS surveys dates from 1957 in the USA, just before the launch of the Russian satellite Sputnik shocked the western world (Withey 1959). These efforts were imitated and adapted across the globe during the 1990s and into the 2000s, first in Asia (Japan, China and India), Russia, Australia and New Zealand, and, in the early 2000s, the survey effort arrived in Latin America sponsored by an Ibero-American network. Brazil and Colombia had earlier surveys in 1987 and 1994. South Africa is constructing a national PUS survey base, after earlier attempts included white-population-only samples. Nigeria is the other African country from where PUS data is emerging (Falade 2014).

Many other surveys are related to specific and often controversial developments in science and technology; they are excluded from this review, not least for reasons of practicality. For example, Eurobarometer ran close to 100 multinational surveys since 1975 on issues such as nuclear power, environment, computers, cyber-society and biotechnology (see Gaskell et al. 2011 on the biotechnology series since 1993). Risk perception research has collected numerous national surveys of various hazards. Also not included here are the occasional science items in international surveys with broader social content. An analysis of this evidence on attitudes to science across countries remains desirable. All in all this amounts to big data on science attitudes, a treasure to be lifted by adventurous prospectors and competent researchers. A corpus of nationally and internationally comparable data has accumulated over the past 50 years which offers opportunities for new analysis, dynamic modelling and comparisons that define a renewed research effort.

Table 11.1 Tables 11.1a to 11.1c list, in three regional groups, all the nationally representative surveys on general attitudes to science known to the authors; Table 11.1d lists the agencies conducting or commissioning there surveys

Table 11.1a shows the surveys for Europe and North America

Year	Europe							N America	
	UK	France	EU	E*	Italy	Iceland	Bulgaria	US	Canada
1957								Michigan	
1970									
1971									
1972		Boy						Harvard	
1973									
1974									
1975									
1976									
1977	EB7	EB7	EB7						
1978	EB10a	EB10a	EB10a						
1979								NSF	
1980									
1981									
1982		Boy							
1983								NSF	
1984									
1985								NSF	
1986	KCL								
1987									
1988	ESRC	Boy/<17						NSF	
1989	EB31	Boy/EB31	EB31	EB31					MST
1990								NSF	
1991									
1992	EB38.1	EB38.1	EB38.1	EB38.1			STS	NSF	
1993									
1994		Boy							
1995								NSF	
1996	OST/Well						STS		
1997								NSF	

(Continued)

Table 11.1a (Continued)

Year	Europe							N America	
	UK	France	EU	E*	Italy	Iceland	Bulgaria	US	Canada
1998									
1999								NSF	
2000	RCENG	EB63.1	EB63.1	EB63.1		EB63.1	EB63.1		
2001	EB55.2	EB55.2	EB55.2	EB55.2				NSF	
2002			EB East	FECYT			EB East		
2003								NSF	
2004	BIS			FECYT					
2005	RCENG	EB63.1	EB63.1	EB63.1		EB63.1	EB63.1		
2006				FECYT				NSF	
2007	(EB)	Boy	(EB)	Ibero	Observa		(EB)		
2008	BIS				Observa			GSS	
2009	WELL			Ibero	Observa			PEW	
2010	EB73.1	EB73.1	EB73.1	FECYT	Observa	UNI/EB	EB73.1	GSS	
2011	BIS	Boy			Observa				
2012	WELL			FECYT	Observa	UNI			
2013	BIS				Observa				MST
2014					Observa				
2015					Observa				

*E refers to European countries outside EU

Table 11.1b for Asia, Africa, Australasia and Russia

Year	Other				Asia					
	Russia	S. Africa	AUS	NZ	Japan	Korea	Malay	India	Taiwan	China
1990										
1991		SAASTA			NISTEP					
1992								NISTED		CAST
1993		SAASTA								
1994										CAST
1995	HSE		STAP							
1996	HSE									CAST
1997	HSE			MST						
1998										
1999	HSE									
2000								STIC		
2001		SAASTA			NISTEP					CAST
2002										
2003	HSE									CRISP
2004								NCAER		
2005										CRISP
2006						Kofac				
2007								NISTED		CRISP
2008									SunYa	
2009								NYRead		
2010		SAASTA	NUA							CRISP
2011										
2012									SunYa	
2013								NISTED		CRISP
2014									SunYa	
2015										CRISP

Table 11.1c shows for Latin America and Caribbean

Year	Brazil	Argentina	Venzuzuela	Costa Rica	Panama	Uruguay	Ecuador	Paraguay	Peru	Chile	Mexico	Trinidad	Columbia
1985													
1986													
1987	CNPq												
1988													
1989													
1990													
1991													
1992													
1993													
1994													ColSci
1995													
1996													
1997											Conacyt		
1998													
1999													
2000													
2001					SENACYT								
2002											Conacyt		
2003	FAPESP	RiCYT									Conacyt		
2004	FAPESP		MCT			RepUni							ColSci
2005											Conacyt		
2006	MCT	SeCyt	MCT								Conacyt	unknown	
2007	Ibero	Ibero	Ibero		Ibero	Ibero				CON/Ibero	Conacyt		Ibero
2008	FAPESP				SENACYT		SENACYT						
2009	Ibero	Ibero				ANII/ibero		Ibero	Ibero		Conacyt		Ibero
2010	MCT												
2011					SENACYT								
2012	PPSUS	MCT/Redes		IDESP							Conacyt		
2013													ColSci
2014	MCT											unknown	
2015													

Table 11.1d National and international institutions sponsoring and conducting PUS surveys

ANII	Uruguay
BAS-IS	Bulgarian Academy of Science, Institute of Sociology, Sofia
BIS	Business, Innovation and Science, UK Ministry
CNPq	Brazilian National Research Foundation
CEVIPOV	Centre for the Study of Political Life, SciencePo, Paris
CONACYT	Mexico
CONARE	Costa Rica
COLCIENCIAS	Columbia
HSE	Higher School of Economics, Russia
ESS	European Social Survey
EB	Eurobarometer, DG-12, later DG Research, Brussels
EVS	European Value Survey
ESRC	Economic and Social Research Council, UK
FAPESP	Research Foundation of the State of Sao Paulo, Brazil
FECYT	Spain
ISSP	International Social Survey Programme
MINCYT	Argentina
MORI	British public opinion research company
MYCT	Venezuela
NSF	US National Science Foundation, Washington, Science Indicators
NISTEP	National Institute of Science and Technology Policy, Japan
NISTED	National Institute of Science, Technology and Development, India
NCAER	National Centre for Applied Economic Research, Delhi, India
MST, MCT	Ministry of Science and Technology (Canada, China, Brazil)
Observa	Non Profit Research Centre, Science in Society, Italy
Wellcome Trust	Research Foundation interested in PUS, Britain
RiCyT	Latin American Network for Science Indicators
CAST	China Academy of Sciences and Technology
OST	Office of Science and Technology, London
PISA	Programme for International Student Assessment, OECD, Paris
SAASTA	South Africa
SECYT	Argentina
SENACYT	Panama and Ecuador
STIC	Strategic Thrust Implementation Committee, Malaysia
WVS	World Value Survey

Paradigms of researching the publics of science

Over the last 40 years the public understanding of science has spawned a field of enquiry that engages, to a greater or lesser degree, sociology, social psychology, history, communication studies and policy analysis. It remains somewhat marginal within social research but vigorous in its output (see Suerdem *et al.* 2013). Table 11.2 gives a schematic overview of three *paradigms* of research into public understanding of science. The table models the relationship between science and the public based on the attribution of deficits (see Bucchi and Trench in this volume). Each paradigm has its prime time and is characterised by a diagnosis of the problem that science faces in its relations with the public. Each paradigm pursues particular research questions through survey research and offers particular solutions to the diagnosed deficit problems. Rather than assuming that the emergence of a new paradigm displaces the older one, more realistically we must assume that these frames of mind co-exist in the present and in different contexts. This is not a model of progress, but one of

Table 11.2 Periods, problems and proposals

Period	Problem Attribution	Proposals for Research
Science Literacy 1960s–1985	Public deficit Knowledge	Measurement of literacy Education
Public Understanding 1985–1995	Public deficit Attitudes	Knowledge drives attitude Attitude change Education Public Relations
Science in Society 1995–present	Trust deficit Expert deficit Notions of the public Crisis of confidence	Participation Deliberation 'Angels', mediators Impact evaluation

Source: modified from Bauer, Miller and Allum 2007

multiplication of discourses. One might argue that the literacy-PUS tradition of research develops in parallel to that of risk perception, with a common function: to deal with a recalcitrant public opinion. It appears that the literacy-PUS line is favoured by the scientific community, while risk perception sits better with engineers and investment managers (see Bauer 2014).

Scientific literacy (mainly 1960s to mid-1980s)

Scientific literacy builds on two ideas. Firstly, science education is essentially part of the secular drive for basic literacy in reading, writing and numeracy. The second idea is that science literacy is a necessary part of civic competence. In a democracy people partake in political decisions in one way or the other, either directly through voting and indirectly via expressions of public opinion. However, the political animal is only effective if it is also familiar with the political process (Althaus 1998). The assumption is that scientific as well as political ignorance breeds alienation and extremism, hence the quest for *civic scientific literacy* (Miller 1998). These ideas highlight the dangers of a cognitive deficit and call for more and better science education through the life cycle. However, it also plays to technocratic attitudes among elites: an ignorant public is disqualified from partaking in policy decisions.

An influential definition of science literacy was proposed by Jon D. Miller (1983, 1992) with four elements: (a) knowledge of basic textbook *facts* of science, (b) an understanding of *methods* such as probability reasoning and experimental design, (c) an appreciation of the *positive outcomes* of science and technology for society, and (d) the *rejection of 'superstitions'*. Miller developed these literacy indicators from earlier work (e.g. Withey 1959) for the NSF in the US. Since 1979, NSF has undertaken a regular audit of the nation's scientific literacy with representative surveys of the adult population. Similar efforts, but less regular, came in the EU in the 1980s, and also elsewhere. This literacy programme is written into a Chinese plan for 2020, which stipulates a measurement of literacy targets for youth, farmers and urban workforce (CAST 2008). The plan ensured that the 2010 China Civic Science Literacy Survey comprised 69,000 interviews, a representative sample of the population in all provinces (He and Gao 2011).

The research agenda

Knowledge is the key problem of this paradigm, and it is measured by quiz-like items (see examples in Table 11.3). Respondents are asked to decide whether a statement giving a scientific textbook fact is true or false.

Table 11.3 Examples of knowledge and attitude items in literacy research

---

Knowledge items (examples)
Item 1. 'Does the earth go around the sun or does the sun go around the earth? (**the earth goes around the sun**; the sun goes around the earth; dk).
Item 2. 'The centre of the earth is very hot' (**true**, false, dk)
Item 3. 'Electrons are smaller than atoms' (**true**, false, dk)
Item 4. 'Antibiotics kill viruses as well as bacteria' (true, **false**, dk)
Item 5: 'The earliest humans lived at the same time as the dinosaurs' (true, **false**, dk)

Attitude items (examples of common items, Likert-type scales)
Item 6: 'Science and technology are making our lives healthier, easier and more comfortable' **(agree = positive)**
Item 7: 'The benefits of science are greater than any harmful effects' **(agree = positive)**
Item 8: 'We depend too much on science and not enough on faith' **(disagree = positive)**
Item 9: 'Science and technology change our lives too fast' **(disagree=positive)**
Scale: 1 strongly agree; 2 agree to some extent; 3 neither/nor; 4 disagree to some extent; 5 strongly disagree; 99 don't know (DK).

Source: e.g. Eurobarometer 31, 1989; see INRA (Europe) and Report International (1993)

Respondents score a point for every correct answer (as highlighted in Table 11.3). It is not trivial to formulate short and unambiguous statements that have an authoritative answer, and to balance easy and difficult items from different fields of knowledge. The responses to these items, often a set of between 10 and 20, must be combined to form a reliable index. Research involves the construction of items and the testing of their scalar value. Finding and testing such items is a bit like laying bricks – they have to stand up in the end. Item response theory is brought into the discussion of the research (Miller and Pardo 2000; Shimizu and Matsuura 2012). In isolation these items have little significance, though the reliability of the scale remains an issue (Pardo and Calvo 2004). Some items are notorious, have travelled far and hit the news headlines, not least because of their scandal value when answered incorrectly and considered in isolation. Public speakers and mass media repeatedly pick out single items as *proof* of public ignorance and as a cause for moral panic. For example, the item about the sun and the earth has seen many citations out of context.

Researchers have explored the 'don't know' ('DK') responses to knowledge items and the ambiguities of ignorance. Variations in DK-responses suggest an index of confidence: women and certain social milieus prefer declaring ignorance rather than guessing; they are less confident to give opinions on science (Bauer 1996). Turner and Michael (1996) describe four types of self-admitted ignorance: embarrassment ('I will go and find a book in the library'); self-identity ('I am not very scientific'); division of labour ('I know somebody who knows'); smugness ('I couldn't care less'). However, sudden changes in DK responses might indicate methodological issues: survey companies can alter their interview protocol (e.g. accept DK as an answer or probe further). Further probing reduces the reported rate of DK. Such changes can also reflect a change of fieldwork contractor, as in the case of the Eurobarometer consortium.

What is to be done?

The literacy research paradigm is fixated on the cognitive deficit, which falls easily into political controversy when one worries about 'misinformation' from the other party (e.g. Lewandowsky *et al.* 2012). Interventions are focused mainly on *education*. Literacy is a matter for continued education, requires attention in the schools' curricula and on the part of a publicly responsible mass media that is called to task to educate the wider public (see e.g. Royal Society 1985). This literacy-cum-education paradigm dominates official Chinese policy as expressed in the 2006 law on the 'popularisation of science' and the related policy statements (see CPI 2006).

Critique

The critique of the literacy paradigm focuses on conceptual as well as empirical issues. Why should science knowledge qualify for special attention? What about historical, financial or legal literacy? The case for *science literacy* needs to be made in competition with other types of literacy. What counts as scientific knowledge? Miller (1983) suggested two dimensions: facts and methods. This stimulated efforts to assess familiarity with scientific procedures such as probability reasoning, experimental design and the importance of theory and hypothesis testing. Others have argued that the essence of science is process, not facts (e.g. Collins and Pinch 1993), and that topics like uncertainty, peer reviewing, scientific controversies and the need for experiments to be replicated should be included in the assessment of literacy. But scales of methodological knowledge are challenging to construct. An open question suggested by Withey (1959) proved insightful: 'Tell me in your own words, what does it mean to study something scientifically'. Respondents' answers can be coded for normative and descriptive awareness (see Bauer and Schoon 1993). Process also includes awareness of scientific institutions and its activities, what Prewitt (1983) called *scientific savvy* and Wynne (1996) called the *body language* of science, which received attention in explorative studies (Bauer et al. 2000; Sturgis and Allum 2004).

Many countries have undertaken audits of adult scientific literacy in order to compare with others. A problem of such comparison remains the fairness of the indicators. The set of knowledge items could be biased towards a particular science base. Countries tend to have a corps of scientific heroes from one rather than the other discipline, and literacy scores are likely to reflect this particular science base. Raza and his colleagues' suggestions for culturally fair indication and analysis deserve more attention (e.g. Shukla 2005; Raza et al. 1996, 2002; Raza 2013).

Other issues arise around the question: is literacy a continuum or a threshold measure? Miller originally envisaged a threshold measure. To qualify as a member of the *attentive public for science* one needs to command some minimal level of science literacy, be interested, command the vocabulary of science and technology, appreciate positive outcomes, and renounce superstitions. However, the definition of this minimal level of literacy changed from audit to audit, and it is unclear whether the reported changes, or for that matter the lack of changes (Miller 2004), reflect shifts in definition or in substance (see Beveridge and Rudell 1998).

Critics have also argued that indicators of textbook knowledge are irrelevant and empirical artefacts. Of real importance is knowledge-in-context that emerges from local controversies and people's concerns (Ziman 1991; Irwin and Wynne 1996). However, what accounts for the consistent correlations between measures of literacy, attitudes and socio-demographic variables? We must recognise a certain intellectual failure to engage with these robust results.

Then there is the question of *superstitions*. For example, does belief in astrology disqualify a member of the public from being scientifically literate, as Miller (1983) suggested and as the Chinese Research Institute for Science Popularisation (CRISP) in China continues to declare? Is the relation of superstition and scientific literacy not more of an empirical matter? Astrology and scientific practice serve different functions in life. To make the rejection of astrology a criterion of literacy bars us from understanding the tolerance-intolerance between science and pseudoscience in everyday life, which is a cultural variable (see Allum and Stoneman 2012; Bauer and Durant 1997; Boy and Michelat 1986).

Knowledge items can be controversial in substance. So, for example, physicists might point out that whether electrons are smaller than atoms cannot be determined in general but depends on circumstances. More problematic is the statement 'The earliest humans lived at the same time as the dinosaurs', which according to biology textbooks is false. This item captures, in particular in the USA, a debate over evolutionary theory which clashes with fundamentalist religious culture. On this, the NSF invited a discussion on knowledge versus belief (Toumey et al. 2010). It is thus not a priori clear whether this item is an indicator of science literacy or of religious belief; this might depend on the wider conversation of science in society.

The very concern with literacy might also respond to a crisis of legitimacy of science in society. But to tackle this crisis by literacy campaigns assumes a gap in the operations of scientists and an illiterate public for which there is little evidence beyond elitist prejudice. If Francis Bacon's late sixteenth-century notion of 'knowledge is power' holds, any attempt to share knowledge without simultaneous empowerment will alienate rather than bring the public closer to science. Literacy is therefore the wrong answer to a question of legitimacy, (Roqueplo 1974) of trust (House of Lords Select Committee on Science and Technology 2000) or of authority more generally (Arendt 1968).

Public understanding of science (after 1985 to mid-1990s)

New concerns emerge under the title public understanding of science, or PUS.[2] In the UK, this is marked by an influential report of the Royal Society (1985) with that very title. PUS inherits the notion of a *public deficit*; however, now it is the attitudinal deficit that is foregrounded (Bodmer 1987). The public is not sufficiently positive about science and technology, sceptical or even outright anti-science. This must be of major concern to scientific institutions like the Royal Society. Old and new good reasons for the public appreciation of science are put forward: it is important for making informed consumer choices; it enhances the competitiveness of industry and commerce; and it is part of national culture (see Thomas and Durant 1987; Gregory and Miller 1998; Felt 2000). The Royal Society of London famously assumed that more knowledge would foster more positive attitudes.

Research agenda

The public positioning vis-à-vis science is mostly measured by Likert-type attitude items. Respondents agree or disagree with statements and thereby express their positive or negative orientation towards science (see Table 11.3 for examples). Some statements, in order to assess a positive attitude, require some respondents to disagree and others to agree, depending on the formulation. A mixed set of items avoids acquiescent response bias, that is the general tendency to agree to most statements in the artificial context of survey interviews. Another issue is how to deal with 'neither/nor' and 'DK' options. Not offering a 'neither/nor' may increase the variance in the data, but this forces people into positions which they do not hold. This would leave no space to express ambivalence, genuinely motivated abstention of judgement or the absence of opinion. There must be space for the *idiot*; that is, in ancient Greece, someone who has not yet formed an opinion (see Lezaun and Soneryd 2007).

Research on science attitudes is concerned with the construction of reliable scales, the variable structure of attitudes (e.g. Pardo and Calvo 2002), the relationship between general and specific attitudes, context effects of previous questions and, most importantly, the relationship between knowledge and attitudes (Sturgis and Allum 2004). The concern for literacy carried over into PUS, as knowledge measures were needed to test the commonplace notion: *The more they know, the more they love it*. However, the emphasis shifted from a threshold measure to that of a continuum of knowledge.

Breakwell and Robertson (2001) found that British girls were less inclined towards science than boys, and this gap did not close between 1987 and 1998. Sturgis and Allum (2001) highlight the knowledge gap between men and women when explaining the attitude gap, controlling for other factors. Crettaz (2004) shows for Switzerland in 2000 that gender does not explain attitudinal gender differences; it is science literacy and general education that make the difference. The persistence of gender gaps for science literacy in rural and urban China, and across age groups, is explored by Chao and Wei (2009), though the gap is closing for those with higher education.

What is to be done?

The practical interventions of the PUS paradigm might be divided into a rationalist and a realist agenda. Both agree with the diagnosis of an attitudinal deficit – the public is insufficiently infatuated with science and technology – but they disagree on what to do about it. For the *rationalist*, public attitudes are a product of information processing with a cognitive-rationalist core. Hence, negative attitudes towards science – or risk perceptions of technology deviating from actuarial assessment – are caused by insufficient information, or they are based on heuristics, such as availability or small sample evidence, that *bias* the public's judgement. It is assumed that, had people all the information and operated without these heuristics, they would display more positive judgements of scientific developments. Thus, they would agree with experts, who do not succumb to these biases as easily as the public does. People thus need more information and training on how to avoid faulty information processing. The battle for the public is thus a battle for rational minds with the weapons of information and training in probability and statistics.

For the *realist,* attitudes express relations with the world. Realists work the emotions and appeal to people's desires, moral stances and gut reactions, and thus follow the logic of modern sophistry in advertising and propaganda. In what is seen as the battle for the hearts of the public, the key question is: how can we make science *sexy*? The consumer public is to be seduced rather than rationally persuaded. According to this logic, there might be little difference between science and washing powder in terms of reaching the audience (see Michael 1998, Toumey *et al.* 2010).

Critique

The critique of deficit models correctly highlighted the pitfalls of reifying knowledge – scientific knowledge is what surveys measure – and insisted on a focus on knowledge-in-context (Ziman 1991) and on how experts relate to the public (Irwin and Wynne 1996). Wynne (1993) suggested the term *institutional neuroticism* to point at prejudices of scientific actors towards the public that create a self-fulfilling prophecy and a vicious circle: the public, cognitively and emotionally deficient, cannot be trusted. This mistrust by scientific actors will be paid back in kind by public mistrust. Negative public attitudes then confirm the assumptions of scientists: the public is not to be trusted. This circularity and confirmatory bias of an *institutional unconscious* calls for *soul-searching,* that is reflexivity among scientific actors, and possibly endorsement of a social epistemology with a plurality of knowledge centres (Jovchelovitch 2007).

The empirical critique of the paradigm focuses on the relations of interest, attitudes and knowledge. The correlation between knowledge and attitudes becomes a focus of research. The results remained inconclusive until recently (see Durant *et al.* 2000; Allum *et al.* 2008). Overall, large-scale surveys show a consistent but small correlation of knowledge and positive attitudes, but they also show larger variance among the knowledgeable. However, on controversial topics, the correlation approaches zero. Thus not all informed citizens are also enthusiastic about all science and technology; in relation to controversial issues, familiarity breeds contempt. In hindsight, it is surprising that anybody ever expected this to be different.

The measurement of attitudes is the remit of social psychology (e.g. Eagly and Cheiken 1993; McGuire 1986). In classical theory, cognitive elaboration is not a factor of positive attitudes but a quality of the attitude: knowledge fortifies the attitude to resist influence and makes it more predictive of behaviour, whatever its direction (Pomerantz et al. 1995). What emerges is that knowledge does matter but not in the way scientists' common sense assumes. Better-informed citizens do not have more positive attitudes, but poorly informed citizens more easily sway their views.

Many surveys measure the people's interest in science. Eurobarometer surveys suggest that self-reported interest is rising and falling over time, while knowledge is increasing (Miller et al. 2002). Shukla and Bauer (2012: 102) confirm this trend and show that *familiarity breeds disinterest*, thus touching upon another naïve assumption of the literacy and PUS paradigms: *The more we know, the more we are interested*. Interest in science deserves to be compared with interest in other matters (Paul 2008).

Science in-and-of Society (mid-1990s to present)

The critique of the public deficit models ushered in a reversal of the attribution. The public deficit of trust is mirrored by a deficit on the part of science and technology and its representatives. The focus shifts to the deficit of the scientific expert: their prejudices of the public.

Diagnosis

Evidence of negative attitudes from large-scale surveys is contextualised with focus group research and in quasi-ethnographic observations and reinterpreted as a 'crisis of confidence' (House of Lords Select Committee on Science and Technology 2000; Miller 2001). Science and technology operate *in* society and therefore stand relative to other sectors of society. The views of the public held by scientific experts come under scrutiny. Prejudices operate in policymaking and communication efforts; and these alienate the public. The decline in trust of the public vis-à-vis science might also indicate the revival of an enlightenment notion of a sceptical but informed public opinion (Bensaude-Vincent 2001). It remains unclear, however, whether there is a long-term decline in public trust or only an institutional anxiety. The existing data has not received a systematic analysis as yet.

What is to be done?

Writings on the new governance of science advise public involvement as part of a new deal between science and society (e.g. Jasanoff 2005). For the science-in-society paradigm the distinction between research and intervention blurs. Many are committed to action research and reject the separation of analysis and intervention. This agenda, academically grounded as it may be, often ends in political consultancy with a very pragmatic outlook. Notions of the public, of public opinion and of the public sphere, are reported back as *theories espoused* and *theories-in-action* (Argyris and Schön 1978) to stimulate reflective change of mind among these scientific actors (see Braun and Schulz 2010; Lezaun and Soneryd 2007).

Advice proliferates on how to rebuild public trust by addressing its paradoxes: trust is relational; once it is on the cards, it is already lost; trust cannot be engineered, it is granted to who deserves it (see Luhmann 1979). Public participation is the Prince's way to rebuild public trust: the House of Lords Select Committee on Science and Technology report (2000) lists many formats such as citizen juries, deliberative opinion polling, consensus conferencing, national debates, hearings etc., or what Jasanoff (2003) calls *technologies of humility*. Academic writings compare and order the virtues, the experience and the know-how of these exercises (Gregory *et al.* 2007; Abels and Bora 2004; Joss and Belluci 2002; Einsiedel *et al.* 2001). Many of these lively discussions merge with a professionalism of public relations for science (see Borchelt and Nielsen in this volume).

Critique

Deliberative activities are time-consuming, require know-how, and they are thus increasingly outsourced to a newly forming private sector of *angels*. These are age-old mediators, in this context, not between Heaven and Earth but between a disenchanted public and the institutions of science, industry and policymaking. But, for the utilitarian spirit, the democratic ethos is not self-sufficient. 'Does the deliberation process pay off?' is a pertinent question. Also a private sector of outsourced PR *angels* makes claims and offers services, often with spurious product differentiations, which require critical consumer testing.

The answer to the audit problem requires process and impact measures. Researchers therefore advocate quasi-experimental evaluations of participatory events and suggest process and outcome indicators (see Rowe and Frewer 2004; Bütschi and Nentwich 2002), including indicators such as changing public literacy and attitudes. It seems that discussions of public engagement have moved from motivation to empirical analysis and a critical assessment of formats, aspiration and reality. A key question that remains to be addressed comparatively is how does the roll-out of public event-making relate to changes in public knowledge and attitudes in the long run? PUS surveys are unlikely to evaluate any one particular event of deliberation or public engagement, but in the long run the surveys should pick up a cumulative shift in science culture, if it happens.

Afterthought on survey research and the deficit concept

PUS survey research has at times been hampered by an association with the *deficit concept*. As the polemic has it, the PUS survey researcher is a *positivist* who constructs the *public deficit* of knowledge, attitude and trust in the service of sponsors in government, business or learned societies. In doing so, surveys bolster existing powers that seek to control public opinion. By contrast the *Critical-constructivist* researcher avoids this ideological entanglement of the PUS survey by mobilising exclusively qualitative data.[3] The reflexivity of qualitative research will open the sponsors to a change of mind. *Critical* qualitative research emancipates the public from the grip of elite prejudice: where doxa was, there will be logos. Problematic in all this is neither the difference in knowledge interests, nor the potential move from prejudice (doxa) to enlightenment (logos), but the identification of knowledge interest and survey method protocol:

> Quantitative survey research = Positivist = anxious control, de-contextual prejudice
> Qualitative research = Critical-constructivist = reflexivity and change

These implicit equations are of unclear origin like any urban myth, but they were catalysed by the reception of the influential British research programme on PUS of the late 1980s. Irwin and Wynne (1996), in summarising that effort, framed a polemic with unfortunate consequences.[4] The *essentialisation* of sample surveys in elite anxiety and a control agenda is a fallacy, historically unfounded and unduly restrictive of research (see Kallerud and Ramberg 2002). These polemics made our Ibero-American colleagues stay away from measuring knowledge, something colleagues now seem to come to regret when constructing indirect measures of information (Polino and Castelfranchi 2012). In Britain, government surveys of attitude only reintroduced literacy items in 2013, after they were dropped in 2000 to avoid the embarrassment of association with the deficit concept. The identification of data protocol and knowledge interest ignores the interpretive flexibility afforded by any instrument.

Where to go from here?

Progress in PUS survey research remains modest, but recently gained a new momentum. None of the new discourses made the previous ones obsolete. PUS adds attitudes to the literacy concept. Science-in-society, whilst rejecting the deficit model, cannot avoid the thorny audit culture, and ironically reinvents the measures of public understanding to evaluate the effects of public participation events; but what do the PUS surveys tell us so far? The evidence can be summarised (see also Bauer *et al.* 2012) in the following points:

- Knowledge of science measured by textbook items is generally improving as a linear trend since 1989 in Europe, Japan, China and the USA; attitudes and interest in science show no such linear trend over the same time period. It is shown for Japan that knowledge is improving in all age groups.
- Knowledge and attitudes are not necessarily positively correlated. The more controversy on the issue, the lower is the correlation. On aggregate, in the European context, the correlation between knowledge and welfare expectation from science is negative: the more knowledgeable the population, the less the welfare expectations from science. In the very different context of India, this correlation remains positive.
- In the USA, younger age cohorts are more prone to endorse pseudoscientific pursuits such as interest in UFOs and astrology despite their rising science literacy.
- Knowledge is not so much a driver of attitudes as a definer of the quality of the attitudes. Cognitively elaborated attitudes are more difficult to change, attitudes based on low level of knowledge are more volatile.
- The gender gap in science literacy is closing in many contexts; the gap is much less in evidence among younger age cohorts than older ones. However, the gender gap in attitudes to science persists: women tend to be more sceptical on the claimed achievements.
- Across European countries, the generation with the highest knowledge of science is the baby-boom generation, born in the 1950s and educated in the 1960s.
- Across Europe, in the youngest age groups born after the mid-1970s, those with only primary education do better in science than those with secondary education.
- While among the older generations, the gap in interest is between those with higher or secondary education and those with only primary education, among the youngest age groups, the new gap in science interest is between primary or higher education and those who left school with only secondary education.
- Different facets of attitudes to science move differently across time and age groups.

These general results need to be more clearly specified and contextualised on the available data. In order to capitalise on the new opportunities that arise from a global corpus of data accumulating over the past 50 years we need to open the research agenda. This includes taking survey research for what it really is: a powerful and *movable immobile*, a format of representation of science culture among others. Achieving this, this field of research will enter its most fertile period yet. Public understanding of science is a process. The occasional survey, media analysis or focus group might have news or scandal value, but it is not yet a valid analysis of the historical dynamic of science culture. Now it is time to be ambitious again. Recent research is developing these efforts in five directions.

1. To integrate national and international surveys into a more global database for purposes of comparative analysis

Such a database of PUS would bring the field into the age of *big data*. The patronage of an international body such as the World Bank, UNESCO or OECD may be needed to consolidate the collaboration with existing social science data archives. A promising model is the OECD assessment of educational achievements (PISA) which focus periodically on *scientific literacy*, albeit only for 15-year-olds. These results create large and detailed databases across many countries. However, according to Sjoeberg (2012), PISA suffers from serious gaps between espoused intentions and achieved measurement, and an economistic bias of privileging the producer/consumer over the literate citizen. The OECD Science Indicator Group may turn its attention to adult attitudes to science, which will catalyse the global data integration advocated here. But an OECD effort will not compensate for the lack of data in Africa and South East Asia; though South Africa is moving ahead with the Social Attitude Survey series (Reddy *et al.* 2009).

2. To produce sophisticated secondary analysis of the growing longitudinal database

PUS campaigns have encouraged research on target group segmentation and consumer-type profiling of the public in existing data. Lebart (1984) worked with correspondence analysis and clustering. The British public has been clustered into six groups (OST 2000): confident believers, technophiles, supporters, concerned, the 'not sure', and the 'not for me'. However, the classification seems to add little to an education gradient. A more useful segmentation was developed for Portugal (da Costa *et al.* 2002). Mejlgaard and Stares (2012) and Stares (2009) develop typologies based on latent trait modelling of participation using interest, informedness and knowledge. PUS researchers might also look across to other forms of literacy. There is a literature emerging on financial literacy (Lusardi and Mitchell 2006) based on concerns about people making poor decisions for retirement savings. There is also a quest for literacy in the arts and humanities (see Liu *et al.* 2005).

3. To construct dynamic models of public understanding of science over time, including cohort analytic and quasi-panel models, and test these in different contexts

PUS research will, in the coming years, capitalise on the longitudinal data that is available in many contexts. Gauchat (2012) finds that the authority of science is seriously challenged in the USA, but this is mostly among Republican voters. Bauer and Howard (2013) offer a view of science culture in modern Spain compared with the rest of Europe based on four waves of Eurobarometer data, 1989–2005, and centred on changes in public attention, enculturation and progressivism across generations.

4. To work towards indicators of the qualitative diversity of 'cultures of science'

The cultural authority of science is being reconsidered (Gauchat 2011) and the quest for indicators of a *science culture* is gaining track (Godin and Gingras 2000; Bauer 2012b). Shukla and Bauer (2012) validate the science culture index to do justice to the diversities of Europe and

India; and Song (2010) develops a system of indicators to compare Korean cities and Korea with neighbouring countries. Vogt (2012) proposes a model of science culture spiralling over the two axes of exoteric/esoteric and dialogical/monological communication. All these indicator ideas take up the challenge of combining *objective* scientific performance and *subjective* perception data. The quest for indicators of science culture will grapple with old and semantically rich conversations, such as the one on 'scientific temper' in India (see Mahanti 2013; Raza 2013; Kumar 2011).

The analysis of science cultures in the plural raises questions on so-called superstitions, traditional beliefs, theory of evolution and people's relations to animals, not only in terms of knowledge but in terms of (in)compatibility of different beliefs (see Allum and Stoneman 2012; Crettaz 2012). Attitudes must be understood in the context of iconic imagination, which comes to the fore when common sense is challenged by scientific research. Representations familiarise the unfamiliar and give attitudes the directions of approach or avoidance (Farr 1993; Wagner 2007). Such an outlook shifts PUS research from rank-ordering people by gradients of knowledge or attitudes to the characterisation of images of science in the function of different lifeworlds (e.g. Boy 1989; Bauer and Schoon 1993; Durant *et al.* 1992). In the twenty-first century, science is a globalised pursuit; but science culture is not, if it ever will be. Studying the social representations of science opens the door widely for mixed methods of enquiry (Bauer and Gaskell 2008; Wagner and Hayes 2005).

5. To develop complementary data streams such as mass media monitoring and longitudinal qualitative research to map the public understanding of science

The PUS paradigm has extended its range of data streams. Mass media monitoring, in particular of print and online contents, is cost-effective and easily extended backward in time and updated into the present. The salience and framing of science in mass media offer indicators of attention and reveal the trends, such as the medicalisation of news (Bauer 1998), the issue cycles (Schäfer 2012), and the waves of attention over the past 150 years (Bauer 2012a; Bauer *et al.* 2006; Bucchi and Mazzolini 2003; LaFollette 1990). Public attention to science is no historical constant, and these fluctuations are far from understood.

Survey indicators are also making an inroad into the study of the mobilisation of scientists (Bauer and Jensen 2011). The mobilisation of scientists for public engagement across times and place invites comparison to the *great awakenings* in American evangelism (Barkun 1985); they are cyclical rather than a historical constant. To get a better sense of the scope of this mobilisation we need investigations of the historical origins of the current mobilisation wave since the 1980s. The very commissioning of PUS survey research is likely to be a part of this mobilisation effort in a particular historical context that deserves to be elucidated (see Gregory and Lock 2008).

Key questions

- What have been the main paradigms for PUS survey research and what are the principal differences between them?
- How do such paradigms relate to broader debates about models of science communication?
- What do surveys indicate about the correlation of knowledge of science with attitudes to science? What implications do such findings have for policymaking?

Notes

1 This new version of the chapter substantially updates and expands the material covered in the first edition of this Handbook (Bauer 2008).
2 PUS was also extended to PUST to include 'T' for technology, PUSTE to include 'E' for engineering, or PUSH to include 'H' for the humanities, the latter indicating a more continental understanding of *Wissenschaft*. The dating of these phases is liberal and follows mainly the influential British experience. In the US, the AAAS had a Standing Committee on 'PUS' through the 1970s (see Kohlstedt et al. 1999).
3 I use the capital 'C' to indicate this posture of foundational critique. It is hard to imagine how a critical mind could be the privilege only on one side of this polemic.
4 In Britain, a research programme on public understanding of science ran 1987–1990. The publication (Irwin and Wynne 1996) that became the summary of this programme 'excluded' the three projects that worked with numerical data: the survey of British attitudes (see Durant et al. 1989, 1992), the survey of British adolescents (see Breakwell and Robertson 2001), and the analysis of mass media reportage (see Hansen and Dickinson 1992). Alan Irwin recalls (personal communication, January 2007) that the book selected contributions in order to counter-balance the publicity which the survey of 1988 generated through its *Nature* piece (Durant et al. 1989) and was not designed to be a statement against survey research.

References

Abels, G. and Bora, A. (2004) *Demokratische Technikbewertung*, Bielefeld: Transcript Verlag.
Allum, N. (2010) 'Science literacy', in S. Hornig Priest (ed.) *Encyclopedia of Science and Technology Communication*, Los Angeles: Sage, 724–727.
Allum, N. and Stoneman, P. (2012) 'Beliefs about astrology across Europe', in M. W. Bauer, R. Shukla and N. Allum (eds) *The Culture of Science – How the Public Relates to Science across the Globe*, London and New York: Routledge, 301–322.
Althaus, S. (1998) 'Information effects in collective preferences', *American Political Science Review*, 92, 3: 545–558.
Arendt, H. (1968) 'What is authority?' in H. Arendt, *Between Past and Future: Eight Essays in Political Thought*, New York: Viking Press, 91–142.
Argyris, C. and Schön, D. A. (1978) *Organizational Learning: A Theory of Action Perspective*, Reading, MA: Addison-Wesley.
Barkun, M (1985) 'The awakening cycle controversy', *Sociological Analysis*, 46, 4: 425–443.
Bauer, M. (1996) 'Socio-economic correlates of DK-responses in knowledge surveys', *Social Science Information*, 35, 1: 39–68.
Bauer, M. (1998) 'The medicalisation of science news: from the "rocket-scalpel" to the "gene-meteorite" complex', *Social Science Information*, 37, 4: 731–751.
Bauer, M. W. (2008) 'Survey research and the public understanding of science' in M. Bucchi and B. Trench (eds) *Handbook of Public Communication of Science and Technology*, London and New York: Routledge, 111–130.
Bauer, M. W. (2012a) 'Public attention to science 1820–2010 – a "longue duree" picture', in S. Rödder, M. Franzen and P. Weingart (eds) *The Sciences' Media Connection: Public Communication and its Repercussions. Sociology of the Sciences Yearbook 28*, Dordrecht: Springer, 35–58.
Bauer, M. W. (2012b) 'Science culture and its indicators', in B. Schiele, M. Claessens and S. Shi (eds) *Science Communication in the World: Practices, Theories and Trends*, Dordecht: Springer, 295–312.
Bauer, M. W. (2014) *Atom, Bytes and Genes: Public Resistance and Techno-Scientific Responses*, London and New York: Routledge.
Bauer, M., Allum, N., and Miller, S. (2007) 'What can we learn from 25 years of PUS survey research? Liberating and expanding the agenda', *Public Understanding of Science*, 16, 1, 79–95.
Bauer, M. and Schoon, I. (1993) 'Mapping variety in public understanding of science', *Public Understanding of Science*, 2, 2: 141–155.
Bauer, M. and Durant, J. (1997) 'Belief in astrology: a social-psychological analysis', *Culture and Cosmos: A Journal of the History of Astrology and Cultural Astronomy*, 1, 1: 55–72.
Bauer, M. W. and Gaskell, G. (2008) 'Social representations theory: a progressive research programme for social psychology', *Journal for the Theory of Social Behaviour*, 38, 4: 335–354.
Bauer, M. W. and Jensen, P. (2011) 'The mobilisation of scientists for public engagement', *Public Understanding of Science*, 20, 1: 3–11.

Bauer, M. W. and Howard, S. (2013) *The Culture of Science in Modern Spain: An Analysis across Time, Age Cohorts and Regions*, Bilbao: Fundación BBVA.

Bauer, M.W., Petkova, K. and Boyadjewa, P. (2000) 'Public knowledge of and attitudes to science: alternative measures', *Science, Technology and Human Values*, 25, 1: 30–51.

Bauer, M. W., Shukla, R. and Allum, N. (eds) (2012) *The Culture of Science: How the Public Relates to Science across the Globe,* London and New York: Routledge.

Bauer, M.W., Petkova, K., Boyadjieva, P. and Gornev, G. (2006) 'Long-term trends in the public representations of science across the iron curtain: Britain and Bulgaria, 1946–95', *Social Studies of Science*, 36, 1: 97–129.

Bensaude-Vincent, B (2001) 'A genealogy of the increasing gap between science and the public', *Public Understanding of Science*, 10, 1: 99–113.

Beveridge, A. A. and Rudel, F. (1988) 'An evaluation of "public attitudes toward science and technology" in Science Indicators: the 1985 report', *Public Opinion Quarterly*, 52, 3: 374–385.

Bodmer, W. (1987) 'The public understanding of science', *Science and Public Affairs*, 2: 69–90.

Boy, D. (1989) *Les Attitudes des Français à l'Égard de la Science*, Paris: CNRS, CEVIPOF.

Boy, D. (2012) *Les représentations sociales de la science et de la technique – rapport de recherche*, Paris: SOFRES and CEVIPOF-CNRS.

Boy, D. and Michelat, G. (1986) 'Croyance aux parasciences: dimensions sociales et culturelles', *Revue Française de Sociologie*, 27, 2: 175–204.

Braun, K. and Schultz, S. (2010) 'A certain amount of engineering involved: constructing the public in participatory governance arrangements', *Public Understanding of Science*, 19, 4: 403–419.

Breakwell, G. M. and Robertson, T. (2001) 'The gender gap in science attitudes, parental and peer influences: changes between 1987/88 and 1997/98', *Public Understanding of Science*, 10, 1: 71–82.

Bucchi, M. and Mazzolini, R. (2003) 'Big science, little news: science coverage in the Italian daily press, 1946–1987', *Public Understanding of Science*, 12, 1: 7–24.

Bütschi, D. and Nentwich, M. (2002) 'The role of participatory technology assessment in the policy-making process', in S. Joss and S. Bellucci (eds) *Participatory Technology Assessement*, London: Centre for the Study of Democracy, 233–256.

CAST (2008) *Outline of the National Scheme for Scientific Literacy*, Beijing: Popular Science Press.

Chao, Z. and Wei, H. (2009) 'Study of gender differences in scientific literacy of Chinese Public', *Science, Technology and Society*, 14, 2: 385–406.

CIP (2006) *Outlines of the National Scheme for Science Literacy*, Beijing: Popular Science Press.

Collins, H. and Pinch, T. (1993) *The Golem: What Everyone Should Know about Science*, Cambridge, Cambridge University Press.

da Costa, A. R., Avila, P. and Mateus, S. (2002) *Publicos da ciencia em Portugal*, Lisbon: Gravida.

Crettaz von Roten, F. (2004) 'Gender differences in attitudes towards science in Switzerland', *Public Understanding of Science*, 13, 2, 191–199.

Crettaz von Roten, F. (2012) 'The human-animal boundary in Switzerland', in M. Bauer, R. Shukla and N. Allum (eds) *The Culture of Science: How The Public Relates to Science across the Globe*, London and New York: Routledge, 323–335.

Durant, J., Evans, G. and Thomas, G. P. (1989) 'The public understanding of science', *Nature,* 340, 6228: 11–14.

Durant, J., Evans, G. and Thomas, G. P. (1992) 'Public understanding of science in Britain: the role of medicine in the popular representation of science', *Public Understanding of Science*, 1, 2: 161–182.

Durant, J., Bauer, M., Midden, C., Gaskell, G. and Liakopoulos, M. (2000) 'Two cultures of public understanding of science', in M. Dierkes and C. von Grote (eds) *Between Understanding and Trust: The Public, Science and Technology*, Reading: Harwood Academics Publishers, 131–156.

Eagly, A. H. and Chaiken, S. (1993) *The Psychology of Attitudes*, Fort Worth: Harcourt Brace College Publishers.

Einsiedel, E., Jelsoe, E. and Breck, T. (2001) 'Publics at the technology table: the consensus conference in Denmark, Canada and Australia', *Public Understanding of Science*, 10, 1: 83–98.

Etzioni, A. and Nunn, C. (1976) 'The public appreciation of science in contemporary America', in G. Holton and W. A. Planpied (eds) *Science and Its Public: The Changing Relationship, Boston Studies in the Philosophy of Science, Volume 33*, Dordrecht: Reidel, 229–243.

European Commission (2012) *Monitoring Policy and Research Activities on Science in Society in Europe (MASIS): Final Synthesis Report*, Brussels: European Commission.

Falade, B. A. (2014) *Vaccination Resistance, Religion and Attitudes to Science in Nigeria*, PhD in Social Psychology, London: London School of Economics.

Farr, R. M. (1993) 'Common sense, science and social representations', *Public Understanding of Science*, 2, 3: 189-204.
Felt, U. (2000) 'Why should the public "understand" science? A historical perspective on aspects of public understanding of science', in M. Dierkes and C. von Grote (eds) *Between Understanding and Trust: The Public, Science and Technology*, Reading: Harwood Academics Publishers, 7–38.
Fuller, S. (2001) 'Science', in T. O. Sloane (ed.) *Encyclopaedia of Rhetoric*, Oxford: Oxford University Press: 703–713.
Gaskell, G., Allansdottir, A., Allum, N., Castro, P., Esmer, Y., Fischler, C., Jackson, J., Kronberger, N., Hampel, J., Mejlgaard, N., Quintanilha, A., Rammer, A., Revuelta, G., Stares, S., Torgersen, H. and Wager, W. (2011) 'Winds of change: the 2010 Eurobarometer on the life sciences', *Nature Biotechnology*, 29, 1: 113–114.
Gauchat, G. (2011) 'The cultural authority of science: public trust and acceptance of organized science', *Public Understanding of Science*, 20, 6: 751–770.
Gauchat, G. (2012) 'Politicization of science in the public sphere: a study of public trust in the United States, 1974 to 2010', *American Sociological Review*, 77, 2: 167–187.
Godin, B. and Gingras, Y. (2000) 'What is scientific and technological culture and how is it measured? A multi-dimensional model', *Public Understanding of Science*, 9, 1: 43–58.
Gregory, J. and Miller, S. (1998) *Science in Public: Communication, Culture and Credibility*, New York: Plenum Trade.
Gregory, J. and Lock, S. J. (2008) 'The evolution of "public understanding of science": public engagement as a tool of science policy in the UK', *Sociological Compass*, 2, 4: 1252–1265.
Gregory, J., Agar, J., Lock, S. and Harris, S. (2007) 'Public engagement of science in the private sector: a new form of PR?', in M. W. Bauer and M. Bucchi (eds) *Journalism, Science and Society: Science Communication between News and PR*, London and New York: Routledge, 203–214.
Hansen, A. and Dickinson, R. (1992) 'Science coverage in the British mass media: media output and source input', *Communication*, 17, 3: 365–377.
He, W. and Gao, H. (2011) 'Twenty years of scientific literacy research in China', in G. Raza, F. Ren, H. J. Khan and W. He (eds) *Constructing Culture of Science: Communication of Science in India and China*, New Delhi and Beijing: CSIR-NISCAIR and CRISP, 21–36.
House of Lords Select Committee on Science and Technology (2000) *Science and Society, Third Report*, London: HMSO.
INRA (Europe) and Report International (1993) *Europeans, Science and Technology – Public Understanding and Attitdues*, Brussels: European Commission, DG 12.
Irwin, A. and Wynne, B. (1996) *Misunderstanding Science? The Public Reconstruction of Science and Technology*, Cambridge: Cambridge University Press.
Jasanoff, S. (2003) 'Technologies of humility', *Minerva*, 41, 3: 223–244.
Jasanoff, S. (2005) *Designs on Nature, Science and Democracy in Europe and the United States*, Princeton: Princeton University Press.
Joss, S. and Bellucci, S. (eds) (2002) *Participatory Technology Assessment*, London: Centre for the Study of Democracy.
Jovchelovitch, S. (2007) *Knowledge in Context: Representation, Community and Culture*, London: Sage.
Kallerud, E. and Ramberg, I. (2002) 'The order of discourse in surveys of public understanding of science', *Public Understanding of Science*, 11, 3: 213–224.
Kohlstedt, S. G., Sokal, M. M. and Lewenstein, B. (1999) *The Establishment of Science in America*, Washington: AAAS.
Kumar, P. V. S. (2011) 'Cultural nature of scientific temper', in G. Raza, F. Ren, H. J. Khan and W. He (eds) *Constructing Culture of Science: Communication of Science in India and China*, New Delhi and Beijing: CSIR-NISCAIR and CRISP, 257–278.
LaFollette, M. C. (1990) *Making Science Our Own: Public Images of Science, 1910–1955*, Chicago, Chicago University Press.
Lebart, L. (1984) 'Complementary use of correspondence analysis and cluster analysis', in A. Greenacre and J. Basius (eds) *Correspondence Analysis in the Social Sciences*, London: Academic, 162–178.
Lewandowsky, S., Ecker, U. K. H., Seifert, C. M., Schwarz, N. and Cook, J. (2012) 'Misinformation and its correction: continued influence and successful debiasing', *Psychological Science in the Public Interest*, 13, 3: 106–131.
Lezaun, J. and Soneryd, L. (2007) 'Consulting citizens: technologies of elicitation and the mobility of publics', *Public Understanding of Science*, 16, 3: 279–297.
Liu, J. H., Goldstein-Hawes, R., Hilton, D., Huang, L., Gastardo-Conaco, C., Dresler-Hawke, E., Pittolo, F., Hong, Y., Ward, W., Abraham, S., Kashima, Y., Kashima, E., Ohashi, M. M., Yuki, M. and Hidaka, Y. (2005)

'Social representations of events and people in world history across twelve cultures', *Journal of Cross-Cultural Psychology*, 36, 2: 171–191.
Luhmann, N. (1979) *Trust and Power*, Chichester: Wiley.
Lusardi, A. M. and Mitchell, O. S. (2006) *Financial Literacy and Planning: Implications for Retirement Planning*, Dartmouth College: Department of Economics.
McGuire, W. J. (1986) 'The vicissitudes of attitudes and similar representational constructs in 20th century psychology', *European Journal of Social Psychology*, 16, 2: 89–130.
Mahanti, S. (2013) 'A perspective on scientific temper in India', *Journal of Scientific Temper*, 1, 1–2: 46–62.
Mejlgaard, N., and Stares, S. (2012) 'Validating survey measures of scientific citizenship', in M. Bauer, R. Shukla and N. Allum (eds) *The Culture of Science: How the Public Relates to Science across the Globe*, London and New York: Routledge, 418–435.
Michael, M. (1998) 'Between citizen and consumer: multiplying the meanings of "public understanding of science"', *Public Understanding of Science*, 7, 4: 313–328.
Miller, J. D. (1983) 'Scientific literacy: a conceptual and empirical review', *Daedalus*, 112, 2: 29–48.
Miller, J. D. (1992) 'Towards a scientific understanding of the public understanding of science and technology', *Public Understanding of Science*, 1, 1: 23–30.
Miller, J. D. (1998) 'The measurement of civic scientific literacy', *Public Understanding of Science*, 7, 3: 203–224.
Miller, J. D. (2004) 'Public understanding of, and attitudes toward, scientific research: what we know and what we need to know', *Public Understanding of Science*, 13, 3: 273–294.
Miller, J.D., and Pardo, R. (2000) 'Civic Scientific Literacy and Attitude to Science and Technology: a Comparative Analysis of the European Union, the United States, Japan and Canada', in M. Dierkes and C. von Grote (eds) *Between Understanding and Trust: the Public, Science and Technology*, Amsterdam: Harwood Academic Publishers, 81–129.
Miller, S (2001) 'Public understanding of science at the cross-roads', *Public Understanding of Science*, 10, 1: 115–120.
Miller, S, Caro, P., Koulaidis, V., de Semir, V., Staveloz, W. and Vargas, R. (2002) *Benchmarking the Promotion of RTD Culture and Public Understanding of Science*, Brussels: Commission of the European Communities.
OST (2000) *Science and the Public: a review of science communication and public attitudes to science in Britain*, London: Office of Science and Technology/Wellcome Trust.
Pardo, R. and Calvo, F. (2002) 'Attitudes toward science among the European public: a methodological analysis', *Public Understanding of Science*, 11, 2: 155–195.
Pardo, R. and Calvo, F. (2004) 'The cognitive dimension of public perceptions of science: methodological issues', *Public Understanding of Science*, 13, 3: 203–227.
Paul, J. S. (2008) 'Interest: the curious emotion', *Current Directions in Psychological Science*, 17, 1: 57–60.
Pion, G. M. and Lipsey, M. W. (1981) 'Public attitudes towards science and technology: what have the surveys told us?', *Public Opinion Quarterly*, 45, 3: 303–316.
Polino, C. and Castelfranchi, Y. (2012) 'Information and attitudes towards science and technology in Iberoamerica', in M. W. Bauer, R. Shukla and N. Allum (eds) *The Culture of Science: How the Public Relates to Science across the Globe*, London and New York: Routledge, 158–178.
Pomerantz, E. M., Chaiken, S. and Tordesillas, R. S. (1995) 'Attitude strength and resistance processes', *Journal of Personality and Social Psychology*, 69, 3: 408–419.
Prewitt, K. (1983) 'Scientific literacy and democratic theory', *Daedalus*, 112, 2: 49–64.
Raza, G. (2013) 'Scientific temper and Indian democracy', in P. Baranger and B. Schiele (eds) *Science Communication Today: International Perspectives, Issues and Strategies*, Paris: CNRS Editions, 59–72.
Raza, G., Singh, S. and Dutt, B. (2002) 'Public, science and cultural distance', *Science Communication*, 23, 3: 292–309.
Raza, G., Singh, S., Dutt, B. and Chander, J. (1996) *Confluence of Science and People's Knowledge at the Sangam*, New Delhi: NISTED.
Reddy, V., Juan, A., Bantwini, B. and Gastrow, M. (2009) *Science and the Public in Stratified Societies: A Scoping Exercise*, Durban: Human Sciences Research Council.
Roqueplo, P. (1974) *Le partage du savoir – Science, culture, vulgarisation*, Paris: Seuil.
Rowe, G. and Frewer, L. (2004) 'Evaluating public participation exercises: a research agenda', *Science, Technology and Human Values*, 29, 4: 512–556.
Royal Society [of London] (1985) *The Public Understanding of Science*, London: Royal Society.
Schäfer, M. S. (2012) 'Taking stock: a meta-analysis of studies of media's coverage of science', *Public Understanding of Science*, 21, 6: 650–663.

Shimizu, K. and Matsuura, T. (2012) 'Knowledge of science and technology in Japan: IRT scores for 1991 and 2001', in M. W. Bauer, R. Shukla and N. Allum (eds) *The Culture of Science: How the Public Relates to Science across the Globe*, London and New York: Routledge, 110–125.

Shukla, R. (2005) *India Science Report – Science Education, Human Resources and Public Attitudes towards Science and Technology*, Delhi: NCAER.

Shukla, R. and Bauer, M. W. (2012) 'The science culture index (SCI): construction and validation', in M. W. Bauer, R. Shukla and N. Allum (eds) *The Culture of Science: How the Public Relates to Science across the Globe*, London and New York: Routledge, 179–199.

Sjoeberg, S. (2012) 'PISA: politics, fundamental problems, and intriguing results', *Recherches en Education*, 14, September.

Song, J. (2010) 'Rethinking the context and social foundation of science education', in U. J. Lee (ed) *The World of Science Education: Science education in Asia*, Rotterdam and Boston: Sense Publishers, 155–169.

Stares, S. (2009) 'Using latent class models to explore cross-national typologies of public engagement with science and technology in Europe', *Science, Technology and Society*, 14, 2, 298–329.

Sturgis, P. J. and Allum, N. (2001) 'Gender differences in scientific knowledge and attitudes towards science: a reply to Hayes and Tariq', *Public Understanding of Science*, 10, 4: 427–430.

Sturgis, P. J. and Allum, N. C. (2004) 'Science in society: re-evaluating the deficit model of public attitudes', *Public Understanding of Science*, 13, 1: 55–74.

Suerdem, A., Bauer, M. W., Howard, S. and Ruby, L. (2013) 'PUS in turbulent times II: a shifting vocabulary that brokers inter-disciplinary knowledge', *Public Understanding of Science*, 22, 1: 2–15.

Thomas, G. P. and Durant, J. R. (1987) 'Why should we promote the public understanding of science?', in M. Shortland (ed.) *Scientific Literacy Papers*, Oxford: Department for External Studies, 1–14.

Toumey, C., Besley, J., Blanchard, M., Brown, M., Cobb, M., Howard Ecklund, E., Glass, M., Guterbock, T. M., Kelly, A. E., Lewenstein, B. (2010) *Science in the Service of Citizens and Consumers – the NSF Workshop on Public Knowledge of Science*, Columbia: USC Nano Center.

Turner, J. and Michael, M. (1996) 'What do we know about "don't knows"? Or, contexts of ignorance', *Social Science Information*, 35, 1: 15–37.

Vogt, C. (2012) 'The spiral of scientific culture and cultural well-being: Brazil and Ibero-America', *Public Understanding of Science*, 21, 1: 4–16.

Wagner, W. (2007) 'Vernacular science knowledge: its role in everyday life communication', *Public Understanding of Science*, 16, 1: 7–22.

Wagner, W. and Hayes, N. (2005) *Everyday Discourse and Common Sense: The Theory of Social Representations*, London: Palgrave Macmillan.

Withey, S. B. (1959) 'Public opinion about science and the scientist', *Public Opinion Quarterly*, 23, 3: 382–388.

Wynne, B. (1993) 'Public uptake of science: a case for institutional reflexivity', *Public Understanding of Science*, 2, 4: 321–338.

Wynne, B. (1995) 'Public understanding of science', in S. Jasanoff, G. E. Markel, J. C. Petersen and T. Pinch (eds) *Handbook of Science and Technology Studies*, London: Sage, 361–388.

Wynne, B. (1996) 'May the sheep safely graze? A reflexive view of the expert-lay knowledge divide', in S. Lash, B. Szerszynski and B. Wynne (eds) *Risk, Environment and Modernity: Towards a New Ecology*, London: Sage.

Ziman, J. (1991) 'Public understanding of science', *Science, Technology and Human Values*, 16, 1: 99–105.

12

Risk, science and public communication

Third-order thinking about scientific culture

Alan Irwin

Introduction

This chapter explores some different ways of thinking about science communication and risk management.[1] In certain contexts, there has been a transition from *first-order* (or deficit) models of science-public relations to a greater emphasis on public engagement and dialogue (or what will be described here as *second-order* thinking). However, and as it will be presented, *third-order* thinking about risk, science and public communication asks more fundamental questions about the underlying relationship between first- and second-order approaches, the changes that have taken place (both in theory and practice) and the future direction of science communication and scientific governance.

Very importantly, this is not a story of one way of thinking giving way to the next and then the next. Instead, the situation in most national and local settings is of these different *orders* being mixed up (or churned) together. Thus, the deficit model coexists with talk of dialogue and engagement. And while some organisations and individuals look for quick and easy solutions to communication problems, others have begun to reflect upon the limitations, complications and contexts of both deficit and dialogue.

In the following account, the public communication of science and technology is presented as much more than a matter of communication style. Instead, we confront basic issues of the shaping and direction of socio-technical change, the frameworks within which communication takes place, cultures of governance and control (especially relating to the institutions of science and technology) and the choices available to citizens within modern democracies.

Within western Europe in particular, something interesting has been happening to the language of science communication and scientific governance (Hagendijk *et al.* 2005). In Britain, a landmark report from the House of Lords tackled the broad topic of 'Science and Society' by emphasising the 'new mood for dialogue' on science and technology (House of Lords Select Committee on Science and Technology 2000, Section 5.3). Along with other British reports from the late 1990s onwards, the Lords Select Committee presented science's relationship with society as being under strain and sought a greater acknowledgement of doubt and uncertainty and a change in the culture of science communication and decision-making, 'so that it becomes normal to bring science and the public into dialogue about new developments at an early stage' (ibid: 13). In 2002, the European Commission published its *Action Plan on Science and Society*, calling for an 'open dialogue' (European Commission 2002: 21) over technological innovation as part of its 'new partnership' (ibid.: 7) between science and society. The Netherlands, with a longer history of dialogue and engagement concerning science and technology, more recently held a major public debate around genetically modified foods (Hagendijk and Irwin 2006). Denmark has a particularly strong tradition in this area too: notably, consensus conferences in which panels of citizens come together to debate and make recommendations with regard to specific areas of socio-technical change (Horst and Irwin 2010). And it is not just in western Europe that public engagement and dialogue have been both advocated and enacted. Related activities and debates can be found in (among other countries) Canada and the United States, Australia and New Zealand, Brazil and Japan (Einsiedel *et al.* 2011; Hindmarsh and Du Plessis 2008; Macnaghten and Guivant 2011; Yamaguchi 2010).

The basic inspiration for all of these national and international discussions is the notion that more active, open and democratic relations between science and citizens are both desirable and necessary. At the same time, they suggest a critique of what has gone before, of what the UK Chief Scientific Adviser described in the Lords report as a 'rather backward-looking vision' (House of Lords Select Committee on Science and Technology: 25) where 'difficulties in the relationship between science and society are due entirely to ignorance on the part of the public' and 'with enough public-understanding activity, the public can be brought to greater knowledge, whereupon all will be well'. This conceptualisation of an ignorant and uninformed public for science (christened the deficit theory by social scientists in the 1990s: see Wynne 1995; Irwin and Wynne 1996) has been a powerful provocation to change, prompting the argument that we now need to move *from deficit to dialogue* (see Irwin 2006). In what follows, we will present this conceptual and institutional shift as a transition (albeit a partial one) from first-order to second-order thinking.

Quite how much things have changed within scientific governance and science-citizen relations can be gauged by a brief excursion back to the early 1990s in one country which has been especially significant in this context. Britain back then was edging towards what would later be seen as the BSE (or mad cow) crisis. The notion that science-citizen relations were badly mishandled by the relevant government department – and a consequent desire to 'avoid another BSE' – has exerted a powerful influence on institutional thinking about risk communication and management.

Mad cows and first-order thinking

When in 1990 the UK Ministry of Agriculture, Fisheries and Food (MAFF) and its government minister found themselves confronting a new category of risk, they responded in what was then conventional fashion. The issue was the risk to consumers from British beef: could eating contaminated meat lead to a human form of BSE (bovine spongiform encephalopathy)? Reassurances were offered by a variety of governmental and industrial groups, all attempting to convey the message that the risks were minimal and that consumers could purchase British beef with confidence. Famously, the British and international media featured the minister feeding a beefburger to his daughter before a cluster of eager photographers. As one advertisement from the Meat and Livestock Commission put it:

> Eating British beef is completely safe. There is no evidence of any threat to human health caused by this animal health problem (BSE).... This is the view of independent British and European scientists and not just the meat industry. This view has been endorsed by the Department of Health.

> (*The Times*, 18 May, 1990)

Writing just a few years later, I picked out several characteristics of this 'first-order' exercise in risk communication (Irwin 1995: 53). First of all, we can identify an authority claim based on the language of *certainty*. Second, *science* is presented as central to the whole issue. The use of *independent scientists* in the above quotation appeals to an apparent faith that science can be absolutely trusted (in a manner that would not necessarily apply to the meat industry or government). Third, these efforts at risk communication do not draw on public engagement in any meaningful way. Consumers are to be protected rather than consulted. This is *top-down* (or *one-way*) communication. Fourth, this *science-centred* approach to risk management and risk communication takes little account of the diversity, nor the possible *knowledgeability*, of publics. This final point became especially apparent when subsequent control measures were brought into play without any meaningful consultation with the abattoir workers responsible for putting them into practice (and who might have pointed out that the operating conditions of the abattoir are rather different from those of the scientific laboratory).

Although this might not have been so apparent back in 1995, to these points should be added the not-insignificant observation that this mode of risk communication was conspicuously *unsuccessful*. Meat sales suffered and government credibility was damaged. Official claims to certainty were substantially undermined by very public disagreements among scientists concerning the scale of risk. When human cases of variant CJD (Creutzfeldt-Jakob disease, linked to BSE) began to appear, previous expressions of official confidence were judged to be inappropriate and irresponsible. All this was summarised by the inquiry into BSE and variant CJD which published its report a decade later:

> The Government did not lie to the public about BSE. It believed that the risks posed by BSE to humans were remote. The Government was pre-occupied with preventing an alarmist over-reaction to BSE because it believed that the risk was remote. It is now clear that this campaign of reassurance was a mistake. When on 20 March 1996 the Government announced that BSE had probably been transmitted to humans, the public felt that they had been betrayed. Confidence in government pronouncements about risk was a further casualty of BSE.
>
> (Phillips *et al.* 2000, Volume 1, section 1)

Terms such as 'unwarranted reassurance' (ibid: 1150) and 'culture of secrecy' (1258) recur throughout the Phillips report (which at one point goes so far as to suggest that the main object of MAFF's communication strategy was 'sedation' of the publics (1179). Recurrent within the report also was the perceived need by civil servants and others to counteract what were anticipated to be 'alarmist' public and media reactions to the existence of risk. In a highly significant section, it is argued that 'the approach to communication of risk was shaped by a consuming fear of provoking an irrational public scare' (1294). In contrast, the then Chief Scientific Adviser is favourably quoted as arguing that the temptation 'to hold the facts close' so that a 'simple message can be taken out into the market place' should be resisted. Instead, 'the full messy process whereby scientific understanding is arrived at with all its problems has to be spilled out into the open' (1297). In this way a 'culture of trust' rather than one of secrecy can be developed. More broadly, the official report stresses several points that, in the wake of BSE, have become important within the language of scientific governance in many countries (for a discussion across eight European nations, see Hagendijk *et al.* 2005; for a treatment of such issues in a global and *developmental* context see Leach *et al.* 2005):

- trust can only be generated by openness;
- openness requires recognition of uncertainty, where it exists;
- the public should be trusted to respond rationally to openness;
- scientific investigation of risk should be open and transparent;
- the advice and reasoning of advisory committees should be made public.

From first to second order?

The story of BSE in the United Kingdom is therefore also the story of a larger movement in risk management and science communication. The public stance adopted by MAFF in 1990 represented an almost classical representation of first-order thinking about risk communication. By 2000, the official inquiry could draw upon this case in order to advocate much greater transparency and openness – especially in terms of acknowledging uncertainty and respecting public reasoning rather than fearing alarmism. Although not especially emphasised within the report itself, the abolition of MAFF and its replacement with a new government department, DEFRA (Department for Environment, Food and Rural Affairs), also marked a greater willingness to engage in a two-way relationship with the wider publics: in other words, an emphasis on dialogue and deliberation rather than deficit. As was noted above, greater transparency, recognition of uncertainty and public engagement have been advocated by a series of British reports from the late-1990s onwards (Royal Commission on Environmental Pollution 1998; Department of Trade and Industry 2000; Royal Society/Royal Academy of Engineering 2004; see also Irwin 2006).

One central argument within this chapter is that the movement between *first-* and *second-*order thinking raises fundamental questions which take us beyond changes in communication style. It is for this reason that I present these issues in terms of different *orders* as opposed to presenting them simply as a matter of *deficit* and *dialogue*. I suggest that each approach draws upon deeper intellectual and political roots – even if this is not always apparent to those advocating either *first-* or *second-*order approaches.

Looked at in more conceptual terms, first-order thinking depends strongly on what many social theorists – notably, Beck (1992), Bauman (1991) and Giddens (1991) – describe as the culture of modernity, a culture within which science is presented as the embodiment of truth and the task of government becomes one of bringing rationality to human affairs. As Bauman and Beck have argued, notions of uncertainty and ambivalence can fit uneasily within such a culture given its characteristic confidence in science-led progress. Indeed, Bauman has suggested that the whole substance of modern politics and modern life is the quest for order and, as he strongly puts it, the 'extermination of ambivalence' (1991: 7). Certainly, the first-order perspective fits well with the view of government as being primarily concerned with bringing rational principles to bear on political and social challenges. At the same time, the conventional (positivistic) understanding that science can speak 'truth to power' (Jasanoff 1990 (17); see also Jasanoff 2005) reinforces the idea that the wider public can, of epistemological necessity, play only a restricted role in deciding about risk issues. In addition, the economic significance of these questions cannot be ignored, as was especially apparent during the BSE crisis of the 1990s when the impact on British agriculture was clearly a significant governmental and industrial concern. The economic case for technological innovation and development and the first-order case for science-led progress appear to have worked reasonably harmoniously with one another, at least until relatively recently (Ezrahi 1990).

What of the second-order approach to risk communication and management? Certainly, it must be acknowledged that second-order approaches do not have a coherent foundation but instead draw upon a variety of intellectual traditions and (at least as importantly) a diversity of institutional experiences and incremental developments. Whilst points of congruity and overlap exist between specific engagement practices and larger intellectual debates, second-order thinking is also extremely heterogeneous – and to a degree experimental – in character. In that manner, what is presented here as second-order thinking can be seen not just as a radical departure from previous forms of risk and scientific governance but also as offering important continuities. This question of whether second-order thinking offers a radical change or else old wine in new bottles has been a major debate among social scientists and science and technology studies (STS) scholars in particular (Irwin et al. 2013; Wynne 2006).

Attempting to bring some coherence to the diverse roots of second-order thinking, one can first of all point to connections between moves towards greater transparency and engagement and larger discussions about the merits of deliberative democracy and the need to revitalise political institutions – discussions associated with Habermas (1978), Rawls (1972) and Dryzek (2000) among other theorists (Hagendijk and Irwin 2006). Second, Beck and his contemporaries emphasise the manner in which conventional political institutions have come under great challenge when dealing with issues of risk. Whilst modernistic institutions might present themselves as being in control, the evidence from public protests over genetically modified foods, nuclear power and road-building programmes suggests that this may not be the case (Beck 1992). In such circumstances, demands for new forms of democratic accountability and engagement become a central characteristic of contemporary political life. Third, the emergent emphasis on trust, transparency and two-way communication has been stimulated by a series of social scientific studies from the 1990s onwards which suggest a more fundamental institutional challenge in dealing with contested areas of risk (Irwin and Wynne 1996). In contrast to the conventional portrayal of public groups as irrational and uninformed, various empirical studies have explored the knowledgeability and resourcefulness of particular publics when encountering science-related issues within the contexts of everyday life (Bloor 2000; Brown 1987; Epstein 1996; Kerr et al. 1998). One practical implication of this research has been that, rather than simply presenting wider society as an impediment to scientific and technological progress, it is important to examine the operational assumptions and practices of scientific institutions: the very form of reflexive scrutiny that first-order approaches generally evade. Seen from this perspective, second-order thinking may have become a practical necessity if public policy is to be made – and justified – in circumstances of social and technical uncertainty (see Stilgoe et al. 2006).

As has already been suggested, and as Mouffe implies (1993, 2000), the intellectual differences between first- and second-order thinking can be exaggerated. In principle, the commitment to science and to science-led progress, and that to transparency and dialogue do not necessarily contradict one another. After all, many areas of science and technology remain relatively uncontroversial and, indeed, enjoy public support. Going further, what is presented as dialogue, engagement and transparency can, in practice, be very close to public relations and political spin. However, the argument of this chapter is that, equally, it cannot be simply assumed that they are always fully *commensurate* with one another (i.e. that first- and second-order perspectives fit easily together without tension or mutual challenge). Furthermore, whilst it might be tempting to suggest that the emphasis on trust, openness and engagement has somehow replaced older, outmoded forms of practice, I argue in the remainder of this chapter that a more complex (and confusing) situation now operates in which first- and second-order approaches operate very often in uneasy coexistence and unconsidered juxtaposition.

Rather than simply being a matter of shifting from one communication style to another (or a straightforward story of first-order approaches giving way to second-order), I want instead to suggest that the relationship between risk, science and public communication raises more profound questions of scientific and political culture. In that sense, the second-order perspective offers an important but still only partial attempt to tackle the challenges raised by first-order approaches. My argument is that these questions are generally neglected amidst the institutional enthusiasm for new risk communication and public engagement approaches. It is this larger discussion that lies at the heart of third-order thinking, to which we return after the next section.

Putting second-order thinking into practice

What can be said about the actual experience of putting second-order thinking into practice within institutional processes? As one illustration of the kinds of activities which have taken place, the Sciencewise website[2] offers information about a host of current and previous dialogue projects on topics ranging from bioscience to climate change and from health care to information management. In addition, there is no shortage of social scientific accounts offering case studies of the practical experience of engagement, even if these regularly reach rather pessimistic conclusions about the degree to which scientific and risk governance has been transformed (Felt and Fochler 2010; Irwin 2001; Kerr et al. 2007; Rothstein 2007).

One of the most considered examples since the turn of the century has been the British GM Nation? debate into the commercialisation of genetically modified crops in the UK (for a comparison with the parallel Dutch debate, see Hagendijk and Irwin 2006). Taking place during the summer of 2003, the debate was designed to be 'innovative, effective and deliberative' but also 'framed by the public'. Its broad aim was to 'provide meaningful information to government about the nature and spectrum of the public's views, particularly at grass roots level, to inform decision-making'.[3] In practical terms, GM Nation? consisted of national and local meetings (over 600 in all with 20,000 individuals estimated to have taken part), focus groups with citizens pre-selected to represent various socio-demographic characteristics, and a debate website. The eventual conclusions of the debate report were, briefly: people are generally uneasy about GM; the more people engage in GM issues, the harder their attitudes and the more intense their concerns; there is little support for early commercialisation; there is widespread mistrust of government and multinational companies. In summary, the report characterises public opinion over the commercialisation of GM as 'not yet – if ever' (GM Nation? 2003: see also Horlick-Jones et al. 2007).

Here we have what was, in terms of British practice, the most developed application of second-order thinking within the policymaking process thus far. However, the evidence suggests that, rather than representing a shift from older ways of thinking, the situation is indeed best described as one of *uneasy coexistence*. As I consider it below, this has less to do with specific aspects of the debate's design (although such matters can be significant in themselves) than with the wider political and institutional framework within which the debate was conducted (Irwin 2006). Thus, there is an apparent tendency for governmental institutions in particular to view second-order discussions as a discrete phase within the larger policy process, an activity to be fed into decision-making at the appropriate time, alongside other forms of evidence, before business as usual can return. Such an approach imposes fundamental constraints on second-order engagement with science and risk and restricts second-order perspectives so that they fit within the frameworks established by first-order understandings.

It was especially apparent in the debate that members of the public typically *framed* the underlying issues much more broadly than did government and industry officials. Whilst for the concerned civil servants this was a matter of deciding about a particular technical issue, for many members of the public the debate was connected to a wider-ranging set of questions about the power of transnational companies, globalisation, the future of British agriculture, and the comparative benefits of innovation to North American industry and British consumers. Whilst policymakers tended to frame the issue as a matter of 'risk' (to humans and the environment), this by no means captured the full spectrum of public assessments and may in itself represent a first-order framing of the underlying issues (Wynne 2002). Equally, when risk did figure as an issue in public debate, this was generally balanced against questions of *need* rather than presented as a rational calculus of relative harm (see also Jones 2004).

It must also be observed that any hope that such an exercise in engagement would lead to social consensus was certainly disappointed. One characteristic outcome of engagement exercises is that they lead to further accusations and arguments; in this case, it was suggested both that the exercise was hijacked by activist groups and that it was far too restricted in participation, depth and coverage (House of Commons, Environment, Food, and Rural Affairs Committee, 2003). This was even more apparent in the Dutch national debate over the same issue when activist groups simply withdrew their co-operation in protest at the manner in which the debate was being framed (Hagendijk and Irwin 2006). Such disputed outcomes also indicate that what one party (typically the institutions of government) might present as engagement can be viewed by another as *old-style* deficit thinking. Like beauty, *engagement* can lie in the eye of the beholder. Of course, this is not necessarily a negative characteristic. Disagreement and controversy may also bring energy, excitement and even passion to debates and in that sense can be viewed as an important societal resource.

From the perspective of more reflexive *third-order* thinking about risk, science and public communication, one clear implication of the GM Nation? debate is that there are many aspects of the relationship between first- and second-order perspectives that remain unexplored and neglected. Certainly, for this example at least, we can dispense with the idea that there has been anything so straightforward as a paradigm shift from one approach to another. It would appear instead that, even in the case of Britain (where criticisms of the deficit theory have been taken relatively seriously) and even in an area such as GM policy (which is especially well developed in this regard) the shift to second-order thinking has been partial, fixed-term and patchy (see especially Horlick-Jones *et al.* 2007).

A similar conclusion can be applied to the iconic example of Danish consensus conferences, mentioned briefly at the start of this chapter. Despite the international representation of consensus conferences as one of the most prominent examples of second-order thinking, within Denmark there has been an academic and political debate over their impact and effectiveness and even their continued operation following a decision in 2011 to stop governmental funding (Blok 2007; Horst 2003; Horst and Irwin 2010; Jensen 2005). Once again, we are reminded that these changes in the language and practice of science communication and scientific governance do not represent an inevitable, once-and-for-all historical sequence but rather a contested focus for discussion, disagreement and societal reassessment.

Since the GM Nation? engagement exercise, there have been many initiatives in public dialogue and consultation, often focusing on new and potentially controversial areas of science and technology such as nanotechnology, stem cell research and synthetic biology. These have taken a variety of forms, although perhaps the most typical has been the bringing together of *ordinary* citizens – aided by background material and sometimes the participation of various experts – to engage with an issue of social and technical importance. However, far from offering an end point to the discussion about the best relationship between risk, science and the wider publics, these typically have served to open up larger issues. It is to some of these larger issues that we turn in the next section of this chapter.

Third-order reflections

Third-order thinking in this context does *not* refer to a new model of scientific governance or science communication which will resolve the problems created by first- and second-order perspectives. Instead, it represents a move away from sloganising about what is best towards more critical reflection – and reflection-informed practice – about the relationship between technical change, institutional priorities and wider conceptions of social welfare and justice. It is very important to stress that this is not simply a matter of categorising individual activities and initiatives into one order or another. As we have suggested, what might appear as dialogue to one party can look remarkably like deficit to another. Equally, this is not about developing a new, improved toolkit (although tools can be very useful) but rather of interrogating the operating assumptions and modes of thought on which individual initiatives depend and considering the practical and conceptual implications of this. Third-order thinking also takes us away from the notion that any approach to communication is necessarily and intrinsically superior. Instead, deciding what is appropriate to any particular situation must be a matter for contextual judgement but also for recognition of the limitations and strengths of all approaches. Put succinctly, third-order thinking invites us to consider what is at stake within societal decisions over science and technology and to build on the notion that different forms of expertise, practice and understanding represent an important resource for change rather than an impediment or burden (Stilgoe *et al.* 2006).

Some third-order thoughts about the discussion so far will help substantiate these general points. In the first place, it is apparent that the expressed institutional enthusiasm for second-order approaches has not been accompanied by systematic and considered attention to the

Table 12.1 Characteristics of first-, second- and third-order thinking about scientific and risk communication

	First Order	Second Order	Third Order
Main focus	Public ignorance and technical education	Dialogue, engagement, transparency, building trust	The direction, quality and need for socio-technical change
Key issues	Communicating science, informing debate, getting the facts straight	Re-establishing public confidence, building consensus, encouraging debate, addressing uncertainty	Setting science and technology in wider cultural context, enhancing reflexivity and critical analysis
Communication style	One-way, top-down	Two-way, bottom-up	Multiple stakeholders, multiple frameworks
Model of scientific governance	Science-led, 'science' and 'politics' to be kept apart	Transparent, responsive to public opinion, accountable	Open to contested problem definitions, beyond government alone, addressing societal concerns and priorities
Socio-technical challenge	Maintaining rationality, encouraging scientific progress and expert independence	Establishing broad societal consensus	Viewing heterogeneity, conditionality and disagreement as a societal resource
Overall perspective	Focusing on science	Focusing on communication and engagement	Focusing on socio-technical/political cultures

policy implementation of such approaches nor to the challenges this might generate. Instead, and as was evident within the GM debate in Britain, the tendency has been to view transparency and engagement as ends in themselves, and as a supplement to conventional procedures. Equally, the assumption that openness and engagement will restore institutional credibility, as opposed to revealing more fundamental antagonisms around social and technical change, remains largely unquestioned.

Second, and linked to that point, we are likely to witness growing criticism of policy approaches which adopt second-order rhetoric without considering the fuller implications of such a perspective. Viewed sceptically, this seems to be an inevitable consequence of second-order thinking – engagement generates the demand for further engagement, and transparency leads to accusations of opacity (Horst 2003). In more immediate terms, the implication is that institutions should not promise more than they can deliver, which implies making explicit the limitations to openness and engagement as well as the constructive possibilities. This is especially important when the contemporary demand for transparency and dialogue must sit alongside the unavoidable requirement for accountability and leadership. A commitment to openness and democracy should not imply an abnegation of institutional responsibility nor that complex issues of socio-scientific decision-making should always be turned over to a referendum. Instead, new forms of leadership are required which are open and transparent but also capable of defending chosen courses of action in full acknowledgement of significant areas of uncertainty and the existence of alternative strategies and perspectives.

Third, the suggestion of this chapter has been that questions of communication strategy with regard to risk and science cannot be uncoupled from larger matters of scientific governance and scientific culture. The language of transparency, two-way exchange, trust and uncertainty can, it is true, be employed in entirely instrumental and superficial terms. Such an approach is likely to generate rather than (as it is intended) appease anxieties. Equally, claims of this sort can signal a short-lived departure (or diversion) within the policy process. This may offer some progress from the old deficit theory but ultimately will operate according to the same principles and assumptions and at best will represent institutional *listening* rather than dialogue. What such instrumental or diversionary approaches to risk, science and communication neglect are the wider questions of the direction, quality and need for socio-scientific change which engagement exercises so typically generate.

Third-order thinking about the character of scientific culture and the possibilities for socio-technical change is not presented here as a panacea to public concerns around technical change nor as a new policy mode. Instead, it suggests that science-public relations need to be placed in wider context and that critical evaluation is required of current approaches to scientific governance and science communication (whether of the first- or second-order variety or any combination thereof). As the case of BSE in Britain suggests, changes have taken place in the institutional treatment of science communication and risk issues in particular, but critical thinking about the significance of such changes – and the need for further change – has been slower in coming. Such thinking will require an attention not simply to the mechanics of science-public relations but also to deeper questions such as the relationship between scientific governance, political economy and innovation strategy, and the operation of national policy processes in an increasingly globalised setting. In recognising the partiality of progress from first- to second-order thinking, we also raise issues that take us to the core of social and scientific *progress* in democratic societies.

One specific manifestation of this way of thinking about science-society relations can be found in a European Science Foundation (ESF) report on 'caring for our futures in turbulent times' (Felt *et al.* 2013). Among other practical points, this report emphasises the importance of valuing the diversity of contemporary society (and especially contemporary Europe), establishing new spaces for science-society interactions, and creating real opportunities within scientific careers for sustained attention to socio-technical concerns. As the report puts it (ibid.: 8):

> Science communication should not mainly aim at persuading citizens and in particular young people to embrace science and technology in a rather unquestioned manner, but rather support them in becoming reflexive members of contemporary knowledge societies through caring for broader science-society issues.

To offer a final example of these issues in practice, nanotechnologies are now being presented both as possessing a huge capacity for social benefit (e.g. cancer-tackling nanobots) and for threat (as all human life is reduced to *grey goo*). Perhaps the dominant political response has been to point to the huge potential of the nanotechnologies and to advocate increased public education in the issues (supported by social surveys which suggest that only a small proportion of the population even recognise the term 'nanotechnology'). However, second-order thinking has also found expression and strong arguments have been made for democratic engagement and scrutiny of the nanotechnologies (Kearnes *et al.* 2006: Royal Society/Royal Academy of Engineering 2004).

Looked at from a third-order perspective, it is hard to deny the benefits of education or the value of democratic discussion. However, what both of these generally neglect is a deeper scrutiny of the possibilities for regional and national autonomy within the worldwide economy, the relationship between nanotechnologies and societal values and preferences (in all their diversity), the strategies being adopted right now by international corporations, and the manner in which current processes of scientific governance serve to assist or hinder the expression of democratic principles. Put differently, while there is a strong tendency to present nanotechnology as a technical issue with social implications it is equally necessary to present it as a fundamentally *social* matter which draws upon scientific and technological knowledge and expertise. This is broadly in the spirit of the model of 'anticipatory governance' developed by researchers associated with the Center for Nanotechnology in Society at Arizona State University: 'Anticipatory governance evokes a distributed capacity for learning and interaction stimulated into present action by reflection on imagined present and future sociotechnical outcomes' (Barben *et al.* 2008: 993). In related work, Arie Rip has noted the increased engagement of social scientists in nanotechnology research programmes, suggesting at least the possibility for greater reflexivity in what he presents as the 'co-evolution of nanoscience/technology and society' (2006: 362; see also Wood *et al.* 2007).

In this situation, the public communication of risk, science and technology both takes on new significance and faces substantial new challenges. More importantly, new possibilities emerge for forms of communication which do not simply trade in the unreflexive language of 'deficit and dialogue' but which open up fresh interconnections between public, scientific, institutional, political and ethical visions of change in all their heterogeneity, conditionality and disagreement.

Concluding remarks

I have tried in the previous sections to give a sense of the state of the art in the emergent field of risk, science and public communication. Some of the discussed areas are well developed. Public engagement initiatives, for example, now have an established history in many countries, even if there is still disagreement about their purposes and outcomes. Other aspects are more tentative.

Inevitably, this relates especially to what I have described above as 'third-order thinking' which by its very nature represents a look forward to an area of discussion and practice still in the margins rather than in the mainstream. This also means that it is harder to give examples of the approach in institutional practice, although certain activities around nanotechnology (among other topics) suggest that changes are taking place. Above all, I want to suggest that this is very much a field in movement rather than steady state, a pathway which is still being defined and not set in concrete. For me, a sense of movement, new possibility and considered alternatives is our best hope for the future.

New research directions are, however, essential in maintaining movement and vitality. Important, but still relatively neglected, issues include science communication as practised by private (rather than public) organisations, the connection between science communication and the operation of the marketplace (including questions of consumer choice and consumption), and the relationship between changing scientific career structures and science communication activities. Important questions can also be asked about, for example, the relationship between science communication and environmental sustainability and about the operation of scientific and risk communication in a changing world order. Certainly, there is still a strong *western* bias to many of our current discussions and this deserves urgent attention. In more academic terms, it seems too that we should broaden the disciplinary foundation of our field, bringing in the contributions of organisational scholars, economists and cultural anthropologists among others. Meanwhile, one major issue concerns the functioning of public debate and engagement not just in a relatively small number of high-profile issues (GM food, nanotechnology, synthetic biology) but also among the much more numerous but somehow neglected areas of socio-technical change (for example, the development of new security technologies, pharmaceuticals and transport systems).

Overall, there seems no shortage of fresh research- and policy-oriented possibilities. Many possible paths lead forward from here. The limits instead seem to be set by a combination of our own socio-technical imagination and the rigidities of societal practice. Looking ahead, we can perhaps find final inspiration in Antonio Machado's famous line: 'There is no road, the road is made by walking.'[4] Within the broad domain outlined in this chapter, there is undoubtedly much walking still to be done.

Key questions

- What are the main criticisms of the first-order (or *deficit*) approach to science-public communication?
- Why has there been disagreement over the practical implementation of public engagement with science and technology?
- What are the main characteristics of third-order thinking and how might these be put into practice?

Notes

1. This chapter has been revised for the second edition of this handbook. Whilst the overall structure has stayed broadly the same, I have brought the discussion up to date, elaborated and developed certain key points (including the relationship between first- and second-order thinking), extended the treatment of third-order thinking and written a new conclusion.
2. www.sciencewise-erc.org.uk/cms/
3. www.gmnation.org.uk/
4. 'No hay camino, se hace camino al andar', quoted in Macfarlane 2012: 236.

References

Barben, D., Fisher, E., Selin, C. and Guston, D. H. (2008) 'Anticipatory governance of nanotechnology: foresight, engagement, and integration', in E. J. Hackett, O. Amsterdamska, M. Lynch and J. Wajcman (eds) *The Handbook of Science and Technology Studies*, Third edition, Cambridge, MA: MIT Press, 979–1000.

Bauman, Z. (1991) *Modernity and Ambivalence*, Cambridge: Polity.

Beck, U. (1992) *Risk Society: Towards a New Modernity*, London, Newbury Park, New Delhi: Sage.

Blok, A. (2007) 'Experts on public trial: on democratizing expertise through a Danish consensus conference', *Public Understanding of Science*, 16, 2: 163–182.

Bloor, M. (2000) 'The South Wales Miners Federation: miners' lung and the instrumental use of expertise, 1900–1950', *Social Studies of Science*, 30, 1: 125–140.

Brown, P. (1987) 'Popular epidemiology: community response to toxic waste induced disease in Woburn Massachusetts', *Science, Technology and Human Values*, 12, 3–4: 76–85.

Department of Trade and Industry (DTI) (2000) *Excellence and Opportunity: A Science and Innovation Policy for the 21st Century*, London: The Stationery Office.

Dryzek, J. S. (2000) *Deliberative Democracy and Beyond: Liberals, Critics, Contestations*, Oxford: Oxford University Press.

Einsiedel, E. F., Jones, M. and Brierley, M. (2011) 'Cultures, contexts and commitments in the governance of controversial technologies: US, UK and Canadian publics and xenotransplantation development', *Science and Public Policy*, 38, 8: 619–628.

Epstein, S. (1996) *Impure Science: AIDS, Activism and the Politics of Knowledge*, Berkeley, CA: University of California Press.

European Commission (2002) *Science and Society Action Plan*, Brussels: European Commission; online at http://ec.europa.eu/research/science-society/pdf/ss_ap_en.pdf

Ezrahi, Y. (1990) *The Descent of Icarus: Science and the Transformation of Contemporary Democracy*, Cambridge, MA and London: Harvard University Press.

Felt, U. and Fochler, M. (2010) 'Machineries for making publics: inscribing and de-scribing publics in public engagement', *Minerva*, 48, 3: 219–238.

Felt, U., Barben, D., Irwin, A., Joly, P. -B., Rip, A., Stirling, A. and Stöckelová, T. (2013) *Science in Society: Caring for our Futures in Turbulent Times*, Science Policy Briefing, Strasbourg: European Science Foundation.

Giddens, A. (1991) *Modernity and Self-identity: Self and Society in the Late Modern Age*, Cambridge: Polity.

GM Nation? (2003) *The Findings of the Public Debate*, London: Department for Business Innovation and Skills.

Habermas, J. (1978) *Towards a Rational Society*, London: Heinemann.

Hagendijk, R. and Irwin, A. (2006) 'Public deliberation and governance: engaging with science and technology in contemporary Europe', *Minerva*, 44, 2: 167–184.

Hagendijk, R., Healey, P., Horst, M. and Irwin, A. (2005) *Report on the STAGE Project: Science, Technology and Governance in Europe*; online at www.stage-research.net

Hindmarsh, R. and Du Plessis, R. (2008) 'GMO regulation and civic participation at the "edge of the world": the case of Australia and New Zealand', *New Genetics and Society*, 27, 3: 181–199.

Horlick-Jones, T., Walls, J., Rowe, G., Pidgeon, N., Poortinga, W., Murdock, G. and O'Riordan, T. (2007) *The GM Debate: Risk, Politics and Public Engagement*, Abingdon: Routledge.

Horst, M. (2003) *Controversy and Collectivity: Articulations of Social and Natural Order in Mass Mediated Representations of Biotechnology*, PhD thesis, Copenhagen Business School, Doctoral School on Knowledge and Management, Department of Management, Politics and Philosophy.

Horst, M. and Irwin, A. (2010) 'Nations at ease with radical knowledge: on consensus, consensusing and false consensusness', *Social Studies of Science*, 40, 1: 105–126.

House of Commons, Environment, Food and Rural Affairs Committee (2003) *Conduct of the GM Public Debate*, Eighteenth Report of Session 2002–2003, London: H. M. Stationery Office.

House of Lords Select Committee on Science and Technology (2000) *Science and Society*, London: The Stationery Office; online at www.publications.parliament.uk/pa/ld199900/ldselect/ldsctech/38/3801.htm

Irwin, A. (1995) *Citizen Science: A Study of People, Expertise and Sustainable Development*, London: Routledge.

Irwin, A. (2001) 'Constructing the scientific citizen: science and democracy in the biosciences', *Public Understanding of Science*, 10, 1: 1–18.

Irwin, A. (2006) 'The politics of talk: coming to terms with the "new" scientific governance', *Social Studies of Science*, 36, 2: 299–320.

Irwin, A. and Wynne, B. (eds) (1996) *Misunderstanding Science? The Public Reconstruction of Science and Technology*, Cambridge: Cambridge University Press.

Irwin, A., Jensen, T. E. and Jones, K. E. (2013) 'The good, the bad and the perfect: criticizing engagement practice', *Social Studies of Science*, 43, 1: 119–136.
Jasanoff, S. (1990) *The Fifth Branch: Science Advisers as Policymakers*, Cambridge, MA and London: Harvard University Press.
Jasanoff, S. (2005) *Designs on Nature: Science and Democracy in Europe and the United States*, Princeton, NJ: Princeton University Press.
Jensen, C. B. (2005) 'Citizen projects and consensus building at the Danish Board of Technology: on experiments in democracy', *Acta Sociologica*, 48, 3: 221–235.
Jones, K. E. (2004) 'BSE and the Phillips report: a cautionary tale about the update of "risk"', in N. Stehr (ed.) *The Governance of Knowledge*, New Brunswick and London: Transaction Publishers, 161–186.
Kearnes, M., Macnaghten, P. and Wilsdon, J. (2006) *Governing at the Nanoscale: People, Policies and Emerging Technologies*, London: Demos.
Kerr, A., Cunningham-Burley, S. and Amos, A. (1998) 'The new human genetics: mobilizing lay expertise', *Public Understanding of Science*, 7, 1: 41–60.
Kerr, A., Cunningham-Burley, S. and Tutton, S. (2007) 'Shifting subject positions: experts and lay people in public dialogue', *Social Studies of Science*, 37, 3: 385–411.
Leach, M., Scoones, I. and Wynne, B. (eds) (2005) *Science and Citizens: Globalization and the Challenge of Engagement*, London and New York: Zed.
Macfarlane, R. (2012) *The Old Ways: A Journey on Foot*, London: Penguin.
Macnaghten, P. and Guivant, J. S. (2011) 'Converging citizens? Nanotechnology and the political imaginary of public engagement in Brazil and the United Kingdom', *Public Understanding of Science*, 20, 2: 207–220.
Mouffe, C. (1993) *The Return of the Political*, London and New York: Verso.
Mouffe, C. (2000) *The Democratic Paradox*, London and New York: Verso.
Phillips, Lord, Bridgeman, J. and Ferguson-Smith, M. (2000) *The BSE Inquiry: The Report*, London: The Stationery Office.
Rawls, J. (1972) *A Theory of Justice*, Oxford: Clarendon Press.
Rip, A. (2006) 'Folk theories of nanotechnologists', *Science as Culture*, 15, 4: 349–365.
Rothstein, H. (2007) 'Talking shop or talking turkey? Institutionalizing consumer representation in risk regulation', *Science, Technology & Human Values*, 32, 5: 582–607.
Royal Commission on Environmental Pollution (RCEP) (1998) *Setting Environmental Standards: 21st Report*, London: The Stationery Office.
Royal Society/Royal Academy of Engineering (RS/RAE) (2004) *Nanoscience and Nanotechnologies: Opportunities and Uncertainties. RS Policy Document 19/04*, London: Royal Society.
Stilgoe, J., Irwin, A. and Jones, K. (2006) *The Received Wisdom: Opening Up Expert Advice*, London: Demos; online at www.demos.co.uk/files/receivedwisdom.pdf
Wood, S., Jones, R. and Geldart, A. (2007) *Nanotechnology: From the Science to the Social*, Economic and Social Research Council: Swindon.
Wynne, B. (1995) 'Public understanding of science', in S. Jasanoff, G. E. Markle, J. C. Petersen and T. Pinch (eds) *Handbook of Science and Technology Studies*, Thousand Oaks, London and New Delhi: Sage, 361–388.
Wynne, B. (2002) 'Risk and environment as legitimatory discourses of technology: reflexivity inside out?', *Current Sociology*, 50, 3: 459–477.
Wynne, B. (2006) 'Public engagement as a means of restoring public trust in science: hitting the notes, but missing the music?', *Community Genetics*, 9, 3: 211–220.
Yamaguchi, T. (2010) 'Discussing nascent technologies: citizens confront nanotechnology in food', *East Asian Science, Technology and Society*, 4, 4: 483–501.

13
Engaging in science policy controversies
Insights from the US climate change debate

Matthew C. Nisbet

Introduction

Nearly 40 years ago, sociologist Dorothy Nelkin began a series of case studies examining the nature of controversies over science and technology (1979, 1984, 1992). In the decades since, the multiple lines of research inspired by these original studies have identified a generalisable set of insights that inform our understanding of contemporary science policy debates and that provide guidance on how to engage decision-makers and the public effectively in such debates.

According to Nelkin, debates that emerged during the 1970s such as those over nuclear energy, environmental pollution and genetic engineering were fundamentally controversies over political control: Who gets to decide the future of these technologies or the actions to address these problems? Which values, interpretations, and world views matter? Are science and technology being deployed in the public interest or on behalf of special interests? In the 1980s, controversies over fetal tissue research, animal experimentation and the teaching of evolution in schools featured a new emphasis on moral absolutes; for combatants in these debates, there could be no compromise. Notably, each case study reflected intensifying tensions in modern society and competing visions for the future, including most notably 'disagreement over the appropriate role of government, the struggle between individual autonomy and community goals', wrote Nelkin (1979: xi).

In these controversies, traditional approaches to science communication that emphasised the translation and dissemination of expert knowledge were unsuited to reduce conflict or promote consensus. In fact, such efforts were more likely to backfire than be successful. The reason, as Funtowicz and Ravetz (1992) explained, was that in these controversies uncertainty and complexity were high, decisions were perceived as morally urgent and, as a result, reaching agreement among a plurality of stakeholders depended on negotiating competing interests and values. In other words, even though science policy controversies featured competing claims to scientific authority, such claims often only obscured underlying values-based differences. Thus, in those cases where the expert community focused its strategy on the dissemination of scientific evidence, this tended to reinforce entrenched positions, since such evidence is often tentative enough to support the values-based arguments of competing sides (Sarewitz 2004).

Frustrated for decades by their inability to resolve political conflicts over science and technology, many scientists blamed public ignorance, irrationality or superstition when a social group ignored their advice or disputed their expertise. Following the 1986 Chernobyl nuclear disaster, Wynne (1992) in a series of studies challenged these dominant assumptions by the expert community. In examining why English sheep farmers doubted government scientist warnings about local soil and livestock contamination from Chernobyl's continent-wide fallout, Wynne proposed that their scepticism about scientific advice was strongly filtered by feelings of distrust and alienation rather than ignorance or irrationality, feelings that were forged by local history, communication mistakes by scientists and, among farmers, a perceived threat to their way of life.

Other scholars have studied the strategic use of language, metaphors, images and cultural allusions as they have appeared in science policy debates and in news coverage of these debates. These researchers studied the process by which advocates and journalists selectively *framed* the social and political relevance of nuclear energy and biotechnology. Were these technologies innovative breakthroughs destined to drive social progress and economic growth, a Frankenstein's monster out of control, or an unstoppable train that had already left the station? Were solutions dependent on holding industry and elected officials accountable, on following the advice of experts, and/or on following majority opinion? Scholars tracked these competing interpretive packages and social representations as they evolved across time, policy arenas, media outlets and countries, noting their influence on public opinion and policy formulation. They concluded that a common set of social meanings could be expected to define the trajectory of public debate, and that these common frames of reference were identifiable across controversies (Bauer and Gaskell 2002; Gamson and Modigliani 1989; Nisbet 2009; Nisbet and Lewenstein 2002).

In combination with this research on the media and on framing, social scientists in the early 2000s began to examine more closely the cognitive and social factors that shape individual attitudes, beliefs and preferences. Among the general public, scientific knowledge was found to be only one factor among several influencing public attitudes; and was only weakly correlated with policy preferences (Allum *et al.* 2008). Studies showed that knowledge was filtered by way of an individual's social and political identity. Under conditions when trusted political leaders disagreed on policy and strategically communicated these differences to the public, highly knowledgeable members of the public who identified with these leaders tended to be the most divided in their opinions. Polarisation among the most knowledgeable members of differing social groups have been observed in studies of debates on stem cell research (Nisbet 2005), nanotechnology (Brossard *et al.* 2009), genetic testing (Allum *et al.*, in press), climate change (Kahan *et al.* 2012) and other topics.

An intense debate over climate change has evolved in recent years in the United States and controversy over food biotechnology has developed in Europe over the same period. In both cases, political leaders and advocates have framed what is at stake in a manner that resonates with the world views and outlooks of differing social groups and segments of the public. The resulting polarisation has been reinforced by the spread of different meanings and divisive interpretations by way of online news, commentary and social media. Moreover, in both the US and Europe, efforts to change the status quo – either to pass climate legislation or to end the ban on genetically modified crops – have been blocked by opponents who benefit from structural advantages within the political system.

In these debates, frustrated advocates (including many scientists), call for ever more aggressive confrontation of their opponents, believing that such strategies are the only way to achieve desired policy outcomes. Yet even though such efforts may be an essential feature of social change, for the expert community and their allied organisations, other strategies are needed if some semblance of consensus or agreement is to be achieved. Navigating the terrain of science policy controversies requires an understanding of the factors that seed polarisation; and the strategies available for restoring co-operation, for decreasing the perception of entrenched group differences and for building broader consensus.

Specific to the debate over climate change, as I review in this chapter, research suggests that a first strategy includes going beyond the polarised, oppositional parties involved and bringing to the conversation a greater diversity of trusted societal leaders who can frame the issue in a manner that resonates with the identity and cultural background of broader segments of the public. A second strategy starts with experts and their institutions who in serving as *honest brokers* must be proactive in expanding the range of technological options and policy choices considered by decision-makers and the public. Finally, a third strategy involves substantive investment in the civic capacity of society to discuss, debate, learn about and participate in policy decisions via localised media and public forums.

Why Americans disagree about climate change

Unlike a conventional environmental threat such as smog or acid rain, climate change can be defined as a *wicked problem*, a problem that is the product of multiple social, ecological and technological systems, difficult to define, with no clear solution, seemingly intractable and often plagued by chronic policy failures and intense disagreement. Wicked problems require almost constant risk reduction, conflict management and political negotiation that seldom bring an *end* or resolution. Like poverty or war, climate change is not something likely to be solved, eliminated or ended, but rather a condition that society will struggle to understand, make sense of, and do better or worse at in managing (Hulme 2009; Rittle and Webber 1973).

The *Oxford Handbook on Climate Change and Society* (Dryzek *et al.* 2011) reflects the difficulties experts face in reaching consensus on the nature of climate change as a social problem and the actions needed. Across 47 chapters and 600 pages, top international scholars responded to the editors' invitation to help them 'lay out the various ways that climate change affects society and what society might do in response' (ibid.: 4). Easy answers, however, were not forthcoming. The contributors to the volume represented:

> substantial differences when it comes to identifying what matters, what is wrong, what is right, how it got to be that way, who is responsible, and not least, what should be done …. Commissioning, reading, and editing these contributions has left us acutely aware of the limitations of human knowledge – and the major constraints on intelligent human action – when it comes to complex socio-ecological systems.
>
> (Ibid.)

The intractable problem of climate change has presented the opportunity for advocates and policy entrepreneurs to promote prescriptions that align with their preferred future. As science policy scholar Roger Pielke Jr. (2011: 62) aptly summarises: 'Climate change is a bit like a policy inkblot on which people map onto the issue their hopes and values associated with their vision for what a better world would look like'.

Bill McKibben in 1989 published *The End of Nature*, recognised as the first popular book about climate change and in this and many subsequent works, warned that humans had become the 'most powerful force for change on the planet' ([1989] 2006: xix), a potentially catastrophic achievement that marked an end to our traditional understanding of nature. Climate change, unlike other environmental problems, was not conventionally solvable; our best hope was to avert the most devastating impacts, McKibben wrote. Yet he was deeply sceptical of technological approaches to the problem such as genetic engineering or nuclear energy (Nisbet 2013).

The only possible path to survival, he argued, was through a fundamental reconsideration of our world views, aspirations and life goals and the creation of a new consciousness that would dramatically reorganise society, ending our addiction to fossil fuels, economic growth and consumerism. In this pastoral future, free of consumerism or material ambition, Americans would rarely travel, experiencing the world instead via the Internet, grow much of their own food, power their communities through solar and wind energy, and divert their wealth to developing countries. Only under these transformational conditions, argued McKibben, would we be able to set a moral example for countries like China to change course, all in the hope that these countries will accept a 'grand bargain' towards a cleaner energy path (Nisbet 2013).

Other climate advocates offered a different outlook and set of prescriptions intended to address climate change. Author Amory Lovins and former US Vice-President Al Gore agree that limits to economic growth should be respected, but they also assume limits can be stretched if the right policies and reforms are adopted, enabling environmentally sustainable development to continue indefinitely. The main policy action endorsed by these advocates is to increase the cost of carbon-based energy through *pricing mechanisms* like a carbon tax or cap and trade system so that solar, wind and other innovative energy technologies become more competitive and industries more energy-efficient. In this, business leaders and industry are viewed as valuable partners, and actions on climate change are defined as potentially profitable (Nisbet 2013).

Presented with these visions for what climate change means for the transformation of society and the economy, the fossil fuel industry and their allies among conservative political leaders have – unsurprisingly – opposed any effort to limit greenhouse emissions, often rejecting outright the conclusions of climate scientists whom they see as aligned with advocates like Gore. To block policy action, the fossil fuel industry and their political allies have manufactured doubt in the news media about the reality of man-made climate change, exaggerated the economic costs of action, ridiculed environmentalists, intimidated scientists and manipulated the use of scientific expertise in policymaking (McCright and Dunlap 2010).

Conservative Oklahoma Senator James Inhofe, for example, has personalised the issue in strongly cultural and geographic terms, using frames of reference that resonate with right-wing to moderate members of the public who live in states or work in industries that strongly depend on fossil fuels. Inhofe casts doubt on the conclusions of the Intergovernmental Panel on Climate Change and other major scientific organisations, selectively citing scientific-sounding evidence. Inhofe takes advantage of the fragmented news media, with appearances on television outlets such as Fox News and on political talk radio. In a February 2007 Fox & Friends segment titled *Weather Wars*, Inhofe argued that global warming was due to natural causes and mainstream science was beginning to accept this conclusion. Unchallenged by host Steve Doocy, Inhofe asserted that 'those individuals on the far left, such as Hollywood liberals and the United Nations' want the public to believe that climate change is man-made (Nisbet 2009).

In a series of 'cultural cognition' studies, Kahan and colleagues (2012) have identified a set of world views, cultural dispositions and social processes that help explain why the competing views for society offered by McKibben, Gore and Inhofe make achieving political consensus on climate change so difficult. Kahan *et al.* employ an index of survey measures that classify members of the public by their respective orientations towards either hierarchical and individualist world views (corresponding more generally with more traditionally right-wing political views) or their opposing orientation towards a communitarian and egalitarian outlook (corresponding more generally with left-wing political views). Members of the public scoring high on hierarchical and individualist values tend to be sceptical of environmental threats like climate change since they intuitively sense that actions to reduce environmental risks will adversely impact commerce and industry, institutions that they deeply value and respect. In contrast, members of the public scoring high on communitarian and egalitarian values see policy actions that restrict commerce and industry as benefiting the broader community and the most vulnerable in society. This segment of the public readily accept the risks posed by climate change since actions to restrict greenhouse gases from industry are consistent with their vision for what a better world would look like (Kahan et al. 2012).

These differences in world views, hopes, values and visions for society are reflected and reinforced by dramatic changes in the media system over the past decade. In the 24-hour political news cycle, commentators and bloggers on the political left and right rely on the latest insider strategy, negative attack or embarrassing gaffe to appeal to ideologically motivated audiences, connecting almost every policy issue to the broader struggle for control of American politics between liberals and conservatives. In this regard, the divisiveness and rancour that typify online commentary about climate change is driven, in part, by what Berry and Sobieraj characterise in a series of studies as the media 'outrage industry' (2008: 4, 2014). This discourse culture specialises in provoking emotional responses from audiences, trading in exaggerations, insults, name calling and partial truths about opponents and reducing complex issues to 'ad hominem attacks, overgeneralisations, mockery, and dire forecasts of impending doom'.

Moral outrage in the media feeds on and foments Americans' face-to-face conversations and exchanges in online social networks. As people have sorted themselves into like-minded residential areas, workplaces and political districts, the similarity of Americans' social, political and geographic enclaves has increased appreciably (Abramowitz 2010). On climate change, many Americans are unlikely to report personally knowing people who hold different views from their own. Instead, the *political other* is a caricature offered on blogs, talk radio or cable news. For Hierarchical Individualists, those who support action on climate change are *eco-fascists*; and for Communitarian Egalitarians, those who express doubts are *denialists*. In each case, the opposing side is viewed as incapable of either reason or compromise.

Editorial and business decisions at prestige news outlets have also unwittingly boosted polarisation on climate change. *The New York Times* and *Washington Post*, most notably, have cut back on news coverage of climate change and other science issues, letting go of many of their most experienced reporters, allowing advocacy-oriented media outlets and commentators to fill the information gap. As a consequence, careful reporting at these outlets on the technical details of science and policy have been replaced by morally framed interpretations from bloggers and advocacy journalists at other outlets. Online news and commentary are also highly socially contextualised, passed along and preselected by people who are likely to share world views and political preferences. If an individual incidentally *bumps* into news about climate change by way of Twitter, Facebook, or Google+, the news item is likely to be the subject of meta-commentary that frames the political and moral relevance of the information. Taking advantage of these self-reinforcing spirals, advocacy groups devote considerable resources to flooding social media with politically favourable comments and purposively selected stories (Scheufele and Nisbet 2012).

Even when individuals, prompted by a focusing event like extreme weather or a major scientific report, do decide to seek out more information about climate change, further selectivity is likely to occur. In this case, liberals might choose to search for information on 'climate change', and encounter one set of differentially framed search results; whereas a conservative searching for information on 'global warming' encounters an entirely different set of search results. Not only does word choice shape the information returned through Google, but so does the past browsing and search history of the individual, adding an additional layer of selectivity and bias to the information encountered (Brossard 2013).

Promoting consensus and empowering the public

Frustrated by the political paralysis on climate change, environmentalists and their political allies have invested in ever bolder, more urgent efforts to build a politically powerful base of support for action on climate change. In a 2012 cover article in *Rolling Stone* magazine that quickly became a social media sensation, Bill McKibben called for a new sense of 'moral outrage' directed at the fossil fuel industry. Given the urgency of climate change, 'we need to view the fossil-fuel industry in a new light', he argued. 'It has become a rogue industry, reckless like no other force on Earth. It is Public Enemy Number One to the survival of our planetary civilisation' (McKibben 2012).

Drawing comparisons with the civil rights and anti-apartheid movements, McKibben urged readers to join his organisation 350.org in protesting against the proposed Keystone XL tar sands oil pipeline across the US and to pressure local universities, colleges, churches and governments to divest their holdings in fossil fuel companies. In *The Nation* magazine, 350.org board member and bestselling author Naomi Klein (2011) had argued the need for climate activists to copy the political strategies of the US conservative movement:

> Just as climate denialism has become a core identity issue on the right, utterly entwined with defending current systems of power and wealth, the scientific reality of climate change must, for progressives, occupy a central place in a coherent narrative about the perils of unrestrained greed and the need for real alternatives.

The same year in *Rolling Stone*, Al Gore (2011) drew parallels with the US civil rights movement, urging readers to 'become a committed advocate for solving the crisis' by speaking up in everyday conversations when people express doubts about the threat. He encouraged readers to join his advocacy group the Climate Reality Project[1] and to contact newspapers and television programmes to 'let them know you're fed up with their stubborn and cowardly resistance to reporting the facts of this issue'. The Climate Reality Project has organised a series of 24-hour web broadcasts that 'bring together artists, scientists, celebrities, economists, and other experts to explore the many ways we're all paying for carbon pollution in our daily lives – wherever we may live – and how we can solve this with a market price on carbon'.

Though these advocacy efforts might bring much-needed political pressure on key elected officials in the short term, in the long term such strategies, if not also balanced by alternative investments by the expert community, may only intensify polarisation and policy gridlock. A key finding of Kahan and colleagues (2012) was that the most knowledgeable and cognitively sophisticated Hierarchical Individualists and Egalitarian Communitarians tend to be the most divided in their views of climate change. A major reason, they argue, is that in comparison with their less-informed counterparts, these individuals are better attuned to what other members of their cultural group think and believe. The desire to remain aligned with the outlook of others in their cultural group strongly shapes their opinion on climate change.

Therefore, the more that those involved in the climate change movement are perceived by the broader public to be predominantly liberal, Democratic-leaning, and Egalitarian Communitarian in their outlook, the more likely those with differing cultural identities are to dismiss climate change as a threat and to view policy actions to address the problem as in conflict with their vision of society and their future. To offset these barriers to consensus building, the expert community and their partners need to recruit opinion leaders from a greater diversity of societal sectors and to encourage these opinion-leaders to *frame* climate change in ways that resonate with their respective cultural groups, activating feelings of concern, responsibility and obligation. The following sections review some of the key strategic options and how they have been tested in a series of studies. These three options are: promoting new frames of reference and cultural voices; diversifying policy options and technological choices; investing in civic capacity and public deliberation.

Promoting new frames of reference and cultural voices

In a series of studies conducted with Maibach and colleagues, we investigated how a diversity of Americans understand the health and security risks of climate change and how they react to information about climate change when it is framed in terms of these alternative dimensions. Our goal was to inform the work of public health professionals, municipal managers and planners, and other trusted civic leaders as they seek to engage broader publics on the health and security risks posed by climate change.

Framing climate change in terms of public health stresses climate change's potential to increase the incidence of infectious diseases, asthma, allergies, heat stroke and other salient health problems, especially among the most vulnerable populations: the elderly and children. In the process, the public health frame makes climate change personally relevant to new audiences by connecting the issue to health problems that are already familiar and perceived as important. The frame also shifts the geographic location of impacts, replacing visuals of remote Arctic regions, animals, and peoples with more socially proximate neighbours and places across local communities and cities. Coverage at local television news outlets and specialised urban media is also generated (Nisbet 2009; Weathers *et al.*, 2013).

Efforts to protect and defend people and communities are also easily localised. State and municipal governments have greater control, responsibility and authority over climate change adaptation-related policy actions. In addition, recruiting Americans to protect their neighbours and defend their communities against climate impacts naturally lends itself to forms of civic participation and community volunteering. In these cases, because of the localisation of the issue and the non-political nature of participation, barriers related to polarisation may be more easily overcome and a diversity of organisations can work on the issue without being labeled as 'advocates', 'activists' or 'environmentalists'. Moreover, once community members from differing political backgrounds join together to achieve a broadly inspiring goal like protecting people and a local way of life, then the networks of trust and collaboration formed can be used to move this diverse segment toward co-operation in pursuit of national policy goals (Nisbet *et al.* 2012; Weathers *et al.*, 2013).

To test these assumptions, in an initial study, we conducted in-depth interviews with 70 respondents from 29 states, recruiting subjects from six previously defined audience segments. These segments ranged in a continuum from those individuals deeply alarmed by climate change to those who were deeply dismissive of the problem. Across all six audience segments, individuals said that information about the health implications of climate change was both useful and compelling, particularly when locally focused mitigation- and adaptation-related actions were paired with specific benefits to public health (Maibach *et al.* 2010).

In a follow-up study, we conducted a nationally representative web survey in which respondents from each of the six audience segments were randomly assigned to three different experimental conditions, allowing us to evaluate their emotional reactions to strategically framed messages about climate change. Though people in the various audience segments reacted differently to some of the messages, in general framing climate change in terms of public health generated more hope and less anger than framed messages that defined climate change in terms of either national security or environmental threats. Somewhat surprisingly, our findings also indicated that the national security frame could 'boomerang' among audience segments already doubtful or dismissive of the issue, eliciting unintended feelings of anger (Myers *et al.* 2012: 1107).

In a third study, we examined how Americans perceived the risks posed by a major spike in fossil fuel energy prices. According to our analysis of national survey data, about half of American adults believe that our health is at risk from major shifts in fossil fuel prices and availability. Moreover, this belief was widely shared among people of different political ideologies and was strongly held even among individuals otherwise dismissive of climate change. Our findings suggest that many Americans would find relevant and useful these communication efforts that emphasised energy resilience strategies which reduce demand for fossil fuels, thereby limiting greenhouse emissions and preparing communities for fuel shortages or price spikes. Examples include improving home heating and automobile fuel efficiency, increasing the availability and affordability of public transportation and investing in government-sponsored research on cleaner, more efficient energy technologies (Nisbet *et al.* 2011).

Among the public interest organisations applying similar research-based principles to their communication strategies is ecoAmerica (with which I have worked as a consultant and adviser). ecoAmerica is collaborating with opinion leaders and organisations recruited from societal sectors new to the climate change debate, including public health, faith communities, business, higher education and local municipalities. Their goal is not only to empower a more nationally representative 'choir' of cultural voices, but also to promote a new framing of the issue. This new narrative features strong themes of national unity, pride and common identity, emphasising the risks to public health, communities and the economy and the possibility of progress if Americans can come together to defend their local communities against the impacts of climate change (ecoAmerica 2013).

Diversifying policy options and technological choices

Apart from recruiting new voices and emphasising new frames of reference, the expert community can also balance the efforts of climate advocates by expanding the range of policy options and technologies considered. As Roger Pielke Jr. (2007) argues, instead of allowing their expertise to be used in efforts to promote a narrow set of policy approaches, experts and their institutions must act independently as 'honest brokers' to expand the range of policy options and technological choices under consideration by the political community. The broader the menu of policies and technologies under consideration, the greater the opportunity for compromise among decision-makers.

Pielke (2011) notes that polls show that for several years the public has favoured action on climate change but at low levels of intensity, suggesting that it is not a lack of public support that limits policy action. Once technologies are available that make meaningful action on climate change lower cost, then much of the political argument over scientific uncertainty will diminish, he argues.

Kahan and colleagues' findings (2011) strongly suggest that perceptions of climate change are policy- and technology-dependent and that consensus is more likely to occur under conditions of a diverse rather than a narrow set of proposed solutions. In these studies, when Hierarchical Individualists read that the solution to climate change was more nuclear power or geoengineering, their scepticism about expert statements related to climate change decreased and their support for policy responses increased. When the solution to climate change was framed as stricter pollution controls, Hierarchical Individualists' acceptance of expert statements on climate change decreased, whereas Communitarian Egalitarians' increased.

If we apply the reasoning of both Pielke and Kahan to the climate debate, it appears that building political consensus on climate change will depend heavily on experts and their institutions calling attention to a broad portfolio of policy actions and technological solutions, with some actions such as tax incentives for nuclear energy, government support for clean energy research, or proposals to defend and protect local communities against climate change impacts more likely to gain support from both Democrats and Republicans. As effective brokers, scientists and their institutions should proactively encourage journalists, policymakers and the public to discuss a broad menu of options, rather than tacitly allow (or sometimes promote) efforts by climate activists, bloggers and commentators to limit debate to just a handful of options that fit with their own ideology and cultural outlook.

The dynamics identified by Pielke, Kahan and others could be observed in my own analysis of a group of major US foundations and funders who bet heavily on the ability of technocratic expertise to overcome political differences on climate change and in the process committed the mistake of investing in an ideologically narrow set of policy goals and technologies (Nisbet 2011). Several of the country's wealthiest foundations hired a consulting firm to survey the scientific literature and to consult more than 150 leading climate change and energy experts, leading to the 2007 report, *Design to Win: Philanthropy's Role in the Fight Against Global Warming*. This recommended that 'tempering climate change' required a strong cap and trade policy in the US and the European Union, and a binding international agreement on greenhouse gas emissions. The report included little or no discussion of the role of government in directly sponsoring the creation of new energy technologies. The report was additionally notable for the absence of any meaningful discussion of social, political or technological barriers. To understand how this planning document shaped the investment strategies of major funders, I analysed records for 1,246 climate change and energy-related grants distributed between 2008 and 2010 by nine foundations who were either sponsors of the *Design to Win* report or described themselves as following its recommendations. The $368 million funding provided by these foundations reflected a pattern of support focused on achieving a clear set of policy objectives as outlined in the *Design to Win* report. Funding patterns also reflected the *Design to Win* report's framing of climate change as a physical threat that required primarily scientific and economic expertise to solve rather than investments in research that would inform communication campaigns, or in investments in public participation and dialogue. There was very limited or no funding focused on the role of government in promoting innovation or on development of technologies favoured by political conservatives like nuclear energy, carbon capture and storage, or natural gas fracking. Nor was there equivalent investment in important human dimensions of the issue, such as adaptation, health, equity, justice or economic development. Similarly, very few grants supported initiatives designed to better understand public opinion, to evaluate communication strategies, and/or to promote media resources across states and regions.

Investing in civic capacity and public deliberation

The expert community will also need to invest in rebuilding America's civic capacity to discuss, debate and participate in collective decisions. In this regard, universities and other research institutions can serve a vital function in facilitating public dialogue about climate change: by working with philanthropic funders and community partners to sponsor local media platforms and public forums, by convening stakeholders and political groups, and by serving as a resource for collaboration and co-operation. In fact, cities and local regions are the contexts where we can most effectively experiment with communication initiatives that challenge how each of us debate, think and talk about climate change as a social problem. In these forums, new cultural voices can be heard, new cultural framings and meanings emphasised and innovative policy approaches and technological options discussed. Stevenson and Dryzek (2012: 207) emphasise providing 'possibilities for contestation and the reflection it can induce'. Rosen (2012) argues that 'what's possible is a world where different stakeholders "get" that the world looks different to people who hold different stakes'.

By building up local and regional communication capacity, the conditions for eventual change in national politics can be set, rewiring expectations and norms relative to public debate and forging relationships and collaborations that span ideological differences and cultural world views. A university-led public engagement initiative on climate change that was successful in overcoming culturally motivated and group-based polarisation surveyed the public in a coastal Maryland county to better understand their risk perceptions related to sea-level rise and coastal flooding. Respondents' world views as measured in terms of Hierarchical Individualism and Communitarian Egalitarianism were among the strongest predictors of risk perceptions. 'In this environment, traditional communication strategies of providing "objective" assessments are unlikely to staunch further issue polarisation', it was argued (CASI 2013: 12). Yet when the project organisers brought together a sample of 40 local residents to participate in a professionally moderated dialogue about sea-level rise and coastal flooding, Hierarchical Individualists' doubts about the risks posed by the threat decreased. Skillfully moderated public deliberation focused on a local threat made community-wide membership a more salient consideration than this group's cultural identity and political outlook (CASI 2013).

Face-to-face dialogue should be complemented by new online media forums that bridge and add context to opposing perspectives on climate change, expanding discussion of policy options and technological solutions and thereby offering an alternative to the moral outrage that dominates most online commentary. On his Dot Earth blog, part of *The New York Times* opinion section, veteran science reporter Andrew Revkin not only functions as an explainer and informed critic of science but also as a convenor, facilitating discussions among a diversity of experts, advocates and various publics, while contextualising the uncertainty relative to specific claims, technologies and policy approaches (Fahy and Nisbet 2011; Nisbet 2013). Rather than frequently advocating for a position, Revkin prefers posing questions, describing answers from experts and others, in an approach that McKibben has criticised as 'relentlessly middle-seeking' (Nisbet 2013: 56). 'I think anyone who tells you they know the answer on some of these complex issues is not being particularly honest', Revkin responds (2009).

The principles that inform Revkin's blogging at *The New York Times* could also shape the design and sponsorship of media forums sponsored by universities and their partners. As regional newspapers suffer financially and cut coverage of public affairs generally and climate change specifically, new forms of non-profit, university-based media platforms can provide regions of the country with the civic capacity to make informed decisions and choices. A leading prototype for such an initiative is *Ensia*, a foundation-funded web magazine[2] launched by the Institute on the Environment at the University of Minnesota. The online magazine's mission is to use news, commentary, and discussion to identify and inspire new approaches to climate change and other environmental problems. To do so, *Ensia* features reporting by top freelancers, commentaries by experts and thought leaders and a TED conference-like event series that is broadcast and archived online.

Concluding remarks

Successfully navigating controversies such as those over climate change and food biotechnology not only requires a sophisticated, research-based understanding of the factors shaping these debates, but also an acceptance that there are strong limits to what even the best-funded and most carefully planned public engagement strategy can accomplish. In this chapter, I have reviewed three such strategies that, research suggests, may be effective at softening disagreement and creating the opportunity for consensus and agreement. They include investing in new frames of reference and cultural voices; proactively widening the menu of policy options and technological options considered; and investing in localised public and media forums that sponsor dialogue, diverse interactions and collaboration around new ideas and solutions.

Despite the evidence supporting the efficacy of these strategies, the application of research-based principles to science policy controversies does not guarantee conflict resolution. Research findings are often messy, complex and difficult to translate into practice. They are also contingent and subject to revision based on new research, changes in the dynamics surrounding an issue or changes applying across issues and social contexts. Moreover, no matter how knowledgeable and adept the expert community might be in applying research to their public engagement efforts, resolution of intensely polarised debates takes years, if not decades, to achieve. It requires the different sides in a debate to give ground, negotiate and compromise. In no case is this more likely to be true than in the debate over climate change. Yet in the case of climate change, the major question therefore is whether resolution will come too late, preventing society from managing the most serious risks.

The main drivers of eventual resolution and agreement are most likely to be deeper changes in the political system, demographic and social trends, external shocks such as natural disasters and/or breakthroughs in technologies. Applying insights from research on science policy controversies can help bend these dynamics more strongly to one side's favour or to accelerate this long-term process in incremental ways. In the meantime, as observers and participants in the debate, we can all benefit from the wisdom, reflection and insight that such research can provide.

Nelkin in her last edited volume of case studies offered the following outlook (1992: xxiv):

> Based on competing social and political values, few conflicts are in reality resolved. Even as specific debates seem to disappear, the same issues reappear in other contexts .... The persistence of controversy suggests that the issues described in this book are hardly unique events. Rather, they are part of a significant tendency in American society to reassess the social values, the priorities, and the political relationships that underlie technical decisions.

Key questions

- What similarities and differences are there between factors shaping disagreement and polarisation over climate change in the United States and those relating to food biotechnology in Europe and elsewhere?
- Besides framing climate change in terms of either environmental risks or public health threats, what other frames potentially exist and which of these meanings might help overcome polarised differences?
- What values or world views do you think influence how people perceive the risks of climate change and evaluate the various possible technological approaches to the problem?
- What potential do activist approaches such as those pursed by Bill McKibben and 350.org have to inadvertently promote greater polarisation? What benefits do you see to these types of social movement building and civil disobedience strategies?

Notes

1. www.24hoursofreality.org/
2. www.ensia.com

References

Abramowitz, A. (2010) *The Disappearing Center: Engaged Citizens, Polarization, and American Democracy*, New Haven: Yale University Press.

Allum N., Sturgis, P., Tabourazi, D., Brunton-Smith, I. (2008) 'Science knowledge and attitudes across cultures: a meta-analysis', *Public Understanding of Science*, 17, 1: 35–54.

Allum N., Sibley, E., Sturgis, P., Storeman, P. (in press) 'Religions beliefs, knowledge about science and attitudes towards medical genetics', *Public Understanding of Science*, forthcoming.

Bauer M. W. and Gaskell, G. (eds) (2002) *Biotechnology: The Making of a Global Controversy*, Cambridge: Cambridge University Press.

Berry, J. M. and Sobieraj, S. (2008) 'The outrage industry', paper presented at *Going to Extremes: The Fate of the Center in American Politics*, conference held by Dartmouth College, Hanover, NH; online at http://ase.tufts.edu/polsci/prospective/OutrageIndustry.pdf; accessed 25 October 2013.

Berry, J. M. and Sobieraj, S. (2014) *The Outrage Industry: Public Opinion, Media and the New Incivility*, New York: Oxford University Press.

Brossard, D. (2013) 'New media landscapes and the science information consumer', *Proceedings of the National Academy of Sciences*, 110, Supplement 3 14096–14101.

Brossard D., Scheufele, D. A., Kim, E. and Lewenstein, B. V. (2009) 'Religiosity as a perceptual filter: examining processes of opinion formation about nanotechnology', *Public Understanding of Science*, 18, 5: 546–558.

CASI (2013) *Final Project Report: Community Adaptation to Sea-Level Rise and Inundation, A Joint Project of the U.S. Naval Academy and George Mason University*; online at http://bit.ly/1c5H8EA; accessed 25 October 2013.

Dryzek, J. S., Norgaard, R. B. and Schlosberg, D. (2011) 'Climate change and society: approaches and responses', in J. S. Dryzek, R. B. Norgaard and D. Schlosberg (eds) *The Oxford Handbook of Climate Change and Society*, Oxford: Oxford University Press, 3–20.

ecoAmerica (2013) *Overview on Momentus Campaign*, Washington, DC: ecoAmerica; online at http://ecoamerica.org/wp-content/uploads/reports/MomentUs_overview.pdf; accessed 25 October 2013.

Fahy, D. and Nisbet, M. C. (2011) 'The science journalist online: Shifting roles and emerging practices', *Journalism: Theory, Practice and Criticism*, 12, 7: 778–793.

Funtowicz, S. O., and Ravetz, J. R. (1992) 'Three types of risk assessment and the emergence of post-normal science', in S. Krimsky and D. Golding (eds) *Social Theories of Risk,* Westport, CT: Praeger, 251–274.

Gamson, W. A., and Modigliani, A. (1989) 'Media discourse and public opinion on nuclear power: a constructionist approach,' *American Journal of Sociology*, 95, 1: 1–37.

Gore, A. (2011) 'Climate of denial', *Rolling Stone*, 2 June; online at www.rollingstone.com/politics/news/climate-of-denial-20110622; accessed 10 November 2013.

Hulme, M. (2009) *Why We Disagree About Climate Change: Understanding Controversy, Inaction and Opportunity*, Cambridge: Cambridge University Press.

Kahan, D., Wittlin, M., Peters, E., Slovic, P., Ouellette, L. L., Braman, D. and Mandel, G. (2011) 'The tragedy of the risk-perception commons: culture conflict, rationality conflict, and climate change', Cultural Cognition Project, Yale University Working Paper No. 89, Social Science Research Network; online at http://papers.ssrn.com/sol3/papers.cfm?abstract_id=1871503; accessed 25 October 2013.

Kahan, D., Wittlin, M., Peters, E., Slovic, P., Ouellette, L. L., Braman, D. and Mandel, G. (2012) 'The polarizing impact of science literacy and numeracy on perceived climate change risks', *Nature Climate Change*, 2: 732–735.

Klein, N. (2011) 'Capitalism vs. the climate', *The Nation*, 29 November; online at www.thenation.com/article/164497/capitalism-vs-climate#; accessed 25 October 2013.

McCright, A. M. and Dunlap, R. E. (2010) 'Anti-reflexivity: the American conservative movement's success in undermining climate science and policy', *Theory, Culture and Society*, 27, 2–3: 100–133.

McKibben, B. ([1989] 2006) *The End of Nature*, Random House.

McKibben, B. (2012) 'Global warming's terrifying new math', *Rolling Stone*, 19 July; online at www.rollingstone.com/politics/news/global-warmings-terrifying-new-math-20120719; accessed 25 October 2013.

Maibach, E., Nisbet, M. C., Baldwin, P., Akerlof, K. and Diao, G. (2010) 'Reframing climate change as a public health issue: an exploratory study of public reactions', *BMC Public Health*, 10: 299.

Myers, T., Nisbet, M. C., Maibach, E. W. and Leiserowitz, A. (2012) 'A public health frame arouses hopeful emotions about climate change', *Climatic Change Research Letters*, 113: 1105–1121.

Nelkin, D. (1979) *Controversy: The Politics of Technical Decisions*, Newbury Park, CA: Sage Publications.

Nelkin, D. (1984) *Controversy: The Politics of Technical Decisions*, second edition, Newbury Park, CA: Sage Publications.

Nelkin, D. (1992) *Controversy: The Politics of Technical Decisions*, third edition, Newbury Park, CA: Sage Publications.

Nisbet, M. C. (2005) 'The competition for worldviews: values, information, and public support for stem cell research', *International Journal of Public Opinion Research*, 17, 1: 90–112.

Nisbet, M. C. (2009) 'Communicating climate change: why frames matter to public engagement', *Environment*, 51, 2: 514–518.

Nisbet, M. C. (2011) *Climate Shift: Clear Vision for the Next Decade of Public Debate*, Washington, DC: American University, School of Communication; online at http://climateshiftproject.org/wp-content/uploads/2011/08/ClimateShift_report_June2011.pdf; accessed 25 October 2013.

Nisbet, M. C. (2013) *Nature's Prophet: Bill McKibben as Journalist, Public Intellectual, and Activist*, Joan Shorenstein Center for Press, Politics, and Public Policy Discussion Paper Series, D-78 March, Cambridge, MA: Kennedy School of Government, Harvard University; online at http://shorensteincenter.org/2013/03/natures-prophet-bill-mckibben-as-journalist-public-intellectual-and-activist/; accessed 25 October 2013.

Nisbet, M. C. and Lewenstein, B. V. (2002) 'Biotechnology and the American media: the policy process and the elite press, 1970 to 1999', *Science Communication*, 23, 4: 359–391.

Nisbet, M. C., Maibach, E. and Leiserowitz, A. (2011) 'Framing peak petroleum as a public health problem: audience research and participatory engagement', *American Journal of Public Health*, 101, 9: 1620–1626.

Nisbet, M.C., Markowitz, E. M. and Kotcher, J. (2012) 'Winning the conversation: framing and moral messaging in environmental campaigns', in L. Ahern and D. Bortree (eds) *Talking Green: Exploring Current Issues in Environmental Communication*, New York: Peter Lang, 9–36; online at http://climateshiftproject.org/winning-the-conversation-framing-and-moral-messaging-in-environmental-campaigns/; accessed 25 October 2013.

Pielke, R. (2007) *The Honest Broker: Making Sense of Science in Policy and Politics*, New York: Cambridge University Press.

Pielke, R. Jr. (2011) *The Climate Fix: What Scientists and Politicians Won't Tell You About Global Warming*, New York: Basic Books.

Revkin, A. (2009) 'My second half', *The Dot Earth blog, The New York Times*, 21 December; online at http://dotearth.blogs.nytimes.com/2009/12/21/my-second-half/; accessed 25 October 2013.

Rittel, H. W. J., and Webber, M. M. (1973) 'Dilemmas in a General Theory of Planning', *Policy Sciences*, 4: 155–169.

Rosen, J. (2012) 'Covering wicked problems: keynote address to the 2nd UK conference of science journalists', *PressThink Blog*, 25 June; online at http://pressthink.org/2012/06/covering-wicked-problems/; accessed 25 October 2013.

Sarewitz, D. (2004) 'How science makes environmental controversies worse', *Environment Science and Policy*, 7, 5: 385–403.

Scheufele, D. A. and Nisbet, M. C. (2012) 'Online news and the demise of political disagreement', in C. Salmon (ed.) *Communication Yearbook 36*, New York: Routledge, 45–54.

Stevenson, H., and Dryzek, J. (2012) 'The discourse democratization of global climate governance', *Environmental Politics*, 21, 2: 189–210.

Weathers, M., Maibach, E. W. and Nisbet, M. C. (2013) 'Using theory and audience research to convey the human implications of climate change', in D.Y Kim, G. Kreps and A. Singhal (eds) *Health Communication: Strategies for Developing Global Health Programs*, New York: Peter Lang, 190–207.

Wynne, B. (1992) 'Misunderstood misunderstanding: social identities and public uptake of science', *Public Understanding of Science*, 1, 3: 281–304.

14
Communicating the social sciences
A specific challenge?

Angela Cassidy

Introduction

Science communication in both research and practice tends to refer to the physical, chemical and biological sciences, sometimes alongside fields such as medicine, mathematics and engineering. Relatively little attention has been paid by PCST (public communication of science and technology) researchers to how other academic fields such as the social sciences, arts and humanities are discussed in the broader public sphere (Schäfer 2012). The research literature on PCSS (public communication of the social sciences) remains relatively sparse and scattered across many disciplinary areas. The historical impetus for research into science communication and the *public understanding of science* came from concerns about the public position of the natural sciences and this limited remit has influenced the subsequent development of the field. However, social scientists and historians of science significantly contributed to research in this field, and it is these academic traditions that generated the classic critiques of *deficit* approaches to PCST. In this light, it is curious that PCST researchers have rarely conducted studies of PCSS, or applied these critiques to communicating with non-specialists about their own findings. While the lack of attention to social sciences in related fields such as science and technology studies may also contribute to the problem (Camic *et al.* 2011; Danell *et al.* 2013), these legacies cannot fully account for the continuing low profile of PCSS as a research topic.

Many media around the world have specialist science output, such as TV and radio programmes about science and science sections in newspapers, and popular science is widely established as a publishing genre. Science journalism is an established journalistic specialism, and such professionals provide content for both specialist science and mainstream media output. Most of this coverage tends to be of natural science disciplines, although social sciences such as psychology do receive some specialist attention. Particularly in English-speaking media, there is little or no corresponding journalistic specialisation for the social sciences or humanities, and the tendency of PCST research to focus on science, medical and environmental *specialists* may also be a factor in the weakness of PCSS research. However, this does not mean that these fields do not attract much media coverage or public attention – in fact, they are covered widely across the broader, non-specialist media and form a major part of the content of specialist areas such as political, economic and lifestyle journalism. Crime figures, demographic census data, opinion polls, educational research, economic analysis, psychological studies and political theory are all examples of social science research which contribute to the core day-to-day content of contemporary media coverage. Social researchers frequently provide policy, personal and lifestyle advice across many fora, and the much-discussed role of the *public intellectual* is one generally occupied by social science and humanities scholars. Therefore, this chapter will also discuss the public role of arts and humanities disciplines.[1]

Social research forms the core activity of many think tanks, active by definition in the public sphere, and of the work of most policymakers in government and in NGOs. In Britain in particular, social scientists have been instrumental in the development of both science communication and public engagement as research fields as well as initiating widespread change in the policy and practice of academic, governmental and scientific institutions in these areas. Yet social scientists, particularly those researching PCST itself, have paid little attention to the public communication of their own work.

Research literature on social sciences and the media

The disparate nature of PCSS research can make it particularly difficult to find relevant studies in citation databases: searches may turn up large numbers of articles on conflating topics such as '*social science* approaches to science *communication*' or 'the *economics* of the *media*'. However, looking at the work that has been done suggests some tentative trends and insights. Research directly addressing PCSS includes quantitative and qualitative analyses of media content; interviews and surveys conducted with academics and journalists; theoretical analysis, personal experiences and material addressing the public promotion of social science, including 'how to' guides for academics interacting with mass media. Of these, the latter few are most commonplace, recalling the literature on PCST prior to critiques of the deficit model.

In English-speaking countries, strong distinctions are drawn between the natural sciences and social sciences/humanities (studies of the human and social), with *science* generally considered to cover the former but not the latter disciplines. Furthermore, popular ideas about the nature of science reinforce the status of subjects which use quantitative, experimental or statistical methods, such as economics and many areas of psychology. In continental Europe and perhaps elsewhere in the world, conceptions of science can include all forms of scholarly research, as conceptualised in the German term *Wissenschaft* (although see Sala, 2012 for further discussion). This, alongside my own language constraints, have led to an Anglocentric bias in this review, which means any conclusions drawn will be inevitably qualified. I have referred to as global a range of studies as possible but, considering the paucity of the literature as a whole, much more work is needed before we can have a coherent understanding of the effect of cross-cultural differences on the public communication of the social sciences.

A great deal of PCSS literature is still written by social scientists drawing upon their own communication experiences and often resembles older PCST literature in the emphasis on how to get the *correct message* across (e.g. Grauerholtz and Baker-Sperry 2007; Stockelova 2012). The public image problems of social science are discussed and strategies for improvement still tend to centre on upbraiding journalists for sensationalism and inaccuracy and/or publics for their incorrect understandings of social science research (Kendall-Taylor 2012; Seale 2010). Social science funding bodies and professional associations are taking relationships with mass media increasingly seriously, with resources and information being made available for both researchers and journalists (e.g. LSE Public Policy Group 2011; ESRC 2013). In the previous edition of this Handbook (Cassidy 2008) I observed that this area of activity was somewhat dominated by psychologists, perhaps due to the discipline's borderline status straddling the natural and social sciences: now professional associations across the social sciences employ media relations professionals and issue press releases on a routine basis. However, the overriding concern remains with the *promotion* of the social sciences, rather than reflective engagement with why this should be done: this has been greatly accentuated by contractions in research funding and pressures for academics to establish the *impact* of their research on society.

A second area of literature, in places closely related to the above, consists of content analysis of social science media coverage. Weiss and Singer (1988) carried out an extensive study of the American news media during the 1980s, comprising parallel content analysis and interview studies. They found that the majority of coverage concentrated on the research topic (e.g. crime, parenting, relationships), with the research itself appearing in an ancillary role. Furthermore, only 7 per cent of the stories found were written by specialist science journalists, with most coverage authored by generalists, or specialists in other areas. Analysed by content theme rather than disciplinary area, the coverage in US media gave economics the largest share. A similar approach, using a broader sweep of methods, was taken in researching the British situation in the following decade (Fenton *et al.* 1997, 1998) and this study reveals an interesting pattern of similarities and differences between the US and Britain. As in the US, social science was rarely covered by science journalists in Britain: in fact, only one such example was found in the entire sample studied. In contrast to the US study, social issues provided the largest proportion of the British coverage, with economics coming next; psychology was the most frequently represented discipline. Social research was the main focus of most stories in the British study, rather than being mentioned in passing in stories on other topics. Most of the social science coverage analysed appeared as features rather than news reports, and social scientists more often appeared reactively as commentators and advisers on specific issues according to the news agenda, rather than being the principal sources of stories.

Both these studies looked at how much, and where, media coverage of social science appeared, and, again, transatlantic differences emerge. In the US, coverage was distributed evenly across all forms of media, and levels of reporting found were far higher than in Britain, where coverage was heavily concentrated in the broadsheet (or *quality*) press. However, without meaningful comparisons, it is difficult to draw useful conclusions from these figures: are they high or low, and in what terms? Similarly, it is difficult to distinguish whether many of the issues raised by these studies are specific to the social sciences or are broader concerns shared in the public communication of all research. A study by Evans (1995) deals with this problem by directly comparing US media coverage of social and natural sciences. Of the total sample of research coverage, 36 per cent was of social science subjects, although this was not broken down into disciplinary groupings. The Science Museum Media Monitor (Bauer *et al.* 1995), one of the largest studies of its kind, applied a continental European definition of *science* as including the social sciences and reported a gradual increase in the proportion of social science coverage over the second half of the twentieth century, eventually reaching similar levels to that found by Evans. A smaller study carried out by Hansen and Dickinson (1992) found only 15 per cent of coverage was of social sciences, but related topics such as market research, human interest and science policy/education were separated out from this, leading to a combined figure of 28 per cent. Overall, these studies suggest that the social sciences provide a substantial proportion of media coverage of research in both the US and Britain, overtaken only by health and biomedicine. Böhme-Dürr (2009) reported that social sciences in the German media were relatively under-represented. By contrast, Šuljok and Vuković (2013) report higher coverage levels and higher-quality reporting of social sciences in Croatian media, which they attribute in part to post-socialist legacies of media bias towards these disciplines.

Despite such variations, these findings do point towards important differences in how natural and social sciences are covered by mass media, particularly between US and Britain. Evans (1995) reported that social science was much less likely to appear in newspaper science sections than natural science and more likely to be in general news coverage, confirming the idea that science journalists rarely cover the social sciences. In interviews, Dunwoody (1986) found that US science journalists typically look down on social science research as less scientific, express little interest in it, and regard it as requiring little specialist training to report. Similarly, both Schmierbach (2005) and Seale (2010) observe that disciplines employing quantitative and/or experimental methods, such as psychology, economics or social statistics, are more likely to be taken seriously by journalists. Evans also found that social scientists were accorded a lower epistemological status in media reports, with natural scientists more often referred to as 'researchers' or 'scientists', and social scientists more likely to be referred to in terms such as 'the authors of the study' (Evans 1995: 172). He notes the lack of credible, centralised journalistic sources for media coverage of social science research, compared to the roles played by major scientific journals such as *Nature* and *Science*. Organisations such as the Science Media Centre (established in 2002) tend to focus on quantitative social science, reinforcing this tendency.

My own study of British newspaper coverage of evolutionary psychology, comparing it with evolutionary biology, showed the field was covered less often by science journalists and more by non-specialists, and appeared more frequently in features, supplements and commentary pieces and rarely in specialist *science* sections (Cassidy 2005). Fenton *et al.* (1997, 1998) also investigated relationships between social scientists and media professionals, noting that social science was not usually covered by correspondents with any in-depth knowledge of research and that it was rarely newsworthy in its own right, covered instead as part of broader news agendas. Furthermore, they describe the relationship between academics and the media in this area as formal, distant and highly reliant on the role of facilitators (Fenton *et al.* 1998: 70). Thirty per cent of the researchers they interviewed had worked with media only via communications professionals, a pattern reflected in interactions between researchers and journalists at academic conferences. However, a more recent study by Peters (2013), looking at social science and humanities academics in Germany, found less strict demarcations between professional and popular communication and much higher rates of interaction with journalists than that of natural science researchers. Several studies have explored these apparent contradictions by looking not just at the appearance and location of social scientists in media, but also the roles they play as experts. Albaek *et al.* (2003) and Wien (2013) both found that social scientists are more likely to act as commentators on pre-existing news stories across a range of topics, rather than be the originators of coverage through the publication of research findings. This suggests that a fruitful point of further enquiry could be to explore the literature on so-called 'soft news' (Reinemann *et al.* 2012) and the roles played by experts therein (Lester and Hutchins 2011). Finally, a study by Sjöström *et al.* (2013) investigated German audiences' views of social science in the context of the 'violent videogames' debate. These audiences recognised the high visibility of social scientists within this coverage and felt that they had made important and legitimate contributions to the debate.

Reflexive sciences?

So why do journalists, editors and audiences seem to have relationships with social science and humanities disciplines so different from those they have with the natural sciences? Paying attention to the subject matter of these disciplines offers important clues towards understanding how and why they are communicated and perceived. Because they investigate the realm of the human – people, their minds, societies, money, politics, histories and so on – the subjects, investigators, communicators and audiences of social science tend to merge into one another. Unlike most natural sciences, where the specialist training, knowledge and equipment of scientists grants them largely uncontested expertise, social scientists' expertise is often about matters of everyday experience and common-sense knowledge. This impacts on how highly that expertise is regarded. According to Evans (1995), US journalists made strong distinctions between natural science and social science, between natural science and lay opinion, but not between social science and lay opinion. As psychologists McCall and Stocking (1982: 988) put it:

> Everyone, including journalists and editors, fancies himself or herself something of a psychologist, but not an astrophysicist. Results from psychology, but not physics, must therefore square with experience to be credible.

British news media audiences were also found to apply these standards in framing their understandings of social science research findings (Fenton *et al.* 1998). This study discusses the overlaps that result between the professional roles of social scientists and journalists and argues that this resulted in further under-reporting of social science, as journalists often felt it was little different from their own work. More recently, Cooper and Ebeling (2007) studied the working practices of financial and science journalists and similarly argued they have a great deal in common with the analytical processes of sociology.

Similar issues of the legitimacy of social science expertise have also been seen in studies of social scientists' role as expert witnesses. Particularly in the US, legal definitions of *science* tend to be heavily traditional, positivist ones, leading at times to non-natural-science expertise being judged as questionable or even inadmissible (Lynch and Cole 2005; Lynch 2009). However, these overlaps between social science, journalism and everyday knowledge, which Fenton *et al.* (1998: 102) refer to as 'epistemological consonance', and which, following historian of psychology Graham Richards (2010), I have described as 'reflexive science' (Cassidy 2003: 236) can also have positive implications for PCSS.[2] The same media news values that result in natural science struggling to gain media coverage can work in favour of social science. Examples include the news values of relevance (to daily life), consonance (with existing beliefs), topicality, controversy and of course human/personal interest (Weiss and Singer 1988: 144–149; Fenton *et al.* 1998: 103–113; Gregory and Miller 1998: 110–114).

These reflexive properties can also help explain media-industry attitudes that journalists do not require specialist training to report social science, ironically also increasing the chances of social science research being reported in the first place. As described above, generalists tend not to have training in either natural or social science, and neither do editors, increasing the chances that social science will make it through the editorial selection process. This could clearly be seen in mass media coverage of popular evolutionary psychology, which tended to be covered by generalist news and *soft* journalists, and gained media coverage by keying into topics of general appeal at the time, such as gender, sexuality, centre-left politics and the role of the biosciences in society (Cassidy 2005, 2007). However, this also meant that evolutionary psychology claims were contested by a range of actors including academics, but also *lay* commentators and journalists, while all sides drew on personal experience and common sense knowledge to support their arguments.

This kind of challenge can come not only via the media but also directly from publics and research participants themselves, leading at times to uncomfortable challenges for social scientists (Breuer 2011). This highlights the double-edged nature of reflexive science, and at times social scientists can take advantage of this to engage in strategic boundary work, emphasising the similarities or the differences between their research and common sense according to particular rhetorical purposes (Derksen 1997; Shapin 2007). Park (2004) compared the contemporary discourses of popular psychiatrists with those of psychoanalysts, arguing that the two groups strategically position themselves against each other as medical or scientific specialists versus broader intellectual authorities. He relates these opposing, yet complementary, strategies to the differing forms of *public intellectual* visible in contemporary popular culture.

Disciplinary status and public expertise

As we have seen, social research is often regarded as less authoritative than natural science research, and social scientists often struggle with the epistemological status of their disciplines, particularly when attempting to communicate about new research findings. However, we can see that social scientists also take on a range of expert roles in society not available to many natural scientists. Social science and humanities academics are often called upon to provide commentary and analysis on the events and news of the day, and they can find easier access to popular audiences. Albaek *et al.* (2003), as well as Bentley and Kyvik (2010), found that researchers in these fields tend to be more active in popular communications than their colleagues in natural science, medical and technical subjects. A good example of such a role is described by the largely US-based literature on popular and self-help psychology. Considering the obvious popularity of these texts evidenced by their vast sales not only in the United States but globally, this work gives an insight into an arena where social science is highly influential in ordinary people's lives. Indeed there is currently a lively debate within psychology about the efficacy of self-help and its adoption as a serious therapeutic technique (Cuijpers *et al.* 2010). Others have taken a more critical approach to self-help, analysing the rhetorical messages embedded in pop-psychology discourses, with a particular focus on the normative regulation of gender and/or sexual relationships in these texts (e.g. Koeing *et al.* 2010). Other work has addressed the social and political contexts of self-help, showing how these ideas relate to social movements such as feminism or the New Age (Askehave 2004) and the broader values of modern liberal democracies (Philip 2009). However, audience studies have shown that, as with many media forms, readers of self-help literature do not simply absorb these messages but instead use them as a starting point for discussing, negotiating and challenging the claims experts make about their life experiences (George 2012).

The literature on the 'public intellectual' – broadly understood as a person of learning, not necessarily an academic, who uses their knowledge to engage with society via the public domain (Small 2002) – has greatly expanded in recent years. This conversation has largely been an academic one located in humanities and social science disciplines and still has had relatively little connection with PCST or the literature on (natural) scientists as public experts (e.g. Peters in this volume). Examples of public intellectuals could include the late Edward Said, Noam Chomsky and, through his participation in debates about religion and society, Richard Dawkins. Most public intellectuals tend to be social science or humanities scholars; this is unsurprising, given their media role as generalised experts and commentators. Debates in this area have moved on from discussions of charismatic individuals to address the role of various disciplines in society, spurred in particular by the sociologist Michael Burawoy and his influential calls for 'public sociology' (2005) and similar calls for public geographies (Ward 2006). Prior to Burawoy, similar conversations have occurred about 'public anthropology' (Borofsky 1999) and preceding the public understanding of science (PUS) movements of the 1980s, 'public history' (Kelley 1978). Burawoy's call involved a vision of sociology as politically engaged, and much of the subsequent debate has turned on whether and how social scientists should contribute to debates about social justice and inequality (Gattone 2012; Jeffries 2009). However, it is noticeable that the massive growth in academic citations in this area has not been reflected in an equivalent increase in public visibility for the social sciences. This may be in part because the literature on public sociology tends to use academic rather than everyday language, highlighting a further issue with PCSS. Social psychologist Michael Billig's (2013) challenging critique of writing traditions in the social sciences argues that these disciplines actively encourage wordiness, neologism and obscurantism, impeding the clear communication of ideas. Stephen Turner's work on the history of American sociology (2012) supports this idea by demonstrating how sociologists actively turned away from public debates during the second half of the twentieth century in order to boost their intellectual status within the academy. Coming at the issue from a very different angle, a recent bibliometric analysis of journal articles (Okulicz-Kozaryn 2013) found much higher proportions of adjectives and adverb usage in social sciences disciplines than in natural sciences.

Other models for public social science include the turn towards more applied modes of research, actively oriented to the needs of social movements, policymakers and industry (Kropp and Blok 2011; Perry 2012). Such approaches also have their drawbacks, as seen in ongoing controversies over the mobilisation of field anthropologists by the US Army in Iraq and Afghanistan (Forte 2011). An alternative version of what it means for a discipline to be public has been offered by researchers drawing on traditions of participatory action research and public engagement. Rather than further advocating social scientists' role as authoritative experts, instead the idea is to undertake research *in public*. This often involves open processes of data collection and analysis, alongside direct collaborations with research participants, local communities or media organisations, and it is often associated with public history, anthropology and geographies research.

The growth in cross-, multi- or inter-disciplinary research across academia has led to social scientists working with natural scientists more than before and this has prompted a debate about how researchers can communicate across the natural/social sciences divide, highlighting a similar set of issues faced when communicating with journalists or publics, particularly around the status and methods of social science (Barry and Born 2013). Similarly, some have suggested taking a more open approach to research practice, making it possible for partners from different disciplines to understand each other's work at an earlier stage of the research process and to generate shared research goals, aims and questions (e.g. Phillips *et al.* 2012).

Public social sciences in the changing academy

To summarise, the literature on PCSS continues to be sparse, scattered across many disciplinary areas and, despite the knowledge gaps outlined here, has rarely been investigated by PCST scholars. With relatively little work done, it is difficult to reach firm conclusions about social science communication, and so any assertions made here are of necessity provisional and subject to further investigation. Despite this, one thing seems clear: social science and humanities research appears to be both *everywhere and nowhere* in public communication. Social sciences have a lower status than natural sciences, are less likely to prompt original news coverage via their findings, do not merit media or journalistic specialisation and at times are seen as little different from journalism itself. At the same time, social science topics constantly generate new coverage, are seen as relevant to audiences and easy to understand, and appear throughout the media rather than being confined to an area of special interest. As such, social scientists play important roles as commentators and advisers in media and public life on a wide range of social, political and personal issues. Beyond these rather broad-brush assertions, it is still difficult to draw any more nuanced conclusions about PCSS. The criteria used to define science in PCST studies tend to be so variable that it is difficult to draw meaningful comparisons across the literature (Schäfer 2012). Studies including social science and/or humanities disciplines either focus on specific cases or have been conducted across different countries and time periods, using a wide range of methodologies. Differences in research findings about PCSS may be due to cross-cultural differences, changes over time or methodological artefacts. However, without further studies taking a more consistent and preferably comparative approach that looks across a broad spread of disciplines, we cannot reach a better understanding of how disciplinary topic affects how research is communicated in public. The widespread reporting of social science by non-specialists highlights the fact that very little work has been done on how generalist and non-science specialist journalists understand and report academic research. The reflexive nature of social science, and the claim that this is what makes PCSS so different from PCST, also requires further investigation and analysis. This may also cast light on what makes communicating natural science so difficult at times, particularly in those topics very far from human experience – as well as in highly controversial and contested scientific issues.

While social scientists have started to discuss the roles that their disciplines can and should play in wider society, via debates over the public intellectual and public social science, these have been driven by bigger changes in the relationship between society and academia in general. Drives towards research assessment via metrics and contractions in research funding have led to increased pressure for all academics to justify the work they do, often in the language of 'impact' (Buchanan 2013; LSE Public Policy Group 2011). At the same time, movements advocating open science and open access publishing may be fundamentally changing research communications in the social science and humanities (Vincent and Wickham 2013). Finally, further movement towards online modes of communication and the uptake of social media have facilitated and accentuated the above trends (e.g. Kitchin 2013).

It is noticeable that, in Britain, efforts to advocate and promote the social sciences (e.g. Brewer 2013; Campaign for Social Sciences 2011; LSE Public Policy Group 2011) became much more prominent following the announcement of contractions in higher education funding affecting those disciplines (Richardson 2010). While there has been some discussion of the 'impact of impact' (Brewer 2011) and advocacy for more collaborative models of public social science (e.g. Flyvjberg *et al.* 2012), much of this debate continues in the framework of what PCST scholars would describe as a deficit or diffusion communications model.

This raises the question of why few thinkers in the social sciences seem to have turned to PCST or STS scholars to learn more about the public role of their research disciplines. In part, this is clearly because PCST scholars have not been that interested in PCSS, and it raises further questions about our own abilities as public communicators. In the previous version of this chapter (Cassidy 2008) I presented a challenge to researchers and practitioners in PCST: how do we communicate about our work on communication, and publicly engage about public engagement? The extra levels of reflexivity introduced in PCST work (communicating about research which is about communicating about research) are hardly compatible with media news values. Some STS scholars have experimented with communicating in the mode of the public intellectual, with variable consequences (e.g. Fuller 2009; Latour and Sánchez-Criado 2007). A few PCST researchers have reflected on these challenges, particularly in terms of the interactions of PCST research with scientific and policy debates (Chilvers 2012; Kahan 2013) and on engaging, interacting and learning from publics as we do it (e.g. Horst 2011; Michael 2011). An urgent challenge for PCST is to start looking for the answers to these questions, both through further research and by communicating clearly and openly about that research. If we aim to advise other researchers, policymakers, journalists and publics about these issues, then surely we must practice what we preach.

Key questions

- How might public communication vary across different academic disciplines?
- How is the content, practice and communication of research affected by interactions with experiential and common-sense knowledge?
- What are the significant differences between the public roles of natural scientists and scholars in humanities and social sciences?
- How can scholars of PCST improve their own communications practice and engage more productively with other disciplines and civil society?

Useful online resources for PCSS

LSE Impact of Social Sciences project: http://blogs.lse.ac.uk/impactofsocialsciences/
Ethnography Matters: http://ethnographymatters.net/
The Sociological Imagination: http://sociologicalimagination.org/
The Conversation (site syndicating popular content by academics regardless of discipline): http://theconversation.com/uk; http://theconversation.com/au
History and Policy: http://www.historyandpolicy.org/
Campaign for Social Science: http://campaignforsocialscience.org.uk/

Notes

1 This chapter updates the review of that is known about PCSS, highlighting areas of change since publication of the earlier version of this chapter (Cassidy 2008).
2 Unlike many philosophers of social science, I do not think that these properties signal a fundamental division between the natural and human sciences. It is clear that some social science disciplines are more profoundly shaped by reflexive overlaps than others, while on the other hand many natural science topics involve important experiential, political and ethical contributions and contestations (e.g. Kent 2003; Moore and Stilgoe 2009; Spence *et al.* 2011).

References

Albaek, E., Christiansen, P. M. and Togeby, L. (2003) 'Experts in the mass media: researchers as sources in Danish daily newspapers, 1961–2001', *Journalism & Mass Communication Quarterly*, 80, 4: 937–948.
Askehave, I. (2004) 'If language is a game – these are the rules: a search into the rhetoric of the spiritual self-help book, *If Life is a Game – These are the Rules*', *Discourse & Society*, 15, 1: 5–31.
Barry, A. and Born, G. (2013) *Interdisciplinarity: Reconfigurations of the Social and Natural Sciences*, London and New York: Routledge.
Bauer, M., Durant, J., Ragnarsdottir, A. and Rudolfsdottir, A. (1995) *Science and Technology in the British Press 1946–1990: A Systematic Content Analysis of the Press*, London: Science Museum.

Bentley, P. and Kyvik, S. (2010) 'Academic staff and public communication: a survey of popular science publishing across 13 countries', *Public Understanding of Science*, 20, 1: 48–63.
Billig, M. (2013) *Learn to Write Badly: How to Succeed in the Social Sciences*, Cambridge: Cambridge University Press.
Böhme-Dürr, K. (2009) 'Social and natural sciences in German periodicals', *Communications*, 17, 2: 140–277.
Borofsky, R. (1999) 'Public anthropology', *Anthropology News*, 40, 1: 6–7.
Brewer, J. D. (2011) 'The impact of impact', *Research Evaluation*, 20, 3: 255–256.
Brewer, J. D. (2013) *The Public Value of the Social Sciences: An Interpretive Essay*, London: Bloomsbury Academic.
Breuer, F. (2011) 'The "other" speaks up. When social science (re)presentations provoke reactance from the field', *Historical Social Research*, 36, 4: 300–322.
Buchanan, A. (2013) 'Impact and knowledge mobilisation: what I have learnt as Chair of the Economic and Social Research Council Evaluation Committee', *Contemporary Social Science*, 8, 3: 176–190. doi:10.1080/21582041.2013.767469.
Burawoy, M. (2005) 'For public sociology', *American Sociological Review*, 70, 1: 4–28.
Camic, C., Gross, N. and Lamont, M. (2011) *Social Knowledge in the Making*, Chicago: University of Chicago Press.
Campaign for Social Science (2011) *Campaign for Social Science Annual Report 2011*, London: Academy of Social Sciences; online at http://campaignforsocialscience.org.uk/wp-content/uploads/2012/12/Annual-Report-2011.pdf
Cassidy, A. (2003) *Of Academics, Publisher and Journalists: popular evolutionary psychology in the UK*, PhD thesis, University of Edinburgh; online at https://www.academia.edu/383966/of_Academics_Publishers_and_Journalists_Popular_Evolutionary_Psychology_in_the_UK
Cassidy, A. (2005) 'Popular evolutionary psychology in the UK: an unusual case of science in the media?', *Public Understanding of Science*, 14, 2: 115–141.
Cassidy, A. (2007) 'The (sexual) politics of evolution: popular controversy in the late twentieth century UK', *History of Psychology*, 10, 2: 199–227.
Cassidy, A. (2008) 'Communicating the social sciences', in M. Bucchi and B. Trench (eds) *Handbook of Public Communication of Science and Technology*, London and New York: Routledge, 225–236.
Chilvers, J. (2012) 'Reflexive engagement? actors, learning, and reflexivity in public dialogue on science and technology', *Science Communication*, 35, 3: 283–310.
Cooper, G., and Ebeling, M. (2007) 'Epistemology, structure and urgency: the sociology of financial and scientific journalists', *Sociological Research Online*, 12, 3, 8; online at www.socresonline.org.uk/12/3/8.html; doi:10.5153/sro.1558.
Cuijpers, P., Donker, T., van Straten, A., Li, J. and Andersson, G. (2010) 'Is guided self-help as effective as face-to-face psychotherapy for depression and anxiety disorders? A systematic review and meta-analysis of comparative outcome studies', *Psychological Medicine*, 40, 12: 1943–1957.
Danell, R., Larsson, A. and Wisselgren, P. (2013) *Social Science in Context: Historical, Sociological, and Global Perspectives*, Lund: Nordic Academic Press.
Derkson, M. (1997) 'Are we not experimenting then? the rhetorical demarcation of psychology and common sense', *Theory and Psychology*, 7, 4: 435–456.
Dunwoody, S. (1986) 'The science writing inner club: a communication link between science and the lay public', in S. L. Friedman, S. Dunwoody and C. L. Rogers (eds) *Scientists and Journalists: Reporting Science as News*, New York: Macmillan, 155–169.
ESRC (2013) *Impact Toolkit*, Swindon: Economic and Social Research Council; online at www.esrc.ac.uk/funding-and-guidance/impact-toolkit/index.aspx
Evans, W. (1995) 'The mundane and the arcane: prestige media coverage of social and natural science', *Journalism and Mass Communication Quarterly*, 72, 1: 168–177.
Fenton, N., Bryman, A., Deacon, D. and Birmingham, P. (1997) 'Sod off and find us a boffin: journalists and the social science research process', *Sociological Review*, 45, 1: 1–23.
Fenton, N., Bryman, A., Deacon, D. and Birmingham, P. (1998) *Mediating Social Science*, London: Sage.
Flyvbjerg, B., Landman, T. and Schram, S. (eds) (2012) *Real Social Science: Applied Phronesis*, Cambridge: Cambridge University Press.
Forte, M. C. (2011) 'The human terrain system and anthropology: a review of ongoing public debates' *American Anthropologist*, 113, 1: 149–153.
Fuller, S. (2009) 'Science studies goes public: a report on an ongoing performance', *Spontaneous Generations: A Journal for the History and Philosophy of Science*, 2, 1: 11–21; online at http://spontaneousgenerations.library.utoronto.ca/index.php/SpontaneousGenerations/article/view/5069

Gattone, C. F. (2012) 'The social scientist as public intellectual in an age of mass media', *International Journal of Politics, Culture, and Society*, 25, 4: 175–186.

George, K. C. (2012) 'Self-help as women's popular culture in suburban New Jersey: an ethnographic perspective', *Participations: Journal of Audience and Reception Studies*, 9, 2: 23–44; online at http://participations.org/Volume%209/Issue%202/3%20George.pdf

Grauerholtz, L. and Baker-Sperry, L. (2007) 'Feminist research in the public domain: risks and recommendations', *Gender and Society*, 21, 2: 272–294.

Gregory, J. and Miller, S. (1998) *Science in Public: Communication, Culture and Credibility*, New York: Plenum Trade.

Hansen, A. and Dickinson, R. (1992) Science coverage in the British mass media: media output and source input, *Communications*, 17, 3: 365–377.

Horst, M. (2011) 'Taking our own medicine: on an experiment in science communication', *Science and Engineering Ethics*, 17, 4: 801–815.

Jeffries, V. (ed.) (2009) *Handbook of Public Sociology*, Rowman & Littlefield Publishers.

Kahan, D. (2013) 'Does communicating research on public polarization polarize the public?' *Cultural Cognition Project Blog*, 2 July; online at www.culturalcognition.net/blog/2013/7/2/does-communicating-research-on-public-polarization-polarize.html

Kelley, R. (1978) 'Public history: its origins, nature, and prospects', *The Public Historian*, 1, 1: 16–28.

Kendall-Taylor, N. (2012) 'Conflicting models of mind: mapping the gaps between expert and public understandings of child mental health', *Science Communication*, 34, 6: 695–726.

Kent, J. (2003) 'Lay experts and the politics of breast implants', *Public Understanding of Science*, 12, 4: 403–421.

Kitchin, R., Linehan, D., O'Callaghan, C. and Lawton, P. (2013) 'Public geographies through social media', *Dialogues in Human Geography*, 3, 1: 56–72.

Koeing, J., Zimmerman, T. S., Haddock, S. A. and Banning, J. H. (2010) 'Portrayals of single women in the self-help literature', *Journal of Feminist Family Therapy*, 22, 4: 253–274.

Kropp, K. and Blok, A. (2011) 'Mode-2 social science knowledge production? The case of Danish sociology between institutional crisis and new welfare stabilizations', *Science and Public Policy*, 38, 3: 213–224.

Latour, B. and Sánchez-Criado, T. (2007) 'Interview: making the "res public"', *Ephemera: Theory & Politics in Organization*, 7, 2: 364–371; online at www.ephemerajournal.org/contribution/making-res-public

Lester, L. and Hutchins, B. (2011) 'Soft journalism, politics and environmental risk: an Australian story', *Journalism*, 13, 5: 654–667.

LSE Public Policy Group (2011) *Maximizing the Impacts of Your Research: A Handbook For Social Scientists (Consultation Draft 3)*, London: London School of Economics and Political Science; online at http://www.lse.ac.uk/government/research/resgroups/LSEPublicPolicy/docs/LSE_Impact_Handbook_April_2011.pdf

Lynch, M. (2009) 'Going public: a cautionary tale', *Spontaneous Generations: A Journal for the History and Philosophy of Science*, 3, 1; online at http://spontaneousgenerations.library.utoronto.ca/index.php/SpontaneousGenerations/article/view/6085; accessed 4 September 2013.

Lynch, M. and Cole, S. (2005) 'Science and technology studies on trial: dilemmas of expertise', *Social Studies of Science*, 35, 2: 269–311.

McCall, R. S. and Stocking, S. H. (1982) 'Between scientists and public: communicating psychological research in the mass media', *American Psychologist*, 37, 9: 985–995.

Michael, M. (2011) '"What are we busy doing?": engaging the idiot', *Science, Technology & Human Values*, 37, 5: 528–554.

Moore, A. and Stilgoe, J. (2009) 'Experts and anecdotes: the role of "anecdotal evidence" in public scientific controversies', *Science, Technology & Human Values*, 34, 5: 654–677.

Okulicz-Kozaryn, A. (2013) 'Cluttered writing: adjectives and adverbs in academia', *Scientometrics*, 96, 3: 679–681.

Park, D. W. (2004) 'The couch and the clinic: the cultural authority of popular psychiatry and psychoanalysis', *Cultural Studies*, 18, 1: 109–133.

Peters, H. P. (2013) 'Gap between science and media revisited: scientists as public communicators', *Proceedings of the National Academy of Sciences*, 110, Supplement 3: 14102–14109; online at http://www.pnas.org/content/110/Supplement_3/14102.full

Perry, R. K. (2012) 'The politics of applied black studies: an historical synthesis for understanding social science impact', *Journal of Applied Social Science*, 6, 1: 53–64.

Philip, B. (2009) 'Analysing the politics of self-help books on depression', *Journal of Sociology*, 45, 2: 151–168.

Phillips, L., Kristiansen, M., Vehviläinen, M. and Gunnarsson, E. (2012) *Knowledge and Power in Collaborative Research: A Reflexive Approach*, London and New York: Routledge.

Reinemann, C., Stanyer, J., Scherr, S. and Legnante, G. (2012) 'Hard and soft news: a review of concepts, operationalizations and key findings', *Journalism*, 13, 2: 221–239.

Richards, G. (2010) *Putting Psychology in its Place, Fourth Edition: Critical Historical Perspectives*, London and New York: Routledge.

Richardson, H. (2010) 'Humanities to lose English universities teaching grant', *BBC News Online*, 26 October; online at www.bbc.co.uk/news/education-11627843

Sala, R. (2012) 'One, two, or three cultures? Humanities versus the natural and social sciences in modern Germany', *Journal of the Knowledge Economy*, 4, 1: 83–97.

Schäfer, M. S. (2012) 'Taking stock: a meta-analysis of studies on the media's coverage of science', *Public Understanding of Science*, 21, 6: 650–663.

Schmierbach, M. (2005) 'Method matters: the influence of methodology on journalists' assessments of social science research science', *Communication*, 26, 3: 269–287.

Seale, C. (2010) 'How the mass media report social statistics: a case study concerning research on end-of-life decisions', *Social Science & Medicine*, 71, 5: 861–868.

Shapin, S. (2007) 'Expertise, common sense, and the Atkins diet', in J. Porter and P. W. B. Phillips (eds) *Public Science in Liberal Democracy*, Toronto: University of Toronto Press, 174–193.

Sjöström, A., Sowka, A., Gollwitzer, M., Klimmt, C. and Rothmund, T. (2013) 'Exploring audience judgements of social science in media discourse: the case of the violent video games debate', *Journal of Media Psychology: Theories, Methods, and Applications*, 25, 1: 27–38.

Small, E. (ed.) (2002) *The Public Intellectual*, London: Blackwell.

Stockelova, T. (2012) 'Social technology transfer? Movement of social science knowledge beyond the academy', *Theory & Psychology*, 22, 2: 148–161.

Spence, A., Poortinga, W., Butler, C. and Pidgeon, N. F. (2011) 'Perceptions of climate change and willingness to save energy related to flood experience', *Nature Climate Change*, 1, 1: 46–49.

Šuljok, A. and Vuković, M. B. (2013) 'How the Croatian daily press presents science news', *Science & Technology Studies*, 26, 1: 92–112.

Turner, S. P. (2012) 'De-intellectualizing American sociology: a history, of sorts', *Journal of Sociology*, 48, 4: 346–363.

Vincent, N. and Wickham, N. (eds) (2013) *Debating Open Access*, London: British Academy; online at www.britac.ac.uk/openaccess/debatingopenaccess.cfm

Ward, K. (2006) 'Geography and public policy: towards public geographies', *Progress in Human Geography*, 30, 4: 495–503.

Weiss, C. H. and Singer, E. (1988) *Reporting of Social Science in the National Media*, New York: Russell Sage Foundation.

Wien, C. (2013) 'Commentators on daily news or communicators of scholarly achievements? The role of researchers in Danish news media', *Journalism*, online first (25 June); doi:10.1177/1464884913490272.

15
Health campaign research
Enduring challenges and new developments

Robert A. Logan

Introduction

This chapter provides an overview of the conceptual landscape upon which health campaign research is based and discusses some recent developments within the field. The chapter notes some of the mixed outcomes from health campaign research and addresses some current challenges.[1] Some of the common outcome variables used in health campaign scholarship are introduced.

Traditionally, a health intervention campaign attempts to therapeutically impact an audience's (and an individual's) health awareness, knowledge, attitudes, behaviours and decisions (Fishbein and Ajzen 1975; Flora 2001). Until recently health campaign research was dominated by initiatives to foster health information and persuasion consistent with public health priorities. However, the field now includes communication activities to enhance patient care and consumer use of health-care delivery services, as well as to support self-determined pathways to better health.

Contemporary health campaign research covers an increasingly diverse field that includes health communication interactive activities (as opposed to intervention) that use mass media tools, including Internet multi-platform services (personal computers, smartphones and tablets). Recipients now include individuals, health-care consumers, patients, caregivers, communities, organisations, the medically underserved and other targeted audiences. Health campaign settings now include public health, clinical (patient) care, health-care delivery services (such as hospitals, clinics), peer-to-peer communication initiatives (such as patient-to-patient health information sharing *via* social media) and medical homes. The goals of a campaign may reflect predetermined public health or clinical priorities as well as more open-ended, community-based and self-determined goals.

Recent overviews and systematic reviews describe the primary findings from health communication campaigns (e.g. Rice and Katz 2001; Hornik 2002a, 2002b; Atkin 2001; Grilli *et al.* 2002; Snyder *et al.* 2004; Dutta-Bergman 2005; Murero and Rice 2006; Noar 2006; Gibbons *et al.* 2009; Brinn *et al.* 2010; Agency for Healthcare Research and Quality 2011; Car *et al.* 2011; Cugelman *et al.* 2011; Lee *et al.* 2012; Liu *et al.* 2012; Stellefson *et al.* 2013; Zhang and Terry 2013). The expansion of health campaign research recently occurred because of the development of the field of health literacy and significant changes in health information technology, such as the widespread diffusion of mobile phones and tablets, mobile phone health monitoring, the widespread availability of social media and other developments. The increased level of activity changes the opportunities for research into an array of diverse settings, such as doctors' offices, hospitals, clinics, medical homes, support groups and communities.

However, as the field moves into personalised decision support, clinical care and healthcare utilisation, the growth of health campaign research brings fresh challenges to triangulate health communication research traditions, health literacy and consumer health informatics research.

Before discussing these more recent issues, two of the field's enduring challenges must be mentioned. First, health campaign research is influenced by a sense of urgency to address public health and health policy issues. The urgency stems from current challenges in public health and health policy on all continents. In many nations, the optimal functioning of health-care delivery and medical cost controls increasingly rely on disease prevention and wellness care as opposed to an after-the-fact clinical response to acute and chronic illness (US Department of Health and Human Services 2000). The Institute of Medicine of the National Academies (2004, 2006) explains that the ultimate success of a health prevention and wellness model and a transition from the prior emphasis on acute care delivery depend on enhancing the knowledge and capabilities of health-care consumers, patients and citizens. Yet, evidence from health literacy research suggests only about 12 per cent of the population have a fundamental grasp of biomedical language and health's biological/genetic, social and individual determinants (Institute of Medicine of the National Academies 2004; Kutner *et al.* 2006; White 2008). As a result, there is a pressing need to boost health literacy as well as health educational outreach and research initiatives (such as targeted campaigns) in order to address existing gaps and enhance clinical outcomes and the cost-effectiveness of health-care delivery.

Second, health campaign researchers retain a sense of mission to demonstrate the effectiveness of non-commercial mass media to enhance patient and public health within a contextually layered mass media and social environment. A prevailing challenge in health campaign research is the degree to which non-commercial mass communication can generate awareness, inform, persuade, influence decisions and change health behaviours in a context where commercial health marketing, faith and socioculturally derived beliefs, as well as questionable health behaviour patterns (e.g. smoking, obesity, drug and alcohol abuse) are ubiquitous.

Non-commercial health promotion and information efforts to target audiences and the public are manifestations of social marketing. Mass-media-based commercial marketing of health products (such as over-the-counter and prescription pharmaceuticals, medical equipment, beauty products, food, beverages, vitamins and food supplements) has the potential to overwhelm social marketing initiatives. While health campaign initiatives do not assess the comparative effectiveness of social versus commercial marketing campaigns (and the success of commercial health marketing is beyond the scope of this chapter), health campaign researchers face a continuing challenge to demonstrate that social marketing, often using many of the same mass media modalities and techniques as commercial marketers, enhances public and patient health (Randolph and Viswanath 2004). Health campaign social marketing efforts also can be overshadowed by sociocultural, faith-based beliefs and resulting health practices that may or may not be consistent with the goals of patients or public health campaigns.

A related, prevailing challenge in health campaign research is to distinguish the degree to which a campaign's influence compares with other potential influences such as commercial health marketing, faith-based and socioculturally derived beliefs, health news coverage, social *mores*, friends, family and health-care providers. For example, to what extent is the decline in smoking and tobacco use in some countries influenced by public health campaigns, doctors' efforts, changes in social *mores*, the impact of health policy decisions (such as changes in tobacco advertising), news framing and coverage, or other factors? Conversely, where smoking and tobacco use remain widespread, to what extent is this a result of all of these factors in reverse? Since potential health information sources provide a reservoir of potentially confounding variables, this places a special burden on campaign researchers to control for external influences, to manage a campaign and to use statistical controls to parse main findings, and to acknowledge research limitations.

This chapter is divided into subsections that discuss: the intent of health campaigns and research topic areas; health campaign research's underlying conceptual frameworks; common outcomes – what health campaign researchers assess; research results and the importance of the perspective provided by basic research; the impacts of health literacy and health information technology on health campaign research and future challenges.

The intent of health campaigns and research topic areas

In a meta-analysis of 48 social-science-based health communication campaigns, Snyder (2001) found most campaigns either attempt to persuade people to stop an existing health behaviour or promote a new therapeutic behaviour. Some newer, additional purposes of health campaigns include: delivering information to persons as they make clinical or health-care utilisation decisions, providing support to affinity groups with a common disease or condition, supplying health monitoring and information assistance to make informed decisions and enhancing a person's quality of life (Nutbeam 2008; Kreps and Neuhauser 2010; Smith 2009, 2011; Smith and Moore 2011).

Snyder (2001) also observes that health campaign research usually addresses a disease/condition or a public health concern. Among the diseases and conditions frequently associated with chronic disease prevention efforts and health communication campaigns are: AIDS, other sexually transmitted diseases, heart disease, stroke, breast cancer and other cancers, hypertension, diabetes, oral health and sudden infant death syndrome. Among health campaigns that focus more broadly on public health are: promoting smoking cessation, reducing alcohol consumption, safe prescription use, eating more fruit and vegetables and safe sex. Some campaigns are targeted for demographic segments where the probability of risky health behaviours are higher and access to medical services is lower than the general population (Piotrow and Kincaid 2001; Gustafson *et al.* 2002; Dutta-Bergman 2005, 2006; Viswanath *et al.* 2006). Health communication campaigns that are designed to change counter-productive behaviours commonly occur in areas such as smoking, illegal drug use, prescription drug abuse, alcohol abuse, infant sleeping positions, sex with risky partners and alcohol sales to minors (Snyder 2001). Among new therapeutic behaviours that health campaigns may promote are: condom use, healthy diet, exercise, seat belt and bike helmet use, mammography and other preventive screenings, child vaccinations, dental visits, regular medical check-ups, hypertension control and pap smears.

Overall, there are few major diseases and conditions, as well as public health concerns, where health campaign research has *not* occurred. While most of the health communication campaign research that Snyder (2001) reviewed occurred in North America and western Europe, health campaign research now is a global activity (Piotrow *et al.* 1997; Nutbeam 2000; Piotrow and Kincaid 2001; Hornik 2002b; Dutta-Bergman 2005; Haider *et al.* 2009; Sorensen *et al.* 2012; Kreps 2012).

Health campaign research's underlying conceptual frameworks

Analysis of the conceptual foundation of health campaign research probably can be divided into three phases. In the first phase a health campaign is conceived as an effort to inform or persuade an audience in a didactic, predetermined, goal-oriented direction. The second phase expands on this concept and the third suggests a more interactive and holistic conceptual framework (Neuhauser et al. 2013a). Reviewing the field's initial conceptual framework, McGuire (2001) outlines these 13 steps to a successful health campaign: 1. tuning in (exposure to the message); 2. attending to the communication; 3. liking it, maintaining interest in it; 4. comprehending its contents; 5. generating related cognitions; 6. acquiring relevant skills; 7. agreeing with the communication; 8. storing this new position in memory; 9. retrieval of this new position from memory when relevant; 10. decision to act on the position; 11. acting on it; 12. post-action cognitive behaviour on position; 13. proselytising others to behave similarly. Overall, these steps illustrate a top-down, didactic approach to overcome users' skill and knowledge deficits, underpinned by a one-way communicator-to-recipient conceptual foundation.

The development of phase two began as McGuire (ibid.) and other health communication researchers noted inherent barriers that undermined the uniformity between an audience's responses to health messages and the original source's intent (Pettegrew and Logan 1987; Logan 2008). Health communication's barriers or constraints were seen as occurring in five different areas: 1. the source of health messages; 2. the content of health messages; 3. the media delivery channel of health messages; 4. the receiver's (recipient's) post-exposure to health messages; 5. the destination of health messages. For example, McGuire (2001) indicates the *source-based* barriers to a desired audience response as including: the number of potentially different sources to which an individual is exposed, the unanimity or consistency of health-care messages, the perceived demographic similarity between the source and the receiver, and the source's perceived appeal and credibility. *Message-based* barriers to a desired audience response include: the message's perceived appeal, the inclusion or omission of pertinent facts or familiar story frames, the arrangement and organisation of text and graphics, a text's readability for its intended audience, the inclusion of appropriate images or diagrams within a text-based story, and the cultural appropriateness of the content (ibid.). Some of the *channel-based* barriers to a desired audience response regarding health messages are: the perceived appropriate use of print, radio, video or interactive channels to convey health content (ibid.). The *receiver-based* barriers include: differences in perception about (and interest in) health as a by-product of gender, age, education, ethnicity, socio-economic status, literacy, personality, lifestyle, values and health literacy (the ability to understand health information regardless of general education); a recipient's access to mass media (e.g. the unavailability or high cost of broadband) and the health-care delivery system (e.g. a recipient's health insurance status and ability to afford care as well as actual access to clinical care facilities) (McGuire 2001; Institute of Medicine of the National Academies 2004). Some of the common *destination-based* barriers are: immediacy or delay in the ability of a targeted audience to receive information, timing of a message with audience needs, whether health messages were prevention- or cessation-oriented and whether messages urged health behaviours characterised by immediate or delayed rewards (McGuire, 2001).

While the overall conceptual foundation within the second phase retained a one-way relationship from source-to-recipient, the identification of communication barriers along a health transmission's pathway added insights into why campaigns were successful or unsuccessful in meeting their initial goals. The research-derived advice about barriers and remedies also helped improve the management and execution of health campaigns.

The expansion of campaign research's conceptual framework in phase two evolved significantly when campaigns also were seen as challenged by broader socioculturally oriented barriers and constraints. The identification of sociocultural dimensions provided new insights about the efficacy of campaigns as well as an array of fresh campaign approaches and strategies. Three of the sociocultural dimensions identified in phase two were:

- social influences, e.g. the influence of peer pressure and commercial advertising on health behaviours (National Cancer Institute 1991; Evans and Raines 1990; Flynn 1992);
- cognitive behavioural, e.g. the degree a person's problem-solving, decision-making and self-control skills influence health behaviours (Kendall and Holon 1979; National Cancer Institute 1991; Flynn 1992);
- life skills, e.g. the degree to which a commitment to a specific health behaviour requires broader skills and individualised training to foster a healthier lifestyle (Botvin et al. 1980; Botvin and Eng 1982; National Cancer Institute 1991; Wollesen and Peifer 2006).

Health campaigns that used a conceptual framework partially based on social influences emphasised *social inoculation*. To overcome perceived barriers to adopting therapeutic health behaviours, campaign recipients were trained about the subtle influences of advertising and peer pressures. To offset peer pressures to smoke, smoking resistance behaviours were modeled, role-played and even turned into theatre productions to encourage rehearsal and reinforcement of smoking cessation behaviours (Evans and Raines 1990; National Cancer Institute 1991; Flynn 1992). The campaigns that used a conceptual framework partially based on a cognitive behavioural approach emphasised improving a recipient's capacity to respond to health knowledge and make reasoned decisions. For example, to encourage smoking cessation, campaign initiatives trained recipients how to manage smoking impulses, improve self-efficacy and how to reward oneself for making appropriate health decisions (Kendall and Holon 1979; National Cancer Institute 1991; Flynn 1992). The campaigns that used a conceptual framework based on life skills partially emphasised how a recipient could live a healthier life. For example, to encourage smoking cessation, a life skills approach provided broader training about the value of diet, exercise, clinical self-examination, alcohol moderation and relaxation (Botvin et al. 1980; Botvin and Eng 1982; National Cancer Institute 1991).

While phase two resulted in an improved understanding of the barriers to communicating health to intended recipients that included enhanced social situational surveillance, the field's conceptual horizons remained somewhat grounded in an expert-to-recipient model where the recipients' knowledge deficits, skills and sociocultural barriers and underpinnings were conceived as obstacles to assess and overcome.

The transition to phase three began in the 1980s and 1990s and was based on criticisms that the field was conceptually one-dimensional. Rakow (1989) and Salmon (1989) explain recipients might infer (and respond negatively to) the underlying paternalism within health campaign's predominant conceptual framework. They each add that campaigns sometimes seemed more interested in fostering recipient compliance with authority than an interactive dialogue between health-care professionals and the public. While these authors acknowledge a need to base the goals of campaigns on public health needs, they imply campaigns could be grounded within more interactive, integrative, holistic conceptual foundations where campaign goals evolved according to changing community participation and needs (Neuhauser et al. 2009, 2013b).

Dutta-Bergman reinforces the need for a more integrative and interactive conceptual framework to undergird health campaign research, especially in international and low-income demographic settings, noting (2005: 119):

The new direction evident in much of the growing research on community-based campaigns puts importance on acknowledging marginalised people's capability to determine their own choices, model their own behaviours, and develop epistemologies based on self-understanding.

Others also suggest health campaign research should be underpinned by more interest in participant-oriented, community involvement approaches (Neuhauser *et al.* 2009, 2013b). More recently, an alternative conceptual paradigm for the health campaign field has been proposed based on design science (Simon 1996; Hevner *et al.* 2004; Green 2011). Neuhauser *et al.* (2013a) suggest that the implementation of design science as a conceptual framework results in a more interactive and collaborative approach to campaigns, which evolve and change as needed. They found a more flexible, less predetermined, open-ended approach to health campaigns fosters participant involvement, activation and self-determination, and that this approach is especially effective among medically underserved audiences and in developing nations.

The expansion of the focus of campaign research and initiatives from audiences to individuals also marks a third conceptual phase in health campaign research. Nelson *et al.* (2009) explain how social media, mobile phones and other interactive mass media channels, as well as unprecedented data sources such as electronic patient records, foster opportunities for campaigns that are tailored to individuals, as well as to groups and audiences. They suggest one of the advances in future health campaigns may be to help individuals use personal and research data to make better decisions about clinical care and health-care utilisation, for themselves and for others. This suggests, further, that one future direction in health campaign research may be away from interventions and towards supporting health decision-making (Nelson *et al.* 2009; Kreps and Neuhauser 2013). Moving towards individuals and providing informational support (instead of adherence-based interventions) partially shifts health campaigns from a one-way expert-to-recipient model to broader interactive activities for health promotion, education, engagement and collaboration using diverse media tools and services. While a focus on individuals and decision-support may be seen as more operational than conceptual, a shift in campaign focus also creates an opportunity for research that is more interactive, individually tailored, and user generated compared with the original emphases within health campaigns. The number of studies framed by more participatory, interactive approaches is comparatively small, but the use of a more holistic conceptual framework is an area to watch within the health campaign research literature.

Common outcomes: what health campaign researchers assess

Health communication campaigns often are a hybrid of social science and public health/clinical intervention approaches regarding design, implementation and evaluation. The most common outcome variables (that represent what researchers assess) are derived from the fields of health communication, consumer health informatics, health literacy, social psychology and public health. But contributions to health campaign research also come from cognitive psychology, mass communication, public understanding of science, strategic communication, risk communication and several other fields (Kreps 2012).

Formal evaluations of the impact of health campaigns often reflect similar outcome variables that are distributed into independent, intermediate and dependent measures. Starting with some *independent* or predictor variables, social demographic variables, such as age, gender, native language, income, educational status, socio-economic status, employment, marital status and race/ethnicity, are common measures in health campaign research. Such measures are statistically differentiated or associated with the efficacy of a campaign's dependent variables. Other commonly used *independent* variables measure the frequency of mass media use and other media exposure, as well as frequency of exposure to other sources of health information and its perceived source trustworthiness or credibility. Independent variables also measure an individual's predispositions about health, including risk perception, and an individual's inclinations to learn about health and medicine, such as health information-seeking behaviours and use of various health information sources.

Meanwhile, *intermediate* variables tend to represent measures that either might predict a campaign's outcomes, or could be an outcome of a health communication activity. For example, the health literacy of recipients before and after (or pre-post) a health intervention might predict a campaign's success, or its improvement may be a desired outcome. Some other commonly used *intermediate* variables measure consumer/patient/audience attitudes about receiving health information and the frequency of seeking health information from informal sources (such as family members and peers). Similar *intermediate* variables measure recipient sharing of health information, as well as measures of the motivation and capacity to integrate new health knowledge within one's life. The latter measures include a recipient's self-perceived ability to implement a campaign's health recommendations (Bandura 1995).

Some commonly used *dependent* variables include measures of clinical outcomes, non-clinical outcomes and health-care utilisation outcomes. For example, the Consumer Health Informatics Research Resource (CHIRR)[2] notes that clinical outcomes in health literacy research include specific indicators of chronic diseases (such as diabetes, arthritis and hypertension); specific clinical indicators of acute diseases (such as stroke, cancer and heart disease); and specific clinical indicators of conditions (such as depression, anxiety, obesity, stress, addiction and tobacco control). CHIRR identifies non-clinical behavioural outcomes in health literacy research such as health beliefs, self-reflection and risky health behaviours. It counts among health-care utilisation outcomes: days spent within a clinical facility, hospitalisation rates, urgent care use and the use of preventive services such as mammograms.

In addition to quantitatively derived variables, health campaign research also has a tradition of using qualitative methods (Dervin and Frenette 2001; Carr *et al.* 2011). For example, Dervin and Frenette's sense-making theoretical model provides an alternative approach to conceiving and assessing how an individual responds to health communication campaigns. Health communication campaigns that use qualitative research approaches are reviewed by Piotrow and Kincaid (2001) and Dutta-Bergmann (2005). The use of the qualitatively based diffusion of innovations tradition in the design and evaluation of health communication campaigns is discussed by Backer *et al.* (1992) and Haider and Kreps (2004). Regardless of whether a health communication campaign seeks to inform, persuade or engage, or if its methodological approach is grounded in qualitative, quantitative or mixed methods, the field is multidisciplinary and often borrows research tools, communication strategies, theoretical frameworks, research methods and outcome variables from diverse disciplines.

Research results and the perspective provided by basic research

A rough summary of the findings from thousands of health campaign studies since the 1970s might be characterised as more mixed than consistent or unequivocal. By mixed, we mean the evidence suggests campaigns have a modest impact on improving some of their targeted health behavioural goals, rarely achieve all their targeted aspirations, and findings are inconsistently robust across similar outcome measures among and between studies. For example, while health campaigns often demonstrate significant increases in a target audience's awareness, knowledge or intentions to adopt more therapeutic health habits, the degree of statistical rigour among these findings tends to vary between studies. The inconsistent findings across aggregate campaigns additionally make it difficult to suggest there is robust evidence for definitive conclusions about the overall effectiveness of campaigns. Instead, the field has some well-developed research directions and traditions, research questions, working hypotheses and findings that are sufficiently grounded to pose some evidence-based inferences.

The pioneering Stanford Heart Health study provided an early model of a quantitatively rigorous, comprehensive population-based campaign. It mimicked a controlled clinical trial and promoted the prevention of heart disease using a mass media campaign reinforced by participating health-care providers and community health-care organisations. This campaign in one Californian county was more successful in influencing audience awareness of heart disease prevention, attitudes about prevention and intentions to obtain screening or care than the media campaign without interpersonal provider reinforcement deployed in another county. Both interventions were more successful in influencing audience awareness, knowledge and intentions to act than in a control community where no interventions occurred (Farquhar et al. 1983, 1984; Flora 2001).

While the Stanford Heart Health campaign suggested audience exposure resulted in significant improvements in some cognitive and behavioural dimensions, the campaign was criticised for not assessing patient clinical outcomes and whether recipient use of local health-care delivery systems changed as a result of the campaign (Farquhar et al. 1983, 1984; Flora 2001). Despite careful efforts to minimise confounding variables, the campaign was also criticised for not providing more controls to distinguish the campaign's influence from other interpersonal and mass media influences (Farquhar et al. 1983, 1984; Pettegrew and Logan 1987; Flora 2001; Logan 2008). However, the assessment of clinical behaviours (such as increased check-ups, screenings, information seeking and interactions with doctors) as well as measuring clinical outcomes (such as cholesterol reduction) and health-care utilisation (such as reducing emergency room use for heart patients) were logistically difficult in the pre-Internet and pre-electronic-health records era. The ability to control for confounding variables remains an ongoing challenge.

Despite important changes in media modalities, campaign goals, audiences, and target diseases and conditions in the latter part of the twentieth and early twenty-first century, the mixed findings reported in the Stanford Heart Health study and criticisms of that study are similar to the summaries of aggregated results reported within contemporary systematic reviews. Reviews of health campaigns, health literacy and consumer health informatics research broadly suggest findings are mixed and somewhat inconsistent (Grilli et al. 2002; Gibbons et al. 2009; Brinn et al. 2010; Agency for Healthcare Research and Quality 2011; Car et al. 2011; Cugelman et al. 2011; Lee et al. 2012; Liu et al. 2012; Stellefson et al. 2013; Zhang and Terry 2013). Despite the promise of new media modalities that provide unprecedented opportunities for data gathering, data provision and recipient interaction, campaigns using these means vary in their success in boosting awareness and knowledge, changing attitudes and behavioural intentions, and impacting recipient decisions (Noar 2011). Campaigns also vary in their success in influencing changes in recipient information access, trust (or source credibility), sharing of health information, and consumer motivation as well as clinical habits, clinical outcomes and utilisation of the health-care delivery system (Chou et al. 2009; Cugleman et al. 2011; Agency for Healthcare Research and Quality 2011; Chou et al. 2013).

Despite mixed findings, there is sufficient evidence from more than 40 years of campaign research to provide some grounded inferences. A succinct description of some of these grounded inferences (which depicts some lessons learned from health campaigns) can be found on a US

National Cancer Institute (NCI) website.[3] The NCI notes that communication campaigns can: increase audience knowledge and awareness of a health issue, problem, or solution; influence perceptions, beliefs, and attitudes that may change social norms; prompt action; demonstrate or illustrate healthy skills; reinforce knowledge, attitudes, or behaviours; show the benefits of behavioural change; advocate a position on a health issue or policy; increase demand or support for health services; refute myths and misconceptions; and strengthen organisational relationships. NCI's website notes health communication campaigns can affect individuals as well as groups, organisations and communities.

Researchers and practitioners are assured that a well-managed and executed campaign should result in sufficient therapeutic outcomes to justify costs and commitment. Moreover, the mixed findings from health campaign research fostered a generation of *basic research* in health communication to help practitioners improve future initiatives. These studies are described here as *basic research* because their frequent intent is to narrowly assess the underlying dimensions that foster a campaign's probable success instead of conducting a comprehensive campaign. CHIRR, mentioned above, describes an array of basic research constructs and measures that influence audience receptivity and acceptance of health promotion messages.[4] These include specific measures/constructs of a recipient's predispositions to receive health promotional messages and information as well as how recipients project specific characteristics onto health promotion messages. Some of the constructs that assess recipient predispositions to receive health promotional messages and information include: perceived susceptibility, sensation seeking, spiritual health, locus of control, self-efficacy and technological acceptance model. Some of the constructs that assess how recipients project characteristics onto health promotional message include perceived message cognitive value, perceived message sensation value and state reactance. Individually and in aggregate, the studies regarding these constructs and measures provide insights about specific elements that may influence the efficacy of campaigns. As we note below, this type of narrower research is important as unprecedented opportunities emerge to assess clinical outcomes, healthcare utilisation and individual decision-making.

The impacts of health literacy and health information technology

The health campaign research field is undergoing a transformation thanks to the availability of new media platforms, new health information technologies and the growing international interest in health literacy research (Parker and Thorson 2009).

Although there are differences regarding its definition and conceptual underpinnings, health literacy has provided a focal point for international research about interpersonal health communication (provider-to-patient and consumer-to-consumer) as well as health's mass communication (Pleasant and Kuruvilla 2008; Berkman *et al.* 2010; Sorensen *et al.* 2012). At its foundation, health literacy seeks to advance the quality of life for patients, caregivers and citizens through communication interventions (American Medical Association 1999; Nutbeam 2008; Paasche-Orlow *et al.* 2010; Smith and Moore 2011). While the field of health campaign research often is defined by mass communication (as opposed to interpersonal communication) efforts, health literacy initiatives frequently use mass media technologies and health information technology-derived decision support tools to advance all types of health interventions and interactive activities.

Health literacy also contributes to health campaign research because the future of well-managed health-care delivery and patient care and prudent health economics is perceived as partially tied to health communication interventions (Institute of Medicine of the National

Academies 2004; Agency for Healthcare Research and Quality 2011). In addition, the improvement of health literacy is seen by doctors, health-care institutions and health insurers in many countries as foundational to addressing core health policy issues such as better patient outcomes, a higher quality of clinical care, increased patient satisfaction and health-care inequities, such as medically underserved audiences (Institute of Medicine of the National Academies 2004; Nutbeam 2008; Paasche-Orlow et al. 2010; Agency for Healthcare Research and Quality 2011; Sorensen et al. 2012). In short, health literacy research provides a beachhead for health campaign research to evolve into new areas and uses.

The expansion of this research is accelerated by significant developments in health information technology and its use, and the transformative impact of unprecedented clinical data on the practice and conceptual foundations of medicine. Current innovations include: smartphone sensors to remotely track physiological metrics such as glucose, heart rates, intraocular pressure, smartphone sensors to monitor patient behaviours (exercise, weight, adherence to medication instructions), a *tricoder* that monitors vital signs and interfaces with smartphones and the Cloud, and a lab-on-a-chip integrated into a smartphone that assays body chemistry (Topol 2013). Accompanying these and other related developments are provider and patient access to electronic health records and health decision-making tools – and the forthcoming availability of aggregating human phenotype and genotype information to provide a new era of *personalised care* (Topol 2013). The prevention and treatment of disease may be derived partially from interventions informed by genotype and phenotype information rather than the condition of a specific human organ. Informatics tools soon may have the capacity to be tailored (by interacting a person's clinical data with prevailing medical evidence) to assist patients as they make clinical and health-care utilisation decisions. Underpinning these resources is an unprecedented Net-centric environment that can monitor as well as aggregate multimodal information from biological, cognitive, semantic and social networks (Topol 2013; Luciano et al. 2013). The symbiosis of the Internet's information-gathering capacity with social networks suggests an evolution of the availability and quality of health and medical information itself and a transformation of the evidence base available to inform health decisions and support self-determined pathways to better health (Topol 2013; Luciano et al. 2013).

Besides artificial intelligence, the interactivity (or recipient-to-recipient, recipient-to-expert, and expert-to-recipient options) that is a hallmark of the evolving health information technology landscape is illustrated by: crowdsourcing feedback from prescription medication users about a drug's side effects; patient Internet use patterns that suggest disease outbreaks; patient-to-patient health support services and patient portals that provide access to personal health records and medications. However, the advances within information technology, medical information and medical care foster a parallel need for initiatives to transform how patients and the public adapt to innovations and understand how health's enhanced capacities are integral to one's personal welfare and quality of life (Dutta 2009; Noar and Harrington 2012). This means the efforts to educate patients – about diseases/conditions and prevention efforts, how to better utilise health-care services and to improve health-care decisions – need to be accompanied by research that seeks to engage patients in health care's new era and provide some meaning to unprecedented aggregates of clinical and health-care utilisation data. For example, a future challenge in health campaign research may be to help persons navigate and understand information from unprecedented health data resources, such as deriving personal meaning from the summaries of findings from clinical trials that recently became publicly available.[5]

Dutta and Kreps (2013) note the transformative potential of health information technology also needs to be applied to help address health disparities or the inequitable distribution of health-care services and community environments that deter access to health care (and medical treatment) in many countries. They consider it is difficult to imagine a future for health campaign research that is not linked to global health challenges. Similarly, the knowledge gap hypothesis suggests the equitable distribution of information and communication support to medically underserved audiences may be inversely related to the level of technological innovation and expansion within the mass media and health-care professions (Viswanath and Finnegan 1995).

In contrast, self-determination theory suggests the adoption of health informatics may be robust because new health information technologies uniquely combine all three of the core components of intrinsic consumer motivation (Deci and Ryan 2002). Unlike legacy media, fully integrated and optimised health information technologies help persons master health-related tasks and provides them with a platform that advances autonomy and individual decision-making and easily connects with others (Deci and Ryan, 2002; Hesse 2013). The rapid global public acceptance of mobile phones suggests some technological characteristics may be so compelling they seem to eclipse socio-economic, sociocultural and other adoption barriers.

Whether this is proven or not, the capacity of health information technologies to assist underserved audiences, provide self-determined pathways for better health and address health literacy provide significant opportunities for future health campaign research. These opportunities join legacy health campaign research interests to boost public awareness about specific diseases/conditions and public health challenges and to enhance the public utilisation of health-care services. Overall, health information technologies and health literacy provide new horizons for health campaign research to engage, educate, empower, entertain and assist patients and the public to adjust to a changing landscape about health, health care, and quality of life.

Concluding remarks

While the expansion of health campaign research's conceptual framework, outcomes and aspirations should be attractive to a new generation of researchers and practitioners, the field's increasing multidimensionality fosters its own set of challenges. These include working towards integrative rather than fragmented multidisciplinary approaches and continuing some of the basic research within the health communication field. As the field of health campaign research adds interrelated fields of inquiry such as health literacy, consumer health informatics, public understanding of science, health-care utilisation, health information technology and clinical decision-making support, it seems more important than ever to avoid the fragmentation of theory and practices that sometimes occurs within sub-disciplines. The growth of health campaign research fosters fresh challenges to triangulate health communication research traditions, health literacy, and consumer health informatics research and integrate ideas from other related disciplines. Moreover, as Harrington and Noar (2012), Keselman et al. (2008) and Logan and Tse (2007) note, it seems prudent to encourage more interoperable theories, hypotheses, constructs and measures, so that campaign research can more systematically compare data sets, propose possible standards to optimise campaign strategies, nurture interdisciplinary graduate programmes and develop a more cross-cutting evidentiary base.

There is a pressing need for basic research as outlined earlier that evaluates some issues linked to health campaign best practices and management. Among these issues are: What communication strategies are best suited for diverse audiences? What media modality best communicates a campaign message? What combinations of persuasive techniques counteract different types of audience resistance?

There is a tradition of basic research in health communication that informs such issues, which potentially improves the efficacy of health campaigns and expedites the process for practitioners and researchers to implement and assess campaigns. This more focused approach to research should not be overlooked as health campaign research's landscape and ambitions evolve.

Key questions

- To what extent is it possible to assess and demonstrate a health campaign is more strongly associated with intended changes in audience awareness, knowledge, attitudes, motivations and behaviours than are other influences?
- Can a health intervention campaign be considered successful if it does not meet all its intended goals? What constitutes compelling evidence of the value and efficacy of a health campaign?
- How do the rapidly expanding capabilities of health information technology to boost health awareness, provide evidence-based information, assist with clinical decisions and foster more personal or tailored medical interventions contribute to health-care practices and the quality of life between and within countries?

Notes

1. The chapter was rewritten from the earlier version (Logan 2008) to incorporate how developments in the field of health literacy and significant changes in health information technology, as well as recent changes in clinical care, health care utilisation and the health-care delivery system have affected health communication campaign research. The chapter also now covers health campaign research's expansion into areas such as enhancing patient care, consumer use of health-care delivery services, and supporting self-determined pathways to better health. In addition, the current chapter considers how the availability of new media platforms and new health information technologies, along with the growing interest in health literacy and participant-oriented and community involvement research, all impact health campaign research.
2. http://chirr.nlm.nih.gov
3. http://www.cancer.gov/cancertopics/cancerlibrary/pinkbook/page3
4. http://chirr.nlm.nih.gov
5. http://clinicaltrials.gov

References

Agency for Healthcare Research and Quality (2011) *Health Literacy Interventions and Outcomes: An Updated Systematic Review. Evidence Report/Technology Assessment 199*, Rockville, MD: Agency for Healthcare Research and Quality: online at www.ahrq.gov/downloads/pub/evidence/pdf/literacy/literacyup.pdf

American Medical Association (1999) 'Health literacy: report of the Council on Scientific Affairs. Ad Hoc Committee on Health Literacy for the Council on Scientific Affairs, American Medical Association', *Journal of the American Medical Association*, 281, 6: 552–557.

Atkin, C. K. (2001) 'Theory and principles of media health campaigns', in R.E. Rice and C.K. Atkin (eds) *Public Communication Campaigns*, third edition, Thousand Oaks, CA: Sage, 49–68.

Backer, T. E., Rogers, E. M. and Sopory, P. (1992) *Designing Health Communications Campaigns: What Works?* Newbury Park, CA: Sage.

Bandura, A. (1995) 'Exercise of personal and collective efficacy', in A. Bandura (ed.) *Self-Efficacy in Changing Societies*, New York: Cambridge University Press, 1–45.

Berkman, N. D., Davis, T. C. and McCormack, L. (2010) 'Health literacy: what is it?', *Journal of Health Communication*, 15, S2: 9–19.

Botvin, G. J. and Eng, A. (1982) 'The efficacy of a multicomponent approach to the prevention of cigarette smoking', *Preventive Medicine*, 11, 2: 199–211.

Botvin, G. J., Eng, A. and Williams, C. L. (1980) 'Preventing the onset of cigarette smoking through life skills training', *Preventive Medicine*, 9, 1: 135–143.

Brinn, M. P., Carson, K. V., Esterman, A. J., Chang, A. B. and Smith, B. J. (2010) 'Mass media interventions for preventing smoking in young people', *Cochrane Database of Systematic Reviews,* 11; doi: 10.1002/14651858.CD001006.pub2.

Car, J., Lang, B., Colledge, A., Ung, C. and Majeed, A. (2011) 'Interventions for enhancing consumers' online health literacy' *Cochrane Database of Systematic Reviews,* 6; doi: 10.1002/14651858.CD007092.pub2.

Carr, S. M., Lhussier, M., Forster, N., Geddes, L., Deane, K. and Pennington, M. (2011) 'An evidence synthesis of qualitative and quantitative research on component intervention techniques, effectiveness, cost-effectiveness, equity and acceptability of different versions of health-related lifestyle advisor role in improving health', *Health Technology Assessment,* 15, 9: iii–iv.

Chou, W. Y., Prestin, A., Lyons, C. and Wen, K. (2013) 'Web 2.0 for health promotion: reviewing the current evidence', *American Journal of Public Health,* 103, 1: e9–e18.

Chou, W. Y., Hunt, Y. M., Beckjord, E. B., Moser, R. P. and Hesse, B. W. (2009) 'Social media use in the United States: implications for health communication', *Journal of the American Medical Informatics Association,* 18, 3: 319–321.

Cugelman, B., Thelwall, M. and Dawes, P. (2011) 'Online interventions for social marketing health behavior change campaigns: a meta-analysis of psychological architectures and adherence factors', *Journal of Medical Internet Research,* 13, 1: e17.

Deci, E. and Ryan, R. (eds) (2002) *Handbook of Self-Determination Research,* Rochester, NY: University of Rochester Press.

Dervin, B. and Frenette, M. (2001) 'Sense-making methodology: communicating communicatively with campaign audiences', in R. E. Rice and C. K. Atkin (eds) *Public Communication Campaigns,* third edition, Thousand Oaks, CA: Sage, 69–87.

Dutta, M. (2009) 'Health communication: trends and future directions', in J. G. Parker and E. Thorson (eds) *Health Communication in the New Media Landscape,* New York: Springer, 59–92.

Dutta, M. J. and Kreps, G. L. (eds) (2013) *Reducing Health Disparities: Communication Interventions,* New York: Peter Lang.

Dutta-Bergman, M. (2005) 'Theory and practice in health communication campaigns: a critical interrogation', *Health Communication,* 18, 2: 103–122.

Dutta-Bergman, M. (2006) 'Media use theory and internet use for health care', in M. Murero and R. E. Rice (eds) *The Internet and Health Care: Theory, Research and Practice,* Mahwah, NJ: Lawrence Erlbaum, 83–105.

Evans, R. I. and Raines, B. E. (1990) 'Applying a social psychological model across health promotion interventions: cigarettes to smokeless tobacco', in J. Edwards, R. S. Tindale, L. Heath, and E. J. Posavac (eds) *Social Influence Process and Prevention: Social Psychological Applications of Social Issues,* New York: Plenum Press, 143–158.

Farquhar, J. W., Maccoby, N. and Solomon, D. (1984) 'Community applications of behavioral medicine', in W. D. Gentry (ed.) *Handbook of Behavioral Medicine,* New York: Guilford Press: 437–478.

Farquhar, J. W., Fortman, S., Wood, P. and Haskell, W. (1983) 'Communication studies of cardiovascular disease prevention', in N. Kaplan and J. Stander (eds) *Prevention of Coronary Heart Disease: Practical Management of Risk Factors,* Philadelphia, PA: Saunders: 170–181.

Fishbein, M. and Ajzen, I. (1975) *Belief, Attitude, Intention and Behavior: An Introduction in Theory and Research,* Reading, MA: Addison-Wesley.

Flora, J. A. (2001) 'The Stanford community studies: campaigns to reduce cardiovascular disease', in R. E. Rice and C. K. Atkin (eds) *Public Communication Campaigns,* third edition, Thousand Oaks, CA: Sage, 192–213.

Flynn, B. S. (1992) 'Prevention of cigarette smoking through mass media intervention and school programs', *American Journal of Public Health,* 82, 6: 827–834.

Gibbons, M. C., Wilson, R. F., Samal, L., Lehmann, C. U., Dickersin, K., Lehmann, H. P., Aboumatar, H., Finkelstein, J., Shelton, E., Sharma, R. and Bass, E. B. (2009) *Impact of Consumer Health Informatics Applications. Evidence Report/Technology Assessment 188,* Rockville, MD: Agency for Healthcare Research and Quality.

Green, N. (ed.) (2011) *Artificial Intelligence and Health Communication: Proceedings of the American Association of Artificial Intelligence Symposium on Artificial Intelligence and Health Communication,* AAAI Press/The MIT Press.

Grilli, R., Ramsay, C. and Minozzi, S. (2002) 'Mass media interventions: effects on health services utilisation', *Cochrane Database of Systematic Reviews,* 1. doi: 10.1002/14651858.CD000389.

Gustafson, D. H., Hawkins, R. P., Boberg, E. W., McTavish, F. M., Owens, B., Wise, M., Berhe, H. and Pingree, S. (2002) 'CHESS: 10 years of research and development in consumer health informatics for broad populations including the underserved', *International Journal of Medical Informatics,* 65, 3: 169–177.

Haider, M., and Kreps, G. L. (2004) 'Forty years of diffusion of innovations: utility and value in public health', *Journal of Health Communication*, 9, Supplement 1: 3–11.

Haider, M., Ratzan, S. C. and Meltzer, W. (2009) 'International innovations in health communication', in J. G. Parker and E. Thorson (eds) *Health Communication in the New Media Landscape*, New York: Springer, 373–394.

Harrington, N. G. and Noar, S. M. (2012) 'Building an evidence base for eHealth applications: research questions and practice implications', in S. M. Noar and N. G. Harrington (eds) *eHealth Applications: Promising Strategies for Behavior Change*, New York: Routledge, 263–274.

Hesse, B. W. (2013) 'Quantifying the health information revolution: triangulation in an era of big data', Better Health: Evaluation Health Communication Series, Bethesda, MD: National Institutes of Health.

Hevner, A., March, S., Park, J. and Ram, S. (2004) 'Design science in information systems research', *Management Information Systems Quarterly*, 28, 1: 75–105.

Hornik, R. C. (ed.) (2002a) *Public Health Communication: Evidence for Behavior Change*, Mahwah, NJ: Lawrence Erlbaum.

Hornik, R. C. (2002b) 'Communication in support of child survival: evidence and explanations from eight countries', in R. C. Hornik (ed.) *Public Health Communication: Evidence for Behavior Change*, Mahwah, NJ: Lawrence Erlbaum, 219–248.

Institute of Medicine of the National Academies (2004) *Health Literacy: A Prescription to End Confusion*, Washington, DC: The National Academies Press.

Institute of Medicine of the National Academies (2006) *Preventing Medication Errors: Quality Chasm Series*, Washington, DC: The National Academies Press.

Kendall, P. C. and Holon, S. D. (1979) *Cognitive-Behavioral Intentions: Theory, Research and Practice*, New York: Academic Press.

Keselman, A., Logan, R. A., Smith, C. A. LeRoy, G. and Zeng-Treitler, Q. (2008) 'Developing informatics tools and strategies for consumer-centered health communication', *Journal of the American Medical Informatics Association*, 15, 4: 475–483.

Kreps, G. L. (2011) Methodological diversity and integration in health communication inquiry, *Patient Education and Counseling*, 82, 3: 285–291.

Kreps, G. L. (2012) 'The maturation of health communication inquiry: directions for future development and growth', *Journal of Health Communication*, 17, 5: 495–497.

Kreps, G. L. and Neuhauser, L. (2010) 'New directions in eHealth communications: opportunities and challenges', *Patient Education and Counseling*, 78, 3: 329–336.

Kreps, G. L. and Neuhauser, L. (2013) 'Artificial intelligence and immediacy: designing health communication to personal engage consumers and providers', *Patient Education and Counseling*, 92, 2, 205–210.

Kutner, M., Greenberg, E., Jin, Y. and Paulsen, C. (2006) *The Health Literacy of America's Adults: Results from the 2003 National Assessment of Adult Literacy* (NCES2006-483), Washington, DC: National Center for Education Statistics.

Lee, T. W., Lee, S. H., Kim, H. H. and Kang, S. J. (2012) 'Effective intervention strategies to improve health outcomes for cardiovascular disease patients with low health literacy skills: a systematic review', *Asian Nursing Research*, 6, 4: 128–136.

Liu, J. J., Davidson, E. Bhopal, R. S., White, M., Johnson, M.R.D., Netto, G., Deverill, M. and Sheikh, A. (2012) 'Adapting health promotion interventions to meet the needs of ethnic minority groups: mixed-methods evidence synthesis', *Health Technology Assessment*, 16, 44.

Logan, R. A. (2008) 'Health campaigns research', in M. Bucchi and B. Trench (eds) *Handbook of Public Communication of Science and Technology*, London and New York: Routledge, 77–92.

Logan, R. A. and Tse, T. (2007) 'A multidisciplinary conceptual framework for consumer health informatics', *Medinfo*, 12, 2: 1169–1173.

Luciano J. S., Cumming, G. P., Wilkinson, M. D. and Kahana, E. (2013) 'The emergent discipline of health web service', *Journal of Medical Internet Research*, 15, 8: e166.

McGuire, W. J. (2001) 'Input and output variables currently promising for construction persuasive communications', in R. E. Rice and C. K. Atkin (eds) *Public Communication Campaigns*, third edition, Thousand Oaks, CA: Sage, 22–48.

Murero, M. and Rice, R. E. (eds) *The Internet and Health Care: Theory, Research and Practice*, Mahwah, NJ: Lawrence Erlbaum.

National Cancer Institute (1991) *Strategies to Control Tobacco Use in the United States: A Blueprint for Public Health Action in the 1990s*, Washington, DC: National Cancer Institute.

Nelson, D. E., Hesse, B. W. and Croyle, R. T. (2009) *Making Data Talk: Communicating Public Health Data to the Public, Policy Makers, and the Press*, New York: Oxford University Press.

Neuhauser, L., Rothschild, B., Graham, C., Ivey, S. L. and Konishi, S. (2009) 'Participatory design of mass health communication in three languages for seniors and people with disabilities on Medicaid', *American Journal of Public Health*, 99, 12: 2188–2195.

Neuhauser, L., Kreps, G.L., Morrison, K., Athanasoulis, M., Kirienko, N. and Van Brunt, N. (2013a) 'Using design science and artificial intelligence to improve health communication: ChronologyMD case example', *Patient Education and Counseling*, 92, 2: 211–217.

Neuhauser, L., Ivey, S. L., Huang, D., Engelman, A., Tseng, W., Dahrouge, D. S., Gurung, S. and Kealey, M. (2013b) 'Availability and readability of emergency preparedness materials for deaf and hard-of-hearing and older adult populations: issues and assessments', *PLoS One*, 8, 2: e55614. doi: 10.1371/journal.pone.0055614.

Noar, S. M. (2006) 'A 10-year retrospective of research in health mass media campaigns: where do we go from here?' *Journal of Health Communications*, 11, 1: 21–42.

Noar, S.M. (2011) 'Computer technology based-interventions in HIV prevention: state of the evidence and future directions for research', *AIDS Care*, 23, 5: 535–533.

Noar, S.M. and Harrington, N.G. (eds) (2012) *eHealth Applications: Promising Strategies for Behavior Change*, New York: Routledge.

Nutbeam, D. (2000) 'Health literacy as a public health goal: a challenge for contemporary health education and communication strategies into the 21st Century', *Health Promotion International*, 15, 3: 259–267.

Nutbeam, D. (2008) 'The evolving concept of health literacy', *Social Science & Medicine*, 67, 12: 2072-2078.

Paasche-Orlow, M. K., Wilson, E. A. H. and McCormack, L. (2010) 'The evolving field of health literacy research', *Journal of Health Communication, 15, Supplement 2: 5–8*.

Parker, J. G. and Thorson, E. (eds) (2009) *Health Communication in the New Media Landscape*, New York: Springer.

Pettegrew, L. and Logan, R. A. (1987) 'Health communication: review of theory and research', in C. R. Berger and S. H. Chaffee (eds) *Handbook of Communication Science*, Beverly Hills, CA: Sage, 675–710.

Piotrow, P. T. and Kincaid, D. L. (2001) 'Strategic communication for international health programs', in R. E. Rice and C. K. Atkin (eds) *Public Communication Campaigns*, third edition, Thousand Oaks, CA: Sage, 249–266.

Piotrow, P.T., Kincaid, D.L., Rimon II, J. G. and Rinehart, W. (1997) *Health Communication: Lessons from Family Planning and Reproductive Health*, Westport, CN: Praeger.

Pleasant, A. and Kuruvilla, S. (2008) 'A tale of two health literacies: public health and clinical approaches to health literacy', *Health Promotion International*, 23, 2: 152–150.

Rakow. L. F. (1989) 'Information and power: towards a critical theory of information campaigns', in C. Salmon (ed.) *Information Campaigns: Balancing Social Values and Social Change*, Newbury Park, CA: Sage, 164–184.

Randolph, W. and Viswanath, K. (2004) 'Lessons learned from public health mass media campaigns: marketing health in a crowded media world', *Annual Review of Public Health*, 25: 419–437.

Rice, R. E. and Katz, J. E. (eds) (2001) *The Internet and Health Communication: Experiences and Expectations*, Thousand Oaks, CA: Sage.

Salmon, C. (ed.) (1989) *Information Campaigns: Balancing Social Values and Social Change*, Newbury Park, CA: Sage.

Simon, H. (1996) *The Sciences of the Artificial*, third edition, Cambridge, MA: MIT Press.

Smith, S. A. (2009) 'Promoting health literacy: concept, measurement and intervention', PhD thesis, Union Institute & University, Cincinnati, Ohio; *Dissertation Abstracts International*, 70: 9.

Smith, S.A. (2011) 'Health literacy and social service delivery', in S.A. Estrine, H. G. Arthur, R. T. Hettenbach and M. G. Messina (eds) *New Directions in Behavioral Health: Service Delivery Strategies for Vulnerable Populations*, New York: Springer Publishing.

Smith, S. A. and Moore, E. (2011) 'Health literacy and depression in the context of home visitation', *Maternal and Child Health Journal*, 16, 7: 1500–1508; online at: www.springer.com/home?SGWID=0-0-1003-0-0&aqId=2111233&download=1&checkval=ffd392d488b3752bf1d2e6a3e258b189

Snyder, L. B. (2001) 'How effective are medicated health campaigns?' in R. E. Rice and C. K. Atkin (eds) *Public Communication Campaigns*, third edition, Thousand Oaks, CA: Sage, 181–192.

Snyder, L. B., Hamilton, M. A., Mitchell, E. W., Kiwanuka-Tonda, J., Fleming-Milici, F. and Proctor, D. (2004) 'A meta-analysis of the effect of mediated health communication campaigns on behavior change in the United States', *Journal of Health Communication*, 9, Supplement 1: 71–96.

Sorensen, K., Broucke, S.V., Fullam, J., Doyle, G., Pelikan, J., Slonska, A. and Brand, H. (2012) 'Health literacy and public health: a systematic review and integration of definitions and models', *BMC Public Health*, 12: 80. doi:10:1186/1471-2458-12-80.

Stellefson, M. B., Chaney, B., Barry, A. E., Chavarria, E., Tennant, B., Walsh-Childers, K., Sriram, P. S. and Zagora, J. (2013) 'Web 2.0 chronic disease self-management for older adults: a systematic review', *Journal of Medical Internet Research*, 15, 2: e35.

Topol, E. (2013) *The Creative Destruction of Medicine: How the Digital Revolution Will Create Better Health Care*, New York: Basic Books.

US Department of Health and Human Services (2000) *Healthy People 2010, Second Edition. Understanding and Improving Health and Objectives for Improving Health*, Washington, DC: Government Printing Office.

Viswanath, K. and Finnegan, J. R. (1995) 'The knowledge gap hypothesis: twenty-five years later', in B. Burlson (ed.) *Communication Yearbook 19*, Thousand Oaks, CA: Sage, 187–228.

Viswanath, K., Breen, N., Meissner, H., Moser, R. P., Hesse, B., Steele, W. R. and Rakowski, W. (2006) 'Cancer knowledge and disparities in the information age' *Journal of Health Communication*, 11, Supplement 1: 1–17.

White, S. (2008) *Assessing the Nation's Health Literacy: Key Concepts and Findings of the National Assessment of Adult Literacy*, Chicago, IL: American Medical Association Foundation.

Wollesen, L. and Peifer, K. (2006) *Life Skills Progression: An Outcome and Intervention Planning Instrument for Use with Families at Risk*, Baltimore, MD: Brookes.

Zhang, N. J. and Terry, A. (2013) 'Effects of health literacy on patients' adherence to prescribed medications: narrative systematic review and meta-analysis', *Value Health*, 16, 3: A36; doi: 10.1016/j.jval.2013.03.206.

16

Global spread of science communication

Institutions and practices across continents

Brian Trench and Massimiano Bucchi, with Latifah Amin, Gultekin Cakmakci, Bankole A. Falade, Arko Olesk, Carmelo Polino[1]

Introduction

When the first large-scale international conferences of science communication practitioners, educators and researchers took place in the early 1990s, they were very largely restricted to western Europe and North America. But the PCST (Public Communication of Science and Technology) series of conferences now attracts more than 600 participants from over 60 countries in all continents; proposals of contributions to the 2012 conference came from over 50 countries. The World Congress of Science Journalists (700-plus participants from 73 countries in 2013) and the professional conferences of science museums and centres attract similar numbers and distributions of participants.

As an inherently international system, modern science has diffused institutional structures and practices across the continents from its earliest days. In an increasingly globalised world, these processes have accelerated and intensified and, with cross-continental collaborations and movement of personnel, ideas and attitudes have also spread. This includes ideas and attitudes on the place of science in society and on scientists' social roles. Partly based on this shared culture of scientists, but also driven by other globalisation factors in politics and economics, science communication has over a relatively short period become a worldwide phenomenon.

This global spread of science communication, its shapes and its meanings, has become a theme in science communication research in recent years. The proliferation of science communication activities and institutions across the globe, but also the differences and similarities between countries and regions in the organisation of these activities and institutions have become an object of specific interest in the worldwide science communication communities. A collection of country profiles and essays (Schiele *et al.* 2012), that grew out of the PCST conferences featured 31 contributors from six continents, presenting national overviews side by side. An edited volume on national and international surveys of public attitudes to science and technology (Bauer *et al.* 2011) sketched a global view of patterns of scientific culture.

In considering science communication comparatively across countries, we are helped by a large-scale assessment of science-in-society practices in Europe, the MASIS project which surveyed 37 countries.[2] The project's final report categorised national science communication cultures according to six parameters, as 'consolidated', 'developing' or 'fragile'. The parameters 'collectively form a framework for analysing science communication culture' (Mejlgaard et al. 2012: 67), which appears valid beyond Europe. These are restated here with slight modification: the degree of institutionalisation of the science communication infrastructure; the level of attention paid by the political system; the number and diversity of actors involved in science communication; the academic tradition for dissemination of research results; public attitudes towards science; the number and qualifications of science journalists.

When we examine the global spread of science communication we are looking in the first instance at its institutionalisation through the policies and programmes of national governments, national academies and research funders, professional networks, intergovernmental organisations, higher education and research institutions, international charities and commercial companies. In different contexts, the strength of the roles that these actors play and the relations between them can vary significantly. But the state, in its various guises, tends to be the main driver of the institutionalisation of science communication. In looking for markers of the institutionalisation of science communication in individual countries across the globe, one of the first, if not *the* first, to observe is the presence of government programmes to boost science awareness. Other markers include the presence of: communication training for scientists; initiatives to support media attention to science; university taught programmes in science communication; and university research in science communication. In the following sections we examine each of these briefly, with particular reference to their appearance in countries and regions outside western Europe and North America.

Key indicators of global spread

Government programmes to boost science awareness. Policy-making for the economy and for research and development have become ever more closely intertwined since the 1990s, as a central role has been ascribed to science and technology in economic development, whether in taking a country from a largely agrarian or from a traditional industrial base to another phase of development. Across the developing countries, as in the industrially and technically more advanced regions, the knowledge economy and/or sustainable development have become central themes of public policy. Under either or both headings, government programmes and policies, with varying degrees of emphasis and explicitness, refer to the public's views of science and technology as a potential constraint on, or support for, economic and social development.

In the world's two most populous countries, China and India, the state's commitment to popularise science has been written into fundamental legislation for several decades. A more recent policy statement, Japan's *Science and Technology Basic Plan* (2011–2015) links knowledge creation and innovation with working towards 'a sound infrastructure of science and technology information and [to] raise awareness and understanding of science and technology-related issues in Japan'.[3] The Japan Science and Technology Agency's divisions include a Centre for Science Communication, which, 'in addition to communication conveying the knowledge and enjoyment of previous achievements in science and technology, also seeks to promote constructive communication by sharing the tentative nature, uncertainty, and latent risks possessed by science and technology with the nation's citizens, its government, its research institutions, and researchers, for a better society and lifestyle'.[4] This represents a more comprehensive view of public communication of science and of its contexts than may be found in many other similar documents.

Perhaps the most frequently shared feature of such policies is a concern – explicit or implicit – about children's and young people's competence in scientific and technical subjects and their attitudes to developments in science and technology. The context of this concern is also competitive: the high ranking of some South East Asian countries in international surveys of school students' abilities in mathematics and science (e.g. PISA – Programme for International Student Assessment of the OECD) is noted with some alarm in western European countries. Government policies in these regions are targeted, respectively, at closing the gap or maintaining the lead. In a largely linear conception of the relations between education and economy, the preparation of young people in *STEM* (science, technology, engineering and mathematics) subjects is seen as assisting skills supply to the economy. Informal education initiatives of the kind typically endorsed in programmes for science awareness are assigned a complementary role in this national effort.

However, government encouragement for public engagement with science in some western European countries has acquired a much wider scope during the past two decades; the public or publics in question specifically include those with an interest and ability to participate in exchange of ideas and in policy formation. The 'turn to dialogue' signalled in national programmes across the developed world reflects the diversity of publics these programmes encompass. In countries where science communication has been institutionalised more recently, the emphasis tends to be more strongly – or, in some cases, exclusively – on children and young people.

This emphasis, represented also in a conception of science communication as informal education, contributes to a common trait of government programmes and policies on science awareness: the commitment to build or support science centres, generally of the kind that has grown up since the late 1960s and the establishment of Exploratorium in San Francisco (see Schiele in this volume). At the turn of the present century, a group of such centres was opened in Britain as a Millennium Project, with support from the government through The National Lottery. Smaller European countries have built their landmark national science centres in the past two decades, generally in or near the capital city, as a representation of their country's openness to science and technology. In more populous Asian countries (e.g. India and South Korea), science centres are counted in their tens or twenties, and the networks have continued to expand through the 2000s and 2010s with support from regional authorities or state governments. But the most ambitious programme by far is that of China, where the number of science and technology centres almost doubled, from 185 to 380, between 2004 and 2008 (Shi and Zhang 2012).

The Chinese network of science centres is firmly integrated into a government programme of public education, as – in an apparently quite different political context – is Taiwan's National Science and Technology Museum which has the mission 'to enrich citizens' knowledge of science and technology; to inspire citizens' interest in doing scientific and technological researches, and to value the development of science and technology; to record and present our national achievement in the development of science and technology, thereby building up our people's confidence; to promote public education of science and technology in southern Taiwan'.[5]

But elsewhere in Asia different models can be found, illustrating that the development of science communication globally is uneven: the ArtScience Museum in Singapore, with its striking architecture, is part of a commercial leisure and entertainment complex; Miraikan, the National Museum of Emerging Science and Innovation, in Japan states, as a founding principle, that 'science and technology are part of our culture. We provide an open forum for all to ponder and discuss the future roles of science and technology';[6] Petrosains in Malaysia (see under 'Reports from five emerging centres of science comunication' in this chapter) 'is a Science Discovery Centre that uses a fun and interactive approach to tell the story of the science and technology of the petroleum industry', housed in some of the world's tallest buildings, built for the energy company, Petronas.[7]

Government programmes for raising awareness about scientific developments also incorporate other common manifestations that are less visible than science museums and science centres. These include, for example, direct or indirect support for national *weeks of science* or similar concentrated efforts in public science; for innovations in science education; and for Internet-based services for news from research, such as are found in Denmark, Netherlands, Norway and Spain. (Other such Internet-based services have also been established independently of government, e.g. www.AfricaSTI.com, operated by a network of African science journalists.)

Comparing government science awareness programmes, Bultitude and colleagues (2012) found that Brazil's and China's were more oriented to development and addressing social inequalities than those of Australia and Britain; emphasis on education was stronger for China and Britain, and emphasis on culture was strongest for Brazil. An earlier study noted that evaluations of the Australian awareness programme of the 1990s did not establish if it 'caused Australians to become more or less aware of science and technology or of the part science plays in stimulating social and economic development' (Gascoigne and Metcalfe 2001: 75); the authors recommended that evaluation needed to be built into such programmes from the start.

Training and other supports for scientists in public communication. Short courses in media and presentation skills are increasingly available to scientists and other academics from research funders, universities, professional societies and – increasingly – private providers. Across western Europe, the number of such courses is growing continuously; the common requirement of national research agencies and of the European Commission that results of projects funded from these sources are *disseminated* publicly is a strong driver of demand for such training. In the present decade, initiatives to provide courses have been taken in several European countries where none existed previously as, for example, in some Italian universities.

Courses are also provided on an international basis, as in the case of a 2011 communication course for scientists in developing countries; this was promoted on the basis that 'communication skills are particularly important for scientists in developing countries, where the infrastructure for science is weak and where science education needs more support at all educational levels'.[8] The course hosts, the International Centre for Theoretical Physics and Third World Academy of Sciences in Trieste, Italy, noted that 'by improving their communication skills, scientists can play an important role in the development of science in their countries'.

A decade ago, an EU benchmark study of activities in the promotion of research, technology and development culture noted that few countries were doing very much to train their scientific research community to communicate with their fellow citizens or to engage with their concerns (Miller *et al.* 2002). More recently, a cross-country survey reported a significant correlation between communication training and confidence among researchers in communicating with the public (Peters *et al.* 2008). But in many countries where there are similar expectations of researchers, that they engage in various ways with the general public, there is little or no provision of relevant training. Although elements of communication training are increasingly found in doctoral and postdoctoral programmes, these are more likely to be directed to communicating with peers in related disciplines or with prospective business users of technologies arising from research than to communicating with broad publics or with policymakers.

As observed in the MASIS reports for several countries in southern and eastern Europe,[9] support for (mainly younger) researchers wishing to become involved in public communication has come from the British Council and UNESCO as much as from local sources. In October 2013 UNESCO organised the First Regional Science Promotion conference in Serbia,[10] bringing together science promotion professionals, practitioners and enthusiasts from south-eastern Europe to 'share experience, network and formulate the next steps towards strengthening the link between science and society'. Among the topics considered was how to 'improve science communication and language of scientists and researchers allowing them to present their work in a comprehensive manner'.

MASIS reports for several central and eastern European countries cited the UK cultural relations agency, British Council, as a primary player in science communication, principally through the Famelab competitions. Famelab has spread to over 20 countries, mainly among the newer member states of the European Union but also to include Egypt, Hong Kong and Israel; and the British Council has provided the associated training, preparing mainly early-career researchers to present a chosen scientific topic in three minutes before non-specialist audiences.

The British Council has also helped organise science cafés in many countries, and this format has also been applied elsewhere to familiarise scientists and others with communicating about science in informal public settings (see Einsiedel in this volume). In Vietnam, Café Khoa Huc was established by Oxford University's Clinical Research Unit and the Hospital for Tropical Diseases aiming to create 'a friendly atmosphere in which everyone feels free to question and offer their ideas'.[11]

Also originating in Britain, the Science Media Centre (SMC) has come to be seen as a model capable of being applied in other countries; as of late 2013, similar centres had been established in Australia, Canada, Denmark, Japan and New Zealand, with 'more on the way in China, Italy and Norway'.[12] The British SMC has operated since 2002 with support from professional societies and private companies. It gives various kinds of support to scientists engaging with mass media, including short workshops under the title Introduction to the News Media, where scientists with media experience and media professionals present 'the realities of the news media'.[13] The Science Media Centre stresses that this does not represent practical media training, but rather offers 'a flavour of the news media'.

A key issue for the design and delivery of such training is the strength of emphasis on technical and formal aspects of communication. A media skills course may, for example, be largely or exclusively focused on the key elements of writing a news release or of doing a radio or television interview. A course on skills for communicating with lay audiences may, in the same way, be largely or exclusively focused on techniques of storytelling. The final report of the EC-funded Messenger project noted that opinions differed among those consulted for the project on the type of training that scientists need, from media skills that are relatively straightforward to impart to developing 'an awareness of social as well as epistemological considerations' (SIRC and ASCOR 2006). As discussed elsewhere (Trench and Miller 2012), an approach to public communication oriented to dialogue requires preparing scientists to consider carefully the needs of their audiences and to listen well to their concerns. Encouraging scientists to take part in informal conversation, as at science cafés, may require specific forms of support.

Among the other supports offered to scientists, sometimes alongside training, sometimes in its absence, are prizes and other awards for outstanding achievements in communicating science. However, the increasingly common requirement of scientists that they engage in public activities is not generally matched by changes in the way scientists are assessed on their performance overall. Despite the increasing attention to the *third mission* (beyond teaching and research) of higher education institutions and to public access to research centres, there are few formal incentives for scientists to be publicly active. In a letter to *Nature*, correspondents from leading research institutions and the national science centre in Japan noted that the government 'has urged the researchers it funds to improve communication with the tax-paying public' but 'time and effort spent on science communication will not help scientists to secure funding, promotion or employment' (Koizumi *et al.* 2013).

Communication training is often focused on early-career researchers or PhD students, as, for example, in the science and communication workshops held in recent years in India, with funding support from The Wellcome Trust and India's Department of Biotechnology. Meanwhile, training for wider groups involved in science communication is spreading, and it is increasingly internationalised: in September 2013 the first Euro-Mediterranean and Middle East Summer School of Science Communication[14] took place in southern Spain, supporting science communication professionals in their efforts 'to drive development of new science communication endeavours'. The participants included staff of existing science centres and museums as well as newcomers to the field from related organisations such as universities, local authorities and associations.

Initiatives to support media attention to science. National and international bodies have become involved in efforts to encourage media interest in science, and to support journalists giving special attention to this *beat*. We can observe a spreading trend of governments encouraging publicly funded broadcasters to increase and maintain levels of science coverage, in some cases providing support through national awareness programmes, or less directly, through state agencies and institutes in the science and technology sectors. High-technology companies also sometimes feature as sponsors of science programming on television.

Intergovernmental organisations, such as UNESCO, also seek to promote media attention to science and to support media professionals working in this field. Alongside the south-east European regional meeting of science communication professionals that UNESCO hosted in October 2013, it also organised a school of journalists 'oriented towards improving the quality and quantity of ethical science reporting by the SEE [south-eastern European] media and will contribute to increasing public awareness on the importance of scientific knowledge and towards developing a critical science journalism culture in SEE'.[15] On a broader European scale, in 2007 the European Commission published a guide to science journalism training, updated in 2010 (European Commission 2010), and hosted a forum in Barcelona to discuss issues in the media coverage of science and in the associated training.

In Africa, global and continental intergovernmental organisations supported a 2012 workshop in Addis Ababa that gathered science and technology journalists from various African countries, heads of key media institutions and scientists to 'discuss how best to communicate scientific issues to the public'.[16] In Asia, the Pakistan Biotechnology Information Center organises media workshops and training courses aimed at enhancing 'the capacity of electronic and print media to objectively cover biotechnology-related issues' (Choudhary and Youssuf 2011: 254).

Among non-governmental organisations supporting the worldwide development of science coverage in mass media, scidev.net and the World Federation of Science Journalists (WFSJ) deserve specific mention. Scidev.net provides an Internet platform for reporting and discussion of scientific developments particularly in – or from the perspective of – less developed countries. The service has the support of the journals *Nature* and *Science* and of development aid agencies and charities with a particular interest in supporting science and technology in developing countries. It has developed a network of correspondents across the world's regions, encourages emerging talent and publishes practical guides on various aspects of reporting science. Similarly, the WFSJ provides experienced mentors for journalists in less developed countries who wish to specialise in science and offers an online course in science reporting. The Federation's biennial conferences have become a focus for discussion of the effects on science coverage of the crisis affecting media industries, particularly in the more developed countries.

University taught programmes in science communication. Over the past 25 years, programmes leading to awards specifically in science communication have come to be recognised as one of the features of a developed science communication infrastructure. From the earliest examples of masters degrees and postgraduate diplomas in science communication established in Australia, Britain, France, Italy and Spain, such programmes are now found in many western European and Latin American countries and India. Among the more recently established masters programmes in science communication are those in Budapest, Hungary and Lisbon, Portugal, and an online programme by Universitat Pompeu Fabra in Barcelona, Spain. In New Zealand, Otago University in 2013 recruited a second professor of science communication for its programmes; over half of its students in this field come from abroad. India's and Korea's several postgraduate diploma and degree programmes in science communication or science journalism have been driven largely by guidance and funding from national or state government. In Brazil, a Masters in Scientific and Cultural Communication was added to the existing offering in science journalism at the University of Campinas (Vogt et al., 2009). At the National Autonomous University of Mexico, the programme in science popularisation, started in 1996 through a close association with a science museum, has been linked to longer-established studies in the philosophy of science (Haynes 2009). Laurentian University, Ontario, Canada, set up a Graduate Diploma in Science Communication as a joint initiative with the Science North science centre, declaring it 'North America's first and only comprehensive Science Communication program',[17] though there are specialisation strands in communication masters at Drexel University, Philadelphia and at the University of Florida, and preparatory work on a new single-subject masters in science communication began in 2013 at another Canadian university.

These programmes show some common characteristics across quite different cultural and educational settings, though the relative emphasis on social studies of science, communication theory and professional skills can vary considerably (Mulder et al. 2008; Trench 2012). North America is under-represented in this sector, reflecting a preference there for more strictly professional programmes in science writing and science journalism. In this respect and in others, the global spread of science communication is not a uniform diffusion of a universal model. Nor is the trend in one direction only: there are also examples of programmes that have been reduced, suspended or cut as part of their host institutions' rationalisation (Trench 2012).

University research in science communication. Individual projects and institutional programmes in science communication research have grown up alongside postgraduate taught programmes. A first wave of doctoral research projects in science communication featured trained scientists who were converting to science communication. More recently, the taught programmes in science communication have been a source of doctoral researchers. The countries that were earliest to establish postgraduate taught programmes have tended also to be the most strongly represented in formal academic research. In many cases in these countries, those who teach on the degree and diploma programmes have completed such programmes themselves and proceed to PhD degrees. In China, however, research in science communication has developed in the absence of postgraduate teaching in this subject area; the China Research Institute of Science Popularisation (CRISP) has facilitated many doctoral research projects, often also including periods of study abroad and, thus, contributing to the exchange of experience across countries and continents. Science communication research outputs in China are at a comparatively high level, and increasing: a report on the development of science popularisation studies found 1,795 papers published between 2002 and 2007 (Ren et al. 2012).

An attempted characterisation of topics, theories and methods in current PhD research in science communication showed wide variation (van der Sanden and Trench 2010). While the pattern may be complex and even contradictory, the trend in numerical terms appears clear from informal evidence gathered in convening a network meeting in 2012 of early-career researchers: there may have been more PhD projects in science communication under way in late 2013 than were completed in the two preceding decades. A study of science communication research in Australia noted the increase in PhD students from 3 in 1997 to 20 in 2012, and a doubling of the output of research papers written by Australian researchers from the 1990s to the 2000s

(Metcalfe and Gascoigne 2012). In the 2000s, research on public attitudes to science, media and science, and policy on science communication commanded roughly equal attention, at 19, 17 and 16 per cent, respectively, of the primary topics of these papers.

A further outgrowth of postgraduate teaching in science communication has been the publication of specialist academic journals in the field. *Public Understanding of Science*, whose title reflects its provenance in Britain in the years following the 1985 report of the Committee on the Public Understanding of Science, emerged in the early 1990s from the same impetus that led to the Masters in Science Communication in Imperial College London. The renaming of another journal, *Knowledge*, as *Science Communication* and the establishment of *JCOM – Journal of Science Communication*, an online, open-access publication from SISSA, Trieste, also reflected the growth of postgraduate teaching and research in the field. Catering to more local markets, *Quark* was published for over a decade by the Science Communication Observatory at Universitat Pompeu Fabra, Barcelona, and the *Japanese Journal of Science Communication* (Kyoto), *Indian Journal of Science Communication, Science Communicator* and *Journal of Scientific Temper* (originating in India) have emerged in more recent years. The pattern of publishing shows both globalisation and localisation trends. This extends to the keywords used in the description and discussion of science communication, some of which have distinctive or exclusive usage in particular countries or regions: 'scientific temper' in India, 'science popularisation' in China, 'social appropriation of science' in Latin America, 'scientific, technical and industrial culture' (generally denoted by the acronym, CSTI) in France.

Reports from five emerging centres of science comunication

To illustrate how the global spread of science communication is manifested in national contexts, we asked correspondents in five countries that have not typically received much attention in the research literature to outline developments in those countries. It will be noted that each of these reports refers to several of the elements outlined above.

Argentina

The global processes affecting scientific and technological practices have been evidenced in Argentina and other Latin American countries, particularly Brazil, Colombia and Mexico. There are many indicators of how science communication is becoming an expanding cultural industry. The environment of institutional science communication is shaped by Argentina's economic growth that, during the last decade, stimulated public policies and consolidated basic science and technology system capacities. The creation in 2007 of a Ministry for Science, Technology and Productive Innovation was a signal of new times. Over recent years, R&D investment – following a regional tendency – grew faster than in Europe, USA and Canada, though behind Asia. Some areas of biotechnology, nanotechnology, information technology and food technology have expanded considerably (Ricyt 2011). The public policy discourse shifted towards a knowledge economy and reducing dependence on commodities production. Within this framework, the importance of social communication of science, including the reinforcement of traditional museums and new science centres, has been emphasised.

There are also signs of stronger relations between scientific institutions and the mass media system, as suggested in the mediatisation process (Weingart 1998; Väliverronen 2001; Peters *et al.* 2008). Science and technology institutions have progressively incorporated media and public opinion operating logics, including practices, values and institutional and technological modes that media use to operate, supported by formal and informal rules. Some indicators are: the creation or consolidation of groups and structures for public communication in universities and science and technology institutions; the intensification of contacts between scientists and journalists; the increasing salience of a rhetoric of engagement, dialogue and public inclusion. Scientists have tended to gain salience in the public sphere and are participating in wider social debates. This is connected with new national political tendencies where intellectuals, scientists and public figures in general have recovered their public role (Polino 2013). On the other hand, the empirical evidence indicates science journalism is also becoming incrementally professionalised and institutionalised (Gallardo 2011; Vara 2007). During the past 15 years, the media have appointed specialist journalists and increased coverage of science-and-technology-related issues. Argentina's research and development has become more prominent in coverage in the main newspapers, public TV and some commercial radio programmes. Science journalists have organised themselves through a network and a professional association[18] and young professionals with new expectations are entering science journalism and science popularisation, as can be seen in data for Argentina from a worldwide survey (Bauer *et al.* 2013). Many of these are coming through new university programmes (Murriello 2011), though the spread of science communication training programmes is still limited.

Increased public demand is reflected in the publication by the main publishing houses of popular science books, some locally produced and some translated materials from Europe and United States. The market for popular science magazines is also growing. The recent creation of competitions and awards for science journalism and popular science that stimulate novel contributions are another effect of the cultural industries' increased interest in science-related matters. These developments can be seen to relate to audience interests and cultural habits as demonstrated by nationally representative surveys on public understanding of science (Mincyt 2012; Secyt 2004, 2007; Polino 2012).

However, the science communication environment in Argentina also has certain constraints which are shared with other Latin American countries. As pointed out elsewhere (Polino 2013), institutional communication shows some structural weaknesses: despite university and federal institutions acknowledging the importance of press offices, funding is scarce: most of these groups have no guaranteed budgets or permanent positions to produce science communication materials, so many of their practices are voluntary. Although things can be seen to change slowly, scientists are not clearly incentivised to engage in public communication. Often, these activities are considered to be *decorative* from the perspective of a scientific career. The consequence of this is evident: the system tends to integrate only those who are already convinced and to reject new talent.

Another problem is the conception of communication and the perception of the public that underlie many institutional initiatives in science communication and science popularisation. Even with strong evidence showing why the deficit model does not work and how it produces a distorted image of science and technology (e.g. contributions in Bucchi and Trench 2008; Dierkes and von Grote 2000) many university efforts in science communication are still inspired by or oriented to that model. Many scientists often deal with journalists in pedagogical terms: they assume that journalists need to be educated (by the scientists). This produces an obvious tension, which is recreated many times in public lectures, talks and media interventions.

Dialogue and social participation are values not clearly translated into institutional practices: there are almost no sponsored mechanisms for citizen participation. We can also observe that media tend to favour descriptive rather than analytical perspectives; science news is often reduced to scientific discoveries, leaving out perspectives on risks, conflicts of interest or the connections between science and economy. Coverage is dominated by a small number of information sources and, consequently, the media offer limited comparative content.

Estonia

The upheavals of the political, economic and social transition of the 1990s left little intact of the ideology-driven and scientist-centred Soviet science popularisation system. The few survivors like the popular science magazine *Horisont* (Horizon) were run by devoted enthusiasts, but the field was marginal and had poor resources and support. By the turn of the millennium Estonia had started to establish itself as an innovative country, strongly oriented to information technologies. Along with that came the realisation that future successes in that area will be seriously undermined by the lack of scientists and engineers and the low public profile of sciences. Since then, the economic benefits of a high number of scientists and engineers in a society and the need to attract young people to science and engineering has been the dominant discourse in Estonian science communication activities, especially those run by the public sector.

Through the Estonian Research Council and other schemes the government funds various initiatives to engage young people with science-related activities. These focus on hands-on activities, like robotics workshops in schools or interactive study programmes in science centres, and on promoting science as a career choice. Many of them stand out as youth-to-youth projects; one of the most notable has been the Science Bus project by the National Physics Society, delivering science theatre shows to schools. To further increase the public visibility of these activities, September 2011–September 2012 was declared the Year of Science in Estonia.

The biggest government scheme to support science communication has been the partly EU-funded TeaMe programme (2009–2015). Its main outcome for the public has been two prime-time TV programmes: a series of portraits of prominent scientists and a students' game show. The programme has also funded training of scientists and journalists and development of various study materials for schools.

In recent years, science centres and museums have become prominent actors in the field of science communication, acquiring new buildings and new exhibits. The opening of the new science centre AHHAA in Tartu and the new exhibition of the Maritime Museum in the Seaplane Harbour in Tallinn have been major events, and these have attracted record numbers of visitors. AHHAA also co-ordinates the annual Researchers' Night, that has grown into a week-long festival and culminates with a live TV show on the national broadcaster's main channel. Besides the festival, the number of events for adults is still small and they are mostly one-offs.

In the mass media, the only specialised science journalist is employed by the national broadcaster for a weekly radio show, though most of the major newspapers have regular contributors specialising in science. There are several online sites dedicated to science and technology news, one of the most prominent ones (www.novaator.ee) hosted by the University of Tartu. The nature of science coverage is mostly informational and promotional with little critical analysis. As issues of public trust in science are not among the main drivers for science communication, there is little discussion about the nature of science or about the need for critical journalism. Public information from research institutions tends to focus on institutional affairs rather than presenting scientific results – however, there are a few individual scientists who skilfully engage with media directly. The Estonian Research Council annually awards national science popularisation prizes.

The rapid changes of the past decade in the Estonian science communication field have been brought about by a combination of internal and outside factors. For example, the national contest for young scientists was launched after Estonia was approached by the European Commission to send entries to the European contest. Many major projects (museums, TV programmes) have received one-off EU grants. However, this followed decisions made in Estonia to dedicate funds to science communication-related projects. Still, regular funding remains the most pressing question for many activities; the government mostly supports activities via yearly open calls.

Malaysia

The importance attached to science and technology has been reflected in several Malaysian government key policies such as Vision 2020, the 10th Malaysia Plan, the National Science and Technology Policy, the National Biotechnology Policy and the National Agricultural Policy. Related efforts to embed science and innovation in Malaysian society include the declaration of 2010 as Malaysia Innovation Year, 2011 as the Year for the Promotion of Science and Mathematics and 2012 as the National Science and Innovation Movement Year.

The National Science Centre (PSN) and Petrosains science centre are at the heart of science promotion in Malaysia. The first is run by the government under the auspices of the Ministry of Science, Technology and Innovation (MOSTI), and the second is the corporate contribution of Petronas, the leading oil and gas company. PSN has various outreach programmes which give opportunities for people outside the area around the federal territory, Kuala Lumpur, to visit the centre, and science camp programmes which give a chance for school students in rural areas to experience the learning of science and technology through interactive hands-on activities. The centre also conducts a special programme for teachers and organises competitions and carnivals to instill interest in science and technology. Petrosains is an interactive science centre that presents the science and technology of the petroleum industry, as well as general science using a hands-on approach. The National Planetarium is also active in educational and outreach programmes on space education (Zainuddin, 2008).

Various organisations in Malaysia have shown a strong commitment to science events. The Academy of Science Malaysia has initiated programmes such as back-to-school lectures by the fellows, quizzes and competitions, science camps, Science and Mathematics Expos, National Science and Technology Month, and exhibitions and publications to enhance public awareness. Universities and schools throughout the country conduct science camps during the school holidays, either on their own or collaboratively. The various government ministries have also played their part through programmes and campaigns related to their particular responsibilities, either on their own or in collaboration with universities, corporate bodies or NGOs.

One effort by MOSTI has been the creation of the MyBiotech@School programme which has exposed nearly 40,000 students throughout the country to biotechnology through hands-on experiments, multimedia shows, demonstrations and talks by scientists and industry experts (Mivil 2013; BIO-BORNEO 2013; Firdaus-Raih et al. 2005). The Ministry of Natural Resources and Environment and the Department of the Environment have conducted various environmental awareness programmes (Pudin et al. 2005) while the Ministry of Health has organised health-related campaigns (MOH 2010; Malaysian Digest 2013; CAP 2011). Non-governmental organisations have also played a prominent role in science awareness in Malaysia with various projects and websites.[19]

The mainstream newspapers play a significant role in the dissemination of science-related information to the public. In January–March 2009, about 300 science-related pieces of news were published (Arujunan and Aziazan 2011), with about two-thirds of these articles focusing on medical- and health-related issues. The remaining one-third dealt with disciplines such as biotechnology, space, biodiversity, the environment, agriculture and others. The coverage of scientific issues tends to follow current interest or controversy at national or international levels, including the focus of government policy at any particular time.

Nigeria

The future of the Nigerian economy is considered to be predicated on the rapid diffusion of science and technology; this is the government's view of the best way forward for accelerated growth. The successes of the technology-driven mobile telephony industry in generating employment and increasing wealth no doubt contributed to this policy direction and to the hope that this success can be replicated in other sectors. The increase in GSM mobile lines from less than a million in 2001 to over 100 million in 2010 showed the highly significant impact of policy changes on economic growth; in 2001 the state monopoly in telecommunications was ended.

A science and technology summit held in Nigeria in 2010 aimed to stimulate the interest of the public in science, technology and innovation, to encourage indigenous researchers, inventors and innovators, and to promote the domestication of modern technologies. Stakeholders at the summit included the federal agencies, Millennium Development Goal (MDG) agencies, small and medium enterprises, chambers of commerce and industry, and international organisations. A new science, technology and innovation (STI) policy was subsequently launched in 2011, the first such policy statement having been produced in 1986, and reviewed in 1997 and 2003. The new policy emphasises innovation and technology transfer and sets out specific objectives for the promotion of STI communication and inculcation of science culture.

The Nigeria Academy of Science provides advisory services for the federal government on STI, of which one was an audit of research and development agencies; one recommendation was for more synergy among the agencies and the institution of an annual national science and technology forum. The 19 science agencies under the Federal Ministry of Science and Technology were at that time running independently, each holding its own public exhibitions and managing its own library and science museum. Without upscaled science awareness activities at federal and state levels, including regular exhibitions of the numerous outputs of the state's research and development agencies, these will remain largely hidden from the wider public and potential users of the technologies.

The Academy of Science is also actively involved in popularising science. In 2012, it held a workshop on effective communication of science research aimed at bridging the gap between scientists and the public and bringing together young scientists and journalists. The Academy also works in partnership with Schlumberger, a private firm, the Nigeria Young Academy, and science, engineering and mathematics teachers and volunteer employees on the Schlumberger Excellence in Education Development (SEED) programme.[20] SEED gives students and teachers the opportunity to work together on a research project. The programme aims to ignite a passion for science and develop the student's technical potential by building critical thinking, creativity and innovation skills.

The Nigerian press regularly features science and technology articles and *The Guardian*, regarded as the flagship of the Nigeria press, has maintained regular science columns for several decades. Media analysis (Falade 2014) has shown that the percentage of science items in the news is similar to that of the United Kingdom and the United States.

Turkey

With a population of over 75 million that is comparatively young, Turkey has acknowledged the value of science communication through investment of large amounts of money to enhance public engagement with science and technology, promote a scientific culture, and develop a dialogical science communication culture in the country. The Scientific and Technological Research Council of Turkey (TUBITAK), in co-operation with local authorities, has been establishing science centres around the country, aiming to complete a science centre in all 16 metropolitan areas by 2016 and in all 81 cities by 2023 (TUBITAK 2013). TUBITAK is also responsible for promoting, funding and carrying out cutting-edge scientific research and making the findings available to the public. It publishes popular science books as well as popular science magazines for children and for the general public.

The Turkish Ministry of National Education has also been working with TUBITAK and the Turkish Radio and Television Corporation on developing effective forms of science communication. The involvement of the education ministry reflects recent changes in science education: creating engaged and scientifically literate citizens has become a focus of the new science curriculum.[21] The new media literacy curriculum[22] specifically endorses public participation in policy debates about science-related social issues; this is seen as essential to maintain a healthy democracy (Cakmakci and Yalaki 2012).

There are several challenges still to be addressed, among them the small number of researchers in science communication and the very limited output of research on science communication. There is no science communication division in any Faculty of Communication or in other faculties. The Turkish press often covers science-and-technology-related issues, but few of the newspapers have a separate science section. Journalists writing about science have limited knowledge and expertise in science communication (Erdogan 2007). But the need and the means to improve the quality of science communication research, education and practice are overlooked in the policy reports. These activities and the establishment and sustainability of a community of practice in science communication are not given the same importance as material outputs such as science centres. Another challenge that Turkey faces is the unsustainable short-term cycle of policies in science outreach. Over a little more than a decade, the minister of education changed five times and each person in that role had different priorities, agendas and favoured different kinds of science communication models (from deficit to dialogue and participation models). This has caused tensions among the public, policymakers, science communication researchers and practitioners.

Concluding remarks

There are striking parallels in these short summaries of conditions for science communication in five countries that have few, if any, bilateral connections. These summaries can be taken as evidence of trends that are not merely born of international diffusion through contagion but represent a global phenomenon. The term 'science communication' is far from being universally recognised nor is it used uniformly, where it does occur. But in disparate countries, with notably different cultural contexts, a similar kind of commitment is being made to promoting science and, with it, to promoting awareness and appreciation of science. Across these examples, there are similar references to science's role in technological and economic development and to the need to encourage interest in science particularly among children and young people.

It is fair to observe that the contributors of these reports are scholars who are aware of the international publications and discussions on models of science communication, also continued elsewhere in this volume. They are thus more likely than others to draw attention to the continuing force of supposedly *old* models in these *newer* regions. However, one conclusion is inescapable: the supposed turn from deficit approaches to dialogue – however valid or not it may be as an observation of regions with longer traditions of institutionalised science communication – does not apply in regions where the science communication culture is, in the terms of the European mapping mentioned above, 'developing' or 'fragile'.

To make this observation is not to make a judgement, nor to apply an evolutionary perspective. It is a reminder that different social conditions shape institutions and practices of communication differently, that trends validly observed in one region of the world do not necessarily apply elsewhere, that discussions of *old* and *new*, or *better* and *best*, in science communication need to be modulated with reference to specific circumstances. We have seen plentiful evidence that didactically oriented programmes of science awareness can coexist with open-forum, interactive and conversational forms of communication in science centres and science cafés. Indeed, the spread of science cafés across the continents is a strong example of a global format, now adapted to local circumstances very widely. The British movement curiously adopted the fully French term, café scientifique, and provides information and advice to the international movement.[23]

As we have seen, several other science communication formats have also spread globally, including science weeks, science festivals, science media centres, short-course communication training for scientists and postgraduate professional education for science communicators. While the international science communication communities have in many cases networked effectively to learn from each other, they may need to develop more sophisticated tools for thinking about and analysing science communication in a global context. Taking a global view draws attention to the patterns of difference as much as to the patterns of similarity.

Key questions

- What are the main policy and other drivers of the global spread of science communication?
- What are the social and cultural factors that shape science communication institutions and practices in particular contexts?
- What analytical criteria might we use to most effectively identify patterns of similarity and difference in science communication cultures of different countries?

Notes

1. The last-named five authors contributed reports on the situations in their respective countries. Latifah Amin is Associate Professor at the Centre for General Studies, Universiti Kebangsaan Malaysia; Gultekin Cakmakci is Associate Professor in science education at Haccetepe University, Ankara, Turkey; Bankole Falade completed PhD research in 2013 at London School of Economics on science communication in Nigeria; Arko Olesk is Head of the Science and Innovation Communication Centre at Tallinn University, Estonia; Carmelo Polino is a senior researcher in science communication at Centro REDES, Buenos Aires, Argentina.
2. The project website, including all reports, is at www.masis.eu
3. See www.jst.go.jp/EN/about/index.html#NOTE2
4. See www.jst.go.jp/EN/operations/operation2_c.html
5. See http://aspacnet.org/ns/membership/list/national-science-and-technology-museum-taiwan
6. See https://www.miraikan.jst.go.jp/en/aboutus/
7. See www.petrosains.com.my
8. Workshop materials in possession of this chapter's lead author as tutor at the event.
9. See www.masis.eu/english/storage/publications/nationalreports/
10. http://sciprom.cpn.rs/#about
11. www.wellcome.ac.uk/stellent/groups/corporatesite/@msh_grants/documents/web_document/wtp052897.pdf
12. See www.sciencemediacentre.net/
13. See www.sciencemediacentre.org/working-with-us/for-scientists/intro/
14. www.ecsite.eu/news_and_events/news/summer-school-science-communication-and-euro-mediterraneanmiddle-eastern-solida
15. See www.unesco.org/new/en/venice/about-this-office/single-view/news/call_for_participation_south_east_european_science_journalism_school_deadline_8_september_2013/#.UnI-RXAmWSo
16. www1.uneca.org/TabId/3018/Default.aspx?ArticleId=1989
17. www.sciencenorth.ca/sciencecommunication/
18. See www.radpc.org/
19. See www.mac.org.my; www.cancer.org.my; http://ensearch.org/global-gateway/environmental-ngos-in-malaysia

20 See www.planetseed.com
21 See Turkish Ministry of National Education, *Science Curriculum*, at http://goo.gl/jSSG5w
22 See www.medyaokuryazarligi.org.tr
23 See www.cafescientifique.org/

References

Arujanan, M. and Aziazan, B. (2011) 'Biotechnology awareness: from the ivory towers to the masses', in M. J. Navarro and R. A. Hautea (eds) *Communication Challenges and Convergence in Crop Biotechnology*, Ithaca, New York, and Los Baños, Philippines: International Service for the Acquisition of Agri-Biotech Applications and SEAMEO Southeast Asian Regional Center for Graduate Study and Research in Agriculture: 180–202.

Bauer, M., Shukla, R. and Allum, N. (eds) (2011) *The Culture of Science: How the Public Relates to Science Across the Globe*, London: Routledge.

Bauer, M., Howard, S., Yulye, J., Ramos, R., Massarani, L. and Amorim, L. (2013) *Global Science Journalism Report: Working Conditions and Practices, Professional Ethos and Future Expectations*, London: LSE, Museo de Vida, Scidev.net.

BIO-BORNEO (2013) Accelerating Biotechnology For Borneo Through Bioeconomy; online at www.mosti.gov.my/index.php?option=com_content&view=article&id=2642&lang=bm

Bucchi, M. and Trench, B. (eds) (2008) *Handbook of Public Communication of Science and Technology*, London: Routledge.

Bultitude, K., Cheng, D., Durant, G., Jackson, R. and Massarani, L. (2012) 'Comparison of national strategies: documentary analysis for Australia, Brazil, China and the United Kingdom', in M. Bucchi and B. Trench (eds) *Quality, Honesty and Beauty in Science Communication – 12th International Public Communication of Science and Technology Conference*, Vicenza: Observa Science in Society: 43–48.

Cakmakci, G. and Yalaki, Y. (2012) *Promoting Student Teachers' Ideas about Nature of Science through Popular Media*, Trondheim: S-TEAM/NTNU.

CAP (2011) 'Health campaigns must be concerted and supported by every government body'; online at www.consumer.org.my/index.php/health/healthcare/474-health-campaigns-must-be-concerted-and-supported-by-every-government-body

Choudhary, M. I. and Youssuf, S. (2011) 'Initiating science communication in the Organization of Islamic Conferences countries', in M. J. Navarro and R. A Hautea (eds) *Communication Challenges and Convergence in Crop Biotechnology*, International Service for the Acquisition of Agri-biotech Applications (ISAAA), Los Baños, Philippines: Ithaca, New York and SEAMEO Southeast Asian Regional Center for Graduate Study and Research in Agriculture (SEARCA), 244–259.

Dierkes, M. and von Grote, C. (eds) (2000) *Between Understanding and Trust: The Public, Science and Technology*, London: Routledge.

Erdogan, I. (2007) *Journalism and Science Communication in Turkey: Structural Features, Problems and Suggested Solutions (in Turkish)*, Ankara: Pozitif Matbaacilik.

European Commission (2010) *European Guide to Science Journalism Training*, Brussels: European Commission; online at http://ec.europa.eu/research/conferences/2007/bcn2007/guide_to_science_journalism_en.pdf

Falade, B. A. (2014) *Vaccination Resistance, Religion and Attitudes to science in Nigeria*, PhD Diss., London: LSE.

Firdaus-Raih, M., Senafi, S., Murad, A. M., Sidik, N. M., Lian, W. K., Daud, F., Ariffin, S. H. Z., Zamrod, Z., Seng, T. C., Othman, A. S., Harmin, S. A., Saad, M.Y. R. and Mohamed, R. (2005) 'A nationwide biotechnology outreach and awareness program for Malaysian high schools', *Electronic Journal of Biotechnology* 8, 1: 10–16.

Gallardo, S. (2011) 'Profesionalización del periodismo científico. Avances y desafíos. ¿Qué se espera hoy de un periodista científico?', in Ministerio de Ciencia, Tecnología e Innovación Productiva, OEA (eds), *Periodismo y comunicación científica en América Latina*, Buenos Aires: Ministerio de Ciencia, Tecnología e Innovación Productiva.

Gascoigne, T. and Metcalfe, J. (2001) 'The evaluation of national programs of science awareness', *Science Communication*, 23, 1: 66–76.

Haynes, E. R. (2009) 'A graduate course for science communicators: a Mexican approach', *Journal of Science Communication*, 8: 1: C04.

Koizumi, A., Yuko, M. and Shishin, K. (2013) 'Reward research outreach in Japan', *Nature* (1 August), 500, 7460: 29.

Malaysian Digest (2013) 'Health Ministry targets "less salt" after "less sugar" campaign' (15 March); online at www.malaysiandigest.com/news/36-local2/283171-health-ministry-targets-less-salt-after-less-sugar-campaign.html

Mejlgaard, N., Bloch, C., Degn, L., Ravn, T. and Nielsen, M. W. (2012) *Monitoring Policy and Research Activities on Science in Society in Europe (MASIS) Final Synthesis Report*, Brussels: European Commission, Directorate-General for Research and Innovation.

Metcalfe, J. and Gascoigne, T. (2012) 'The evolution of science communication research in Australia', in B. Schiele, M. Claessens and S. Shi (eds) *Science Communication in the World: Practices, Theories and Trends*, Dordrecht: Springer: 19–32.

Miller, S., Caro, P., Koulaidis, V., de Semir, V., Staveloz, W. and Vargas, R. (2002) *Report from the Expert Group, Benchmarking the Promotion of RTD Culture and Public Understanding of Science*, Brussels: European Commission; online at ftp://ftp.cordis.europa.eu/pub/era/docs/bench_pus_0702.pdf

Mincyt (2012) *Tercera Encuesta Nacional de Percepción Pública de la Ciencia*, Buenos Aires: Ministerio de Ciencia, Tecnología e Innovación Productiva.

Mivil, O. (2013) '40,000 students get a taste of biotechnology', *New Straits Times* (22 February); online at www.nst.com.my/nation/general/40-000-students-get-a-taste-of-biotechnology-1.222852

MOH (2010) 'Launch of the 1 Malaysia MDA-GSK oral health awareness campaign'; online at http://imuoralhealth.blogspot.ie/2010/03/launch-of-1malaysia-mda-gsk-oral-health.html

Mulder, H., Longnecker, N. and Davis, L. (2008) 'The state of science communication programs at universities around the world', *Science Communication*, 80, 2: 277–287.

Murriello, S. (2011) 'Especialización en divulgación de la ciencia, la tecnología y la innovación. Universidad Nacional de Río Negro', in Ministerio de Ciencia, Tecnología e Innovación Productiva, OEA (eds) *Periodismo y comunicación científica en América Latina*, Buenos Aires: Ministerio de Ciencia, Tecnología e Innovación Productiva.

Peters, H. P., Brossard, D., de Cheveigné, S., Dunwoody, S., Kallfass, M., Miller, S. and Tsuchida, S. (2008) 'Interactions with the mass media', *Science*, 321, 5886: 204–205.

Polino, C. (2012) 'Información y actitudes hacia la ciencia y la tecnología en Argentina y Brasil. Indicadores seleccionados y comparación con Iberoamérica y Europa', in Ricyt, *El Estado de la Ciencia. Principales indicadores de ciencia y tecnología Iberoamericanos/Interamericanos*, Buenos Aires: Ricyt.

Polino, C. (2013) 'Science communication in Latin American countries: some comments on its current strengths and weaknesses', in P. Barenger and B. Schiele (eds) *Science Communication Today: International Perspectives, Issues and Strategies*, Paris: CNRS Éditions, 263–280.

Pudin, S., Koji, T. and Ambigavathi, P. (2005) *Environmental Education in Malaysia and Japan: A Comparative Assessment*, paper presented at the Education for Sustainable Future International Conference, Ahmedabad, India, 18–20 January; online at www.ceeindia.org/esf/download/paper20.pdf

Ren, F., Lin, Y. and Honglin, L. (2012) 'Science popularisation studies in China', in B. Schiele, M. Claessens and S. Shi (eds) *Science Communication in the World: Practices, Theories and Trends*, Dordrecht: Springer: 65–80.

Ricyt (2011) *El Estado de la Ciencia. Principales Indicadores de Ciencia y Tecnología Iberoamericanos /Interamericanos*, Buenos Aires: REDES/OEI.

Secyt (2004) *Los argentinos y su visión de la ciencia y la tecnología. Primera Encuesta Nacional de Percepción Pública de la Ciencia*, Buenos Aires: Ministerio de Educación, Ciencia y Tecnología.

Secyt (2007) *La Percepción de los Argentinos sobre la Investigación Científica en el País. Segunda Encuesta Nacional*, Buenos Aires: Ministerio de Educación, Ciencia y Tecnología.

Schiele, B., Claessens, M. and Shi, S. (eds) (2012) *Science Communication in the World: Practices, Theories and Trends*, Dordrecht: Springer.

Shi, S., and Zhang, H. (2012) 'Policy perspectives on science popularisation in China', in B. Schiele, M. Claessens and S. Shi (eds), *Science Communication in the World: Practices, Theories and Trends*, Dordrecht: Springer, 81–94.

SIRC and ASCOR (2006) *Media, Science and Society: Engagement and Governance in Europe, Final Report*; online at www.sirc.org/messenger/Final_Report_Draft_1.pdf

Trench, B. (2012) 'Vital and vulnerable: science communication as a university subject', in Schiele, B, Claessens, M. and Shi, S. (eds) *Science Communication in the World: Practices, Theories and Trends*, Dordrecht: Springer, 241–258.

Trench, B. and Miller, S. (2012) 'Policies and practices in supporting scientists' public communication through training', *Science & Public Policy*, 39, 6: 722–731.

TUBITAK (The Scientific and Technological Research Council of Turkey) (2013) *4003 National Support Program: Establishment of Science Center Support Program*.

Väliverronen, E. (2001) 'From mediation to mediatization – the new politics of communicating science and biotechnology', in U. Kivikuru and T. Savolainen (eds) *The Politics of Public Issues*, Helsinki: University of Helsinki, 157–178.

van der Sanden, M. and Trench, B. (2010) 'Analysis of doctoral research in science communication', in *Science Communication without Frontiers: Proceedings of 11th International Conference on Public Communication of Science and Technology*, New Delhi: National Council for Science and Technology Communication, 92–95.

Vara, A. M. (2007) 'Periodismo científico: ¿preparado para enfrentar los conflictos de interés?', *Revista CTS*, 9, 3: 189–209.

Vogt, C., Knobel, M., Camargo, V. R. T. (2009) 'Master's degree program in scientific and cultural communication: preliminary reports on an innovative experience in Brazil', *Journal of Science Communication*, 8, 1: C06.

Weingart, P. (1998) 'Science and the media', *Research Policy*, 27, 8: 869–879.

Zainuddin, M. Z. (2008) 'Perspective of space science education and awareness in Malaysia', *Thai Journal of Physics*, 3: 1–2.

17
Assessing the impact of science communication
Approaches to evaluation

Federico Neresini and Giuseppe Pellegrini

Introduction

In its everyday usage, the term evaluation is employed in a wide variety of contexts to denote the act of expressing judgement on some activity. This is a natural exercise which seems to require neither careful thought nor particular skills: we evaluate when we decide whether to use a particular technology, whether a colleague has done his or her work well, whether the person to whom we are talking has understood what we want to say.

Leaving aside the variety of objects to which it is applied, evaluation therefore means establishing the extent, at least approximately, to which a given action has produced effects which match the purposes for which it was undertaken. This definition is still valid when evaluation leaves the everyday domain and becomes a set of activities performed to assess the results of actions that are more complex and more structured – as well as more ambitious – than those of everyday routine. We thus have evaluation of educational performance, of the social impact of a particular political decision, of the quality of a public service, of the claimed advantages of a technological application or of the effects of a communication campaign. In this chapter we review approaches to the evaluation of initiatives in the public communication of science and technology.[1]

The demand for evaluation of science communication processes is increasing as a result of several trends. Many institutions have changed their communicative strategies, from 'communicating predefined institutionally validated scientific facts to assisting in the creation of spaces for the examination of the underlying value-laden assumptions and interests in competing risk definitions of the various actors involved in techno-environmental controversies' (Maeseele 2012: 69). There are also demands for improvements in public communication of science and for training to develop knowledge and skills to get messages from science across to the public more effectively (Wehrmann and De Bakker 2012). There are proposals for the enhancement of quality in communication through evaluation as part of 'a reflective cycle' (Stevenson and Rea 2012; cf. McKenzie 2012).

In this perspective, evaluation becomes a structured and formal activity, a systematic inquiry which applies specific procedures in gathering and analysing information on the content, structure and results of a project, a programme or a planned intervention (Guba and Lincoln 1989).

Evaluation may also perform a political role in supporting decision-making processes and choices on programmes and activities by trying to reduce the level of uncertainty (Patton 1986: 14). The task of evaluation, therefore, is to yield systematic evidence which informs experience and judgement, furnishing an array of options available to the actors involved in a programme (Weiss 1998). In short, evaluation relates to what determines or explains the success or failure of an action in regard to the goals for which it was first conceived and then undertaken.

Phases and choices in evaluation

The organisation of a communication campaign, the planning of an exhibition, the media publication of a news story, the construction of a website, the establishment of a science centre: these and all other activities related to public communication of science and technology (PCST) develop through time. Their evaluation can be divided into three phases: design (*ex-ante*), implementation (*in itinere*) and conclusion (*ex-post*). Although this is an obvious oversimplification, this distinction is nevertheless a useful means to focus on certain crucial aspects of evaluation (see Grant 2011).

Focusing on the design phase, evaluation concentrates on the adequacy of the resources available with respect to the objectives to be pursued. To this end, assessment is made not only of financial and time aspects, and of human resources, but also of whether the communicative strategy adopted will be able to reach the target audience. Knowledge of the principal characteristics of the interlocutors to be involved in the communication is therefore crucial (Storksdieck and Falk 2004) and it may also be important for subsequent evaluation of the results.[2]

Evaluation *in itinere* largely corresponds to formative evaluation, that is, evaluation intended to establish what is working, and what is not, in the ongoing activity, and to adjust the activity accordingly (Scriven 1991). This requires analysis of patterns of interaction among the actors, identification of obstacles and unforeseen effects, and monitoring of the uses made of the available resources. Evaluation *in itinere* often also uses content analysis and ethnographic observation. The evaluation focus can be, for example, on what happens at a public meeting between scientists and members of the public – as in the case of a consensus conference[3] – or how a visit to a museum or a science centre develops as an experience involving not only learning, but also entertainment, social relationships and emotions (Kotler and Kotler 1998; Storksdieck and Falk 2004).[4] Information may also be collected during an ongoing communicative process to conduct the kind of 'theory-based evaluation' proposed by Weiss:

> a mode of evaluation that brings to the surface the underlying assumptions about why a programme will work. It then tracks those assumptions through the collection and the analysis of data at a series of stages along the way to final outcomes. The evaluation then follows each step to see whether the events assumed to take place in the programme actually take place.
>
> (Weiss 2001: 103; see also Gascoigne and Metcalfe 2001)

An example of effective formative assessment in communication through the Web concerned the integration of technology in university coursework (Bowman 2013). A semester-long wiki project was planned and evaluated using a formative assessment plan. The assessment indicated future structural and communicative modifications, taking into account students' suggestions. An open-ended questionnaire was used to verify content learning, 'the effectiveness of producing a better understanding of real-world applications through the wiki project and their perception of the wiki project's shortcomings and/or challenges' (ibid.: 4).

A good example of formative evaluation can also be found in the Large Hadron Collider (LHC) Communication Project promoted by the British Particle Physics and Astronomy Research Council. The LHC Communication Project aimed to 'engage the public with particle physics, developing a four-year programme with the twin aims of increasing public knowledge of, and support for, particle physics and inspiring young people to choose physics courses at 16-years-old and subsequent decision points' (PSP 2006). The evaluation was carried out to assess levels of knowledge and understanding of particle physics and the basic scientific questions behind the LHC project. It sought to gauge public perception and ask relevant questions to help improve the communication process. These objectives were studied for two target groups, members of the general public who are interested in science and students and their teachers. Formative evaluation, conducted through focus groups, interviews, questionnaire and discussions, gave results that helped refine the methods and content of communication in the course of the project.

Evaluation finds perhaps its most obvious place at the end of the communicative process (i.e. *ex-post*) because it aims to determine and explain the success or failure of an action with respect to the goals that it was intended to achieve. When evaluation is realised at the conclusion of the PCST initiative and is focused on its outcomes, it is also called *summative evaluation*, as opposed to *formative evaluation*. An instrument of summative evaluation for assessing the writing skills of scientists trained in the public communication of science was used in a recent study (Baram-Tsabari and Lewenstein 2013). This instrument assessed written communication in an analytical framework; the study contributed to a better understanding of scientists' skills, revealing that to some extent they need to learn a new language of science, that is 'the discourse of public communication of science' (ibid. 80) as a means of engaging with the public.

In summative evaluation it is common to distinguish between *output* and *outcome*. In the former case, the results are defined as the effective accomplishment of what the initial design envisaged, thus privileging the point of view of the promoters of the communication. In the latter case, the results are instead viewed as changes produced by the communication so that the attention focuses – at least potentially – on all the actors involved in the process. Evaluation may yield contrasting judgements in the two cases but, even more importantly, good results in terms of output offer no guarantees in terms of outcome.

Before tackling the problems involved in conducting the evaluation of a PCST initiative, we focus attention on the operating conditions which actually make it possible. We can observe a certain discrepancy between the widely held conviction that evaluation is useful and necessary on the one hand and the fact that evaluation is not always carried out on the other.

In order to evaluate a PCST initiative, we need to: define the objectives of the initiative from the very beginning, as precisely as possible; devote sufficient resources (money, time, personnel) to it; ensure that the evaluation process and the people in charge of it are guaranteed sufficient legitimacy.

It is quite normal for those juggling a thousand tasks, constantly grappling with daily emergencies, to consider time spent on clarifying what one hopes to achieve with a PCST initiative as a fruitless investment. However, the very opposite may apply: the time initially *lost* on accurately defining the objectives can be rewarded in various ways, not least by laying a solid basis for learning from the experience through evaluation.

Defining an initiative's objectives from the very beginning and planning its evaluation also allow us to set aside the necessary resources for the evaluation process, avoiding the risk of having it become an activity left to people's goodwill and spare time. Like other activities, evaluation has its costs in finance, time and personnel.

One should not underestimate the fact that evaluation can turn out to be a tiresome activity, which confronts us with a realisation of – and a responsibility for – failure, especially when it comes to complex and diverse initiatives such as PCST programmes. If, in addition, an evaluation is perceived to be a mere test exercise that someone imposes on someone else – for example, the initiative's owners testing their target audience, or its funders testing those who carried it out – then the risk of it never being completed increases. Evaluation is a delicate process, thus there are myriad ways of sabotaging it.

Given these characteristics, it is very important to decide who should carry out the evaluation of communication processes. The use of internal or external evaluators is a particular consideration in carrying out effective evaluation work. Several studies assert that an in-house evaluation allows the development of improved self-awareness, enriching and boosting the skills of operators who develop communication programmes (Irwin 2009, Jensen 2011). With this in mind, project managers and operators are privileged players who can carry out an analysis of processes and identify suitable indicators and target groups. An evaluation carried out by in-house staff leads to the efficient use of resources, instantly making the most of the expertise available and taking advantage of the desire to improve, and experiment with, forms of communication.

On the other hand, the use of external evaluators is often recommended by specialised evaluation research publications. Standard texts such as that of Rossi *et al.* (1999) stress the importance of using external personnel so as to ensure the neutrality of an evaluation: external evaluators will not be influenced by the relationship with stakeholders and will maintain a detached approach which will allow them to see what the players involved cannot grasp. Of course, external evaluators will need more time to draft an evaluation plan, identifying the size of the study, the target groups involved, and the processes and the results which should be examined.

The decision to evaluate a communication activity with internal or external operators should in any case be taken with a focus on the evaluators' ability to analyse the communication process and its results in both formative and summative terms and their critical competence in expressing judgements aimed at improving programmes. In short, evaluation should be recognised as an integral and essential part of the programme by all those involved, and it should be agreed among the players involved – rather than imposed from above – as much as possible. Its value should be obvious to those involved, everyone should contribute to its implementation and everyone should benefit from it. By allowing an evaluation to be formative (and not merely summative), producing ongoing feedback which is not simply focused on the outcomes, we increase its feasibility and its value.

Evaluation as a particular kind of social research

It is no coincidence that evaluation and social research share a number of key terms: for instance *analysing, understanding, measuring, explaining*. Evaluation, in fact, is nothing other than social research applied for the purposes just mentioned.

Evaluation must therefore deal with the epistemological and methodological issues which concern social research in general. It would obviously be beyond the scope of this chapter to examine these issues in detail. Nevertheless, they should be briefly discussed in order to highlight important aspects of particular importance for evaluative research.

Consider, for example, *GM Nation?*, an initiative promoted by the British government between 2002 and 2003 and involving activities of various kinds – preliminary workshops, a series of different kinds of public meetings, a dedicated website, focus groups and a survey for collecting participants' feedback (36,553 completed questionnaires) – intended 'to promote an innovative and effective programme of public debate on issues around GM in agriculture and the environment, in the context of the possible commercial growing of GM crops in the UK' (PDSB 2003: 11). The initiative's principal purpose was to involve the public in important decisions concerning the regulation of biotechnologies in the agri-food sector, but it also sought to provide the public with the information that it needed to participate in the debate.

The initiative required a large commitment of resources (approximately €1 million), and when it ended debate began on evaluation of its results. The discussion covered the ability of *GM Nation?* to generate real participation, to give voice to all the positions present in society, and to furnish guidelines for legislation.[5] A number of issues arose in the course of the debate; for example: What is meant by *participation*? Who are the *public*? How can *representativeness* be defined? How can it be established whether, and to what extent, the initiative has been successful?

The fact that no agreement had been reached on these questions on conclusion of the initiative highlights the lack of a sufficiently clear *ex ante* definition of its objectives. It is obvious that the actors involved in *GM Nation?* pursued different goals, or at any rate gave different meanings to the terms used to formulate them (Rowe *et al.* 2005). In other words, each actor observed – and therefore evaluated – the process and its results from a different point of view.

A single criterion with which to solve the point-of-view problem does not exist, for every solution has its pros and cons. But precisely for this reason, although it is still the result of choices and negotiations, evaluative research should be able to rely on some reference parameters which make it possible to establish which perspective has been adopted. Evaluation produces results of value only in relation to the context in which they have been obtained – starting from the aims of the project being observed – rather than in absolute terms.

Another methodological problem that evaluation shares with social research has to do with the opposition between quantitative methods – principally, surveys with standardised questionnaires – and qualitative ones (discursive interviews, ethnographic observation, focus groups, etc.). The former methods tend to privilege a type of interaction between researcher and the phenomenon observed which is characterised by detachment, neutrality and separation; the latter emphasises involvement, a direct relation and a sort of constitutive reciprocity.[6] Nevertheless, if we look at the record of evaluative research we can see clearly that making a rigid opposition between quantitative and qualitative research methods is not very useful. For example, the researchers who, albeit as outsiders, evaluated the 2009 Cambridge Science Festival used both quantitative and qualitative methods (Jensen 2009). In this case, the complexity of the context was tackled using a 'methodological triangulation that aims to compensate for the strengths and weaknesses inherent in any one method of data collection by employing overlapping methods on the same topic' (ibid.: 4). Using different tools is, in fact, a good strategy for improving an evaluation's appropriateness with regard to the nature and the context of the communication programme (Joubert 2007).

In the case of the Blast! project, the Open University and the BBC produced a wide range of scientific programmes aimed at improving pubic engagement in science. Each initiative (podcast, event, video, etc.) was evaluated using different tools to assess its success in terms of science outreach. Mixed methods contributed to assessing and developing these activities (BLAST! 2005).

The Museum and Gallery Strategic Commissioning Programme in Britain offers another interesting experience of mixed-method evaluation. The programme activities were assessed by external evaluators using a combination of questionnaires, interviews, visits and desk research, collecting quantitative and qualitative data. The outcomes were presented in a report showing quantitative data and qualitative results, supported by 13 illustrative case studies (RCMG 2007).

We used a similar research design in an exploratory study at European level of a sample of research institutions asking whether the spread of public engagement activities had led to organisational changes in research institutions, incorporating the public engagement perspective into *routine* institutional activities (Neresini and Bucchi 2011). In the first phase, a quantitative questionnaire was used, followed by a thorough assessment of the organisations' websites. In the second phase, qualitative in-depth interviews were conducted with four types of informants in each institution. This research design permitted a broad analysis of resources, processes and ouputs of public engagement activities.

What is specific in evaluating PCST initiatives?

Evaluative research's dependence on an initiative's objectives and on the type of activity involved in reaching these objectives raises questions as to whether evaluating PCST activities is different from evaluating other activities and what is the relation between the way in which PCST is interpreted and the evaluation of the activity which follows it. It is not the same evaluating a PCST activity whose purpose is transmitting knowledge as it is evaluating one whose purpose is promoting discussion between different social actors about a certain issue. In the first case the results are largely predetermined and in the second the activity will be more open-ended. If PCST activities on different issues can be carried out with the aim of achieving attitude, behaviour and knowledge changes, their evaluation should take this into account. There are many different methodologies available and their selection depends on the type of transformation which is expected as a result of a PCST activity. Evaluating the quantity and quality of the changes produced in terms of knowledge entails methodological choices and evaluation techniques which are different from the evaluation of changes in attitudes or behaviour.

An interesting case of evaluation in context concerned the use of an interactive tablet at the Harvard Museum of Natural History (Horn *et al.* 2012). Video recordings and observations were used to study the interaction between visitors and media. The development of a suitable evaluation process made it possible to judge visitors' interaction and their ability to co-operate. The process also allowed the very evaluation tools employed to be improved.

By focusing the attention on the *object* of the evaluation, it is possible to connect the different aims of a PCST initiative, the different kinds of communication promoted through it and its evaluation. An activity mainly devoted to transferring knowledge or to persuading the public will be evaluated in terms of the changes produced in the public. If the aim is also to promote dialogue or participation, the promoters of the communication initiative themselves will have to be under observation. This happened, for example, with the BIOPOP project in which young European biotechnologists sought to develop innovative models of communication of life science through public events in Bologna (Italy) and Delft (The Netherlands). Under a tent in the main square of these cities, young researchers talked with people about biotechnologies, starting from very simple activities such as looking into a microscope, doing PC games and moulding biological molecules by hand (see www.biopop-eu.org). Since the project's aims were focused on dialogue, the evaluation had to take into account not only the effects on the public participating in the events but also the effects on the young biotechnologists conducting the project.

Evaluation of a series of consensus conferences in the Netherlands in 1994–1995 tried to assess, on the one hand, the impacts on participants – with respect to knowledge and attitude – and, on the other, those on policymakers (Mayer *et al.* 1995). In this perspective, evaluation makes it possible also to collect evidence of unexpected effects, as in the case of the citizen conference on genetic diagnostics carried out in Dresden by the German Hygiene Museum in 2001:

> The results were generally positive, though policy-makers and scientists were dissatisfied with the outcome of the conference: participants grew more critical toward some of the diagnostic techniques rather than more accepting.
>
> (Storksdieck and Falk 2004: 100–101)

However, evaluating PCST activities aimed at dialogue and participation – even if we are dealing with a wide range of different activities – often presents interesting challenges focused not only on the public side (Joss and Durant 1995; Rowe and Frewer 2004).

Evaluating the effects of communication

Communication has paradoxical features also displayed by other aspects of everyday life when examined from a social science perspective. Although communication is a practice constitutive of social relations, as well as of individual experience, and although it has been the subject of numerous studies, it is still a rather obscure phenomenon: for example, a generally agreed definition of communication has not yet been produced, certainly not in relation to PCST specifically (Bucchi 2004). However communication is defined, there is some agreement that it is a process able to engender change in those who take part in it.[7] This raises an intriguing question: when can we say that communication has come about? The answer is apparently rather simple: when we can say that something has changed for those involved in the communication process. This is particularly important as regards evaluation. For if communication produces change, then the aim of evaluation must be to establish the extent and nature of this change.

The intrinsic capacity of communication to produce change in those who take part in it makes understanding its effects especially relevant, as well as interesting. It is not by chance that the problem of the effects of communication has received close attention from scholars of the mass media: '[i]f any one issue can be said to have motivated media studies, it is the question of "effects"' (Jensen 2002: 138). Indeed, all researchers interested in social interaction – from the micro to the macro levels – have sooner or later had to confront this problem.

If there is to be some hope of success in evaluating the effects of communication, one must have an idea – general and therefore somewhat imprecise – of what changes communication can be plausibly expected to produce. As indicated above, these are commonly grouped under knowledge, attitudes or behaviour.

Distinguishing between these three types of change is anything but straightforward. It may be rather obvious that a change at the level of knowledge has to do with learning (I now know that a molecule of water consists of two hydrogen atoms and one oxygen atom); at the same time it is clear that altering the way we conceive a given aspect of our experience – in substance, changing the way that we express judgements – pertains to the level of attitudes (contrary to what I thought before, I now believe that scientists are reliable) and that if I begin to do something that I did not do before (watch science programmes on television, attend a science café, encourage my son or daughter to enrol on a science degree course), this concerns the level of behaviour. However, learning cannot be reduced to the acquisition of information alone; it also concerns change in our interpretative schemas, or cognitive models, meaning the criteria on which we base our judgements. Expressing a judgement is likewise a form of behaviour and involves a certain body of knowledge; while performing an action involves a motivational dimension made up of beliefs, competences and knowledge (Michael 2002).

Although these qualifications raise serious doubts as to the validity of the above classification, distinguishing changes at the level of knowledge, attitudes and behaviour is still useful when addressing the evaluation of communication.

Detecting and understanding these three types of change raise somewhat diverse problems of methodology. Rather than seeking the optimal solution, it is important to be aware of the pros and cons of each technique available for this purpose. For example, whilst the acquisition of information can be measured by means of a standardised questionnaire, change in interpretative patterns could be better observed using more flexible techniques like the in-depth interview. Ethnographic observation may be better suited to documenting the behaviour of visitors to a science centre, perhaps using the shadowing technique (see Fletcher 1999; Sachs 1993). But if we are interested in a change in attitudes, the focus group seems to offer greater advantages.

However, there are various aspects that should be borne in mind. The rigidity of the standardised questionnaire, together with the well-known dependency of respondents on the context in which the questionnaire is applied and on the wording of questions, may introduce considerable bias into the data used for evaluation. The development of a discursive interview depends largely on the characteristics of the interviewer and the setting in which the interview takes place. The data collected through focus groups may reflect, among other things, the way in which the discussion group has been composed. Such considerations also arise in ethnographic research.

Whether the concern of a PCST initiative is to transmit information or to promote interaction between scientists and the public, the principal problem of evaluation is still the same: how can the change produced by a communicative process be observed? The apparently simplest and most obvious solution goes by the name of *experimental design*, that is, comparing the situation before the communicative event with the situation after and assuming that any changes observed are due to the communication which has taken place. Taking this approach, a group has to be involved in a communication process (for example viewing a TV programme on science) and another one has not; if the two groups have been made up through a randomised selection, their possible differences observed *ex-post* are attributed to the communication experience. The same result can be obtained by comparing certain characteristics of, say, a group representing the target audience of an information campaign or the visitors to a science centre with those of a control group not involved in the process. Here too, any changes may be attributed to the communication. In the case of a communication campaign devoted to informing about the risk from salmonella, for example, this kind of comparison was applied to measure the effects on a sample of citizens to which the campaign was addressed (Tiozzo *et al.* 2011).

Although attractive at first sight, the experimental design solution proves to be fraught with difficulties arising from the impossibility of complying with the requirement underlying every kind of experimental research – or quasi-experimental – namely the *other things being equal (ceteris paribus)* condition. For example, if it is decided to compare the knowledge – or the attitudes or the behaviour – of the visitors to a science museum before and after their visits, the activity itself of *ex-ante* data collection will produce changes in the subjects, maybe by prompting them to pay closer attention to certain contents of the exhibition than they would do otherwise. For this reason, what one observes as the visitors leave the museum may not be the difference between what the subjects knew before and what they know now due to the exhibition but rather the difference between what the subjects knew before and what they know now due to their involvement in a communicative process stimulated by the exhibition *and* the questions put to them on entering. The administration of a questionnaire or the conduct of an interview is also a communicative process and can produce a change in the participants.

If one seeks to deal with this difficulty by comparing with a control group in a quasi-experimental design there arises another, almost insoluble, problem of social research: the impossibility of obtaining two groups that are completely identical, or in the more attenuated version of the problem, ones that are sufficiently similar, where *sufficiently* means equal in the characteristics most relevant to the changes that one wants to observe. Visiting the website of one scientific institution rather than another, or watching YouTube science videos may not have a great deal to do with attitudes towards science or with knowledge of evolutionary theory but being pestered by parents to get good marks at school most certainly does, even if one might not think of collecting information about it.

The problem also concerns more refined methodological solutions, such as those depicted in Table 17.1 below as quasi-experimental design with crossing groups, for which the difficulty of obtaining homogeneity between the groups compared is not resolved. According to this research design, the target audience of an initiative in science communication – for example, the students at a particular school – in whom changes in characteristics A and B are expected to occur, are divided into two homogeneous groups consisting of students attending the fourth year (ALPHA and BETA). The situation with regard to characteristics A and B is analysed before the event begins but group ALPHA only for A and group BETA only for B. Then, after the event, the situation of characteristic B is surveyed for the ALPHA group, and the situation of A for the BETA group. It is thus possible to compare the characteristics surveyed *ex ante* and *ex post* with respect to A of two different but homogeneous groups, thereby reducing the problem of the conditioning exerted by the *ex-ante* data collection on the *ex-post* one. However, in this case too, the requirement of perfect homogeneity between the two groups compared is still difficult to fulfil. A good example of this research design applied to evaluation of PCST activities is the IN3B project, realised with the aim of understanding the effects of visits to laboratories at large research centres (Neresini *et al.* 2009).

Once again, the optimal solution does not seem to exist, not even if one abandons the (quasi-)experimental approach and adopts '*ex-post* observation'. Rather than comparing the *ex-ante* and the *ex-post* situations, in fact, the researcher now can examine only the characteristics of the *ex-post* one, interpreting them as indicators of changes produced by the

Table 17.1 Evaluation and experimental design

Approach	Problems
(quasi-)experimental *ex-ante/ex-post* design	the *ex-ante* data collection activity conditions the *ex-post* survey
(quasi-)experimental design with a control group	homogeneity between the experimental group and the control group
(quasi-)experimental design with crossing groups	homogeneity between the groups compared
ex-post (self-)evaluation by the subjects	self-deception and deference to the interviewers
ex-post observation	strong hypotheses are necessary on the characteristics of the groups segmented according to the different behaviours observed

communication. In this case, the target audience – for example, teachers attending an in-service training course – performs self-evaluation of communication initiatives by means of both standardised questionnaires and in-depth interviews. In both cases, however, there are potential problems of self-deception and of deference to the interviewers. Moreover, the researcher is obliged to proceed without the necessary benchmark (i.e. the *ex-ante* situation) to determine any changes that may have occurred.

Alternatively, the researcher may try to survey the change indicators indirectly and then verify whether they assume different values in various segments of the sample defined by variables that relate to the expected change. For example, following a communication campaign, the researcher could test indicators of interest in nanotechnologies, and then determine whether their values differ significantly between subjects with lower and higher levels of education, if other evidence has shown that education level is a discriminating variable with respect to interest in science. If the less educated group continues to show little interest in science, while the more educated one is still very interested, one may conclude that the communication campaign has not produced major changes, or that the changes have simply reinforced already existing attitudes. In this case, the main limitation exists in the need to possess quite detailed knowledge about the relations between certain characteristics of the target audience and other characteristics that the campaign is intended to modify.

We should not underestimate the difficulty arising from the fact that the majority of communicative events are brief, while we often expect them to produce great changes with respect to knowledge, attitudes or behaviour. In other words, there is an evident disproportion between a meeting of citizens and scientists, even if it lasts for an entire day, and the acquisition of new knowledge, new interpretative patterns and new habits. There is an even greater disproportion when the communicative process consists of the viewing of a television programme, the reading of an article, sharing information in a social network or visiting a science centre. Evaluation should take account of this aspect (Storksdieck and Falk 2004, Robillard *et al.* 2013), and indeed awareness of the *time factor* has led to the development of various research strategies. Among the best known are the following:

- analysis of short-term effects, bearing in mind that they are highly unstable (after I have visited an exhibition on quarks I can remember a great many things, but how many will I remember some months later?);
- study of the effects generated by repeated involvement in numerous communicative events of the same type (this is the idea at the basis of numerous research studies on the long-term effects produced by the mass media);
- research designs which envisage at least one follow-up survey to determine which of the changes recorded in the short term are still observable some time later;
- analysis of results and outcomes in particular communicative contexts: informal groups, plays, social media.

The main concern in this discussion is not to identify a social research method without contra-indications for use in evaluation. What evaluation requires is adequate awareness of the advantages and disadvantages of the methodology adopted to observe the change generated by communication. In the end, what matters is not isolating the effects of communication on the one hand and its causes on the other; rather, it is understanding the ways in which they combine. If one seeks to find a non-existent optimal solution, there is a serious risk that final evaluation will be abandoned because it is believed to be impracticable.

In this regard, distinguishing between *sequential causation* and *generative causation* may be particularly useful. In the former case – following Pawson and Tilley (1997) – the evaluation seeks to verify whether a particular result can be legitimately ascribed to a given input, assuming that the cause/effect relation thus identified can be generalised to other situations. In the latter case, it is instead important to understand how the result observed may have ensued from a given input, emphasising the role that the context – obviously in combination with the input – plays in producing the result.

Evaluation of attendance at an open day held by a scientific institution will not establish a causal nexus between what the visitors have seen and any changes in their knowledge, attitudes or behaviour. Rather, it may relate the type of communication experience made possible in that context with the effects observed, taking account of the inevitable shortcomings of such observation. Indeed, it may even be found that the activity of observation itself, in that it is part of the overall communicative process experienced during the open day, has contributed directly to producing the effects ascertained.

Concluding remarks

In spite of its limitations, difficulties, and perhaps even its contradictions, evaluation of the public communication of science is useful and necessary. As we have already stressed, evaluation implies that the objectives of the public communication of science be made clear, with specification – especially by its promoters – of the results that a proposed initiative can be legitimately expected to produce. This means bringing to the surface the assumptions which tend to remain hidden, so that they can be scrutinised.

These assumptions are both necessary because they constitute the motivational basis of PCST activities, and unverifiable because they cannot be demonstrated as true or false, only regarded as more or less convincing. The assumptions of the deficit model (knowledge generates attitudes, and these determine behaviour) can be easily identified in the background of numerous communication initiatives, even if their declared objectives state otherwise, and they can be subjected to more detailed scrutiny. This is not to say that, for instance, public communication of science should not seek to transmit knowledge but instead stimulate interest in science; this assumption has its own shortcomings, and it is not absolutely better than the deficit model. Nor will evaluation settle the matter once and for all. Just as happens to science in public debate (Von Schomberg 1995; Grove-White *et al.* 2000), evaluation cannot cut the Gordian knot tying truth and falsehood together but rather can heighten our awareness of the often paradoxical effects of our activities (Boudon 1977, 1995).

This is not a constraint but an opportunity, another good reason to put the logic of evaluation into practice: for evaluating is nothing other than learning from experience more systematically and more efficiently than can be done spontaneously. In recent years, this pragmatic approach has led institutions and practitioners to rapidly follow developments in the public communication of science, building new relations with the media. The so called 'medialisation' effect (Weingart 2012) has produced new ways of communication and scientists have developed new roles in the public sphere. These transformations require new approaches and a specific assessment.

Including adequate evaluation in PCST initiatives will address these challenges and highlight both their strengths and weaknesses, not only by reducing the risk of persisting in the wrong direction, but also by refining our understanding of what those initiatives seek to achieve (Gammon and Burch 2006). PCST tends to proceed by imitation following the fashion of the moment, neglecting that rapid proliferation of similar initiatives does not necessarily indicate success and that what has worked in one context will not necessarily produce the same results in another. Evaluation may prove very useful in insulating against the allure of simple replication. This should be particularly effective in contexts in which there is a growing demand for *scientific citizenship*, where public communication of science is crucial to the definition of public goods (e.g. climate change issues or nuclear plants).

Clarifying the boundaries between credible expectations and unfounded hopes, reflecting carefully on what has been achieved, scrutinising our initial assumptions – only these can ensure that the present euphoria concerning the public communication of science will not be followed by disappointment over unfulfilled expectations.

Key questions

- What are the conditions necessary to ensure that the evaluation of a PCST initative is actually carried out?
- What factors render the evaluation of a PCST initiative different from that of other types of communication initiatve?
- Take an example of a PCST initiative known to you. Can you determine what kind of change is expected and who is expected to change as a consequence of the initiative?
- What are the main methodological problems associated with the summative evaluation of a given PCST initiative? How can these problems be solved or minimised?

Notes

1. The new version of the chapter includes references to research of the last five years and draws attention to current debates on relevant topics such as different levels of involvement in the communication activity. The explanation of key terms and evaluation's phases has been brought forward, so that readers can follow more easily the discussion about advantages and disadvantages of different methods and approaches. An outline of the conditions required for efficient evaluation has been also inserted before discussion of methodological issues.
2. In this regard useful guidelines have been proposed by Eng *et al.* (1999). These guidelines are partial adaptations of those set out by the National Cancer Institute (1989).
3. See, for example, Barnes (1999) and Mayer *et al.* (1995). The general implications of evaluation of participation initiatives are discussed by Rowe and Frewer (2004).
4. For a survey see Piscitelli and Anderson (2000), Persson (2000), Garnett (2002). On the relevance of the context in which the communication process takes place see, among others, Falk and Dierking (2000), Schiele and Koster (1998).
5. See e.g. Barbagallo and Nelson (2005); Horlick-Jones *et al.* (2006); Irwin (2006).
6. As a consequence, the survey tools used by quantitative methods are therefore more standardised; qualitative methods, by contrast, have greater margins of flexibility, which enables them to adapt to the situation at hand. On this see, for example, Silverman (2004) and Strauss and Corbin (1998).
7. See e.g. Bateson (1972); Maturana and Varela (1980); Von Foerster (1981); Watzlawick *et al.* (1967).

References

Baram-Tsabari, A. and Lewenstein, B.V. (2013) 'An instrument for assessing scientists' written skills in public communication of science', *Science Communication*, 35, 1: 56–85.
Barbagallo, F. and Nelson, J. (2005) 'Report: UK GM dialogue', *Science Communication*, 26, 3: 318–325.
Barnes, M. (1999) *Building a Deliberative Democracy: An Evaluation of Two Citizens' Juries*, London: Institute for Public Policy Research.
Bateson, G. (1972) *Steps to an Ecology of Mind*, New York: Ballantine Books.
BLAST! (2005) *Broadcast-Linked Activities in Science and Technology*; online at www.open.ac.uk/science/main/citizen-science/projects/blast

Boudon, R. (1977) *Effets pervers et ordre social*, Paris: PUF.
Boudon, R. (1995) *Le juste et le vrai. Études sur l'objectivité des valeurs et de la connaissance*, Paris: Fayard.
Bowman, S. W. (2013) 'A formative evaluation of WIKI's as a learning tool in a face to face juvenile justice course', *Educational Technology Research and Development*, 61, 1: 3–24.
Bucchi, M. (2004) 'Can genetics help us rethink communication? Public communication of science as a "double helix"', *New Genetics and Society*, 23, 3: 269–283.
Eng T. Gustafson, D. H., Henderson, J., Jimison, H. and Patrick, K. (1999) 'Introduction of evaluation of interactive health communication applications', *American Journal of Preventive Medicine*, 16, 1: 10–14.
Falk, J. H. and Dierking, L. D. (2000) *Learning from Museums*, Walnut Creek: Altamira Press.
Fletcher, J. K. (1999) *Disappearing Acts: Gender, Power and Relational Practice at Work*, Cambridge, MA: MIT Press.
Gammon, B. and Burch, A. (2006) 'A guide for successfully evaluating science engagement events', in J. Turney (ed.) *Engaging Science: Thoughts, Deeds, Analysis and Action*, London: Wellcome Trust, 80–85.
Garnett, R. (2002) *The Impact of Science Centers/Museums on Their Surrounding Communities*; online at www.astc.org/resource/case/Impact_Study02.pdf
Gascoigne, T. and Metcalfe, J. (2001) 'The evaluation of national programs of science awareness', *Science Communication*, 23, 1: 66–76.
Grant, L. (2011) 'Evaluating success: how to find out what worked (and what didn't)', in D. J. Bennett and R. C. Jennings (eds) *Successful Science Communication*, Cambridge: Cambridge University Press, 403–422.
Grove-White, R., Macnaghten, P. and Wynne, B. (2000) *Wising Up: The Public and New Technologies*, Research report, CSEC, Lancaster: Lancaster University.
Guba, E. and Lincoln, Y. (1989) *Fourth Generation Evaluation*, Newbury Park CA: Sage.
Horlick-Jones, T., Walls, J., Rowe, G., Pidgeon, N., Poortinga, W. and O'Riordan, T. (2006) 'On evaluating the GM Nation? public debate about the commercialisation of transgenic crops in Britain', *New Genetics and Society*, 25, 3: 265–288.
Horn, M. S., Leong, Z. A., Block, F., Diamond, J., Evans, E.M., Phillips, B. and Shen, C. (2012) 'Of BATs and APEs: an interactive tabletop game for natural history museums', in Proceedings of the ACM Conference on Human Factors in Computing Systems CHI'12. ACM Press, 2059–2068.
Irwin, A. (2006) 'Public deliberation and governance: engaging with science and technology in contemporary Europe', *Minerva*, 44, 2: 167–184.
Irwin, A. (2009) 'Moving forwards or in circles? Science communication and scientific governance in an age of innovation', in R. Holliman, E. Whitelegg, E. Scanlon, S. Smidt and J. Thomas (eds) *Investigating Science Communication In The Information Age: Implications For Public Engagement And Popular Media*, Oxford: Oxford University Press, 3–17.
Jensen, K. B. (2002) *The Qualitative Research Process. A Handbook Of Media And Communication Research: Qualitative And Quantitative Methodologies*, London: Routledge.
Jensen, E. (2009) *2009 Cambridge Science Festival: External Evaluation Report*, Cambridge: Anglia Ruskin University.
Jensen, E. (2011) 'Evaluate impact of communication', *Nature*, 469, 7329: 162.
Joss, S. and Durant, J. (1995) *Public participation in science: the role of consensus conferences in Europe*, London: Science Museum.
Joubert, M. (2007) 'Evaluating science communication projects', Scidev.net; online at www.scidev.net/ms/sci_comm/
Kotler, N. and Kotler, P. (1998) *Museum Strategy and Marketing: Designing Missions, Building Audiences, Generating Revenue and Resources*, San Francisco: Jossey Bass.
McKenzie, M. J. (2012) 'Science communication e(value)ation: understanding best practice', in M. Bucchi and B. Trench (eds) *Quality, Honesty and Beauty in Science and Technology Communication*, PCST 2012 Book of Papers, Vicenza: Observa Science in Society, 264–267.
Maeseele, P. (2012) 'Science communication and democratic debate: friends or foes?' in M. Bucchi and B. Trench (eds) *Quality, Honesty and Beauty in Science and Technology Communication*, PCST 2012 Book of Papers, Vicenza: Observa Science in Society, 67–71.
Maturana, H. and Varela, F. (1980) *Autopoiesis and Cognition: The Realization of Living*, Dordrecht: Riedel.
Mayer, I., de Vries, J. and Guerts, J. (1995) 'An evaluation of the effects of participation in a consensus conference', in S. Joss and J. Durant (eds) *Public Participation In Science: The Role Of Consensus Conferences in Europe*, London: The Science Museum, 201–223.

Michael, M. (2002) 'Comprehension, apprehension, prehension: heterogeneity and the public understanding of science', *Science, Technology and Human Values*, 27, 3: 357–378.
National Cancer Institute (1989) *Making Health Communication Programs Work*: NIH Publication No. 89-1493, Bethesda, MD: National Institutes of Health.
Neresini, F. and Bucchi, M. (2011) 'Which indicators for the new public engagement activities? An exploratory study of European research institutions', *Public Understanding of Science*, 20, 1: 64–79.
Neresini, F., Dimopoulos, K., Kallfass, M. and Peters, H. P. (2009) 'Exploring a black box: cross-national study of visits effects on visitors to large physics research centers in Europe', *Science Communication*, 30, 4: 506–533.
Patton, M. Q. (1986) *Utilization-Focused Evaluation*, Newbury Park, CA: Sage.
Pawson, R. and Tilley, N. (1997) *Realistic Evaluation*, London: Sage.
PDSB (2003) *GM Nation? The Findings of the Public Debate: Final Report of the GM Public Debate Steering Board*, London: Clarity.
Persson, P. E. (2000) 'Science centers are thriving and going strong', *Public Understanding of Science*, 9, 4: 449–460.
Piscitelli, B. and Anderson, D. (2000) 'Young children's learning in museums settings', *Visitor Studies*, 3, 3: 3–10.
PSP (2006) *Formative Evaluation of the LHC Communication Project*; online at www.peoplescienceandpolicy.com/projects/hadron_collider.php
RCMG (2007) *Inspiration, Identity, Learning: The Value of Museums. Second study*, Leicester: University of Leicester; online at www2.le.ac.uk/departments/museumstudies/rcmg/projects/inspiration-identity-learning-2/IIL.pdf
Robillard, J. M., Whiteley, L., Johnson, T. W., Lim, J., Wasserman, W. W. and Illes, J. (2013) 'Utilizing social media to study information-seeking and ethical issues in gene therapy', *Journal of Medical Internet Research*, 15, 3: e44.
Rossi, P. H., Freeman, H. E. and Lipsey, M. W. (1999) *Evaluation: A Systematic Approach*, sixth edition, Thousand Oaks: Sage.
Rowe, G. and Frewer, L. J. (2004) 'Evaluating public-participation exercises: a research agenda', *Science Technology & Human Values*, 29, 4: 512–557.
Rowe, G., Horlick-Jones, T., Walls, J. and Pidgeon, N. (2005) 'Difficulties in evaluating public engagement initiatives: reflections on an evaluation of the UK GM Nation? public debate about transgenic crops', *Public Understanding of Science*, 14, 4: 331–352.
Sachs, P. (1993) 'Shadows in the soup: conceptions of work and nature of evidence', *The Quarterly Newsletter of the Laboratory of Human Cognition*, 15, 4: 125–132.
Schiele, B. and Koster, E. (eds) (1998) *La revolution de la muséologie des sciences: vers le musée du XXIe siècle?*, Lyon, France: Presses universitaires de Lyon.
Scriven, M. (1991) *Evaluation Thesaurus*, fourth edition, Thousand Oaks, CA: Sage.
Silverman, D. (2004) *Qualitative Research: Theory, Method and Practice*, second edition, London: Sage.
Stevenson, E. and Rea, H. (2012) 'What does quality mean in public engagement of science?', in M. Bucchi and B. Trench (eds) *Quality, Honesty and Beauty in Science and Technology Communication, PCST 2012 Book of Papers*, Vicenza: Observa Science in Society, 207–211.
Storksdieck, M. and Falk, J. H. (2004) 'Evaluating public understanding of research projects and initiatives', in D. Chittenden, G. Farmelo and B. Lewenstein (eds) *Creating Connections*, Walnut Creek, CA: Alta Mira Press, 87–108.
Strauss, A. L. and Corbin, J. M. (1998) *Basics of Qualitative Research: Techniques and Procedures for Developing Grounded Theory*, second edition, Thousand Oaks, CA: Sage.
Tiozzo B., Mari, S., Magaudda, P., Arzenton, A., Capozza, D., Neresini, F. and Ravarotto, L. (2011) 'Development and evaluation of a risk-communication campaign on salmonellosis', *Food Control*, 22, 1: 109–117.
Von Foerster, H. (1981) *Observing Systems: Selected Papers of Heinz von Foerster*, Seaside, CA: Intersystems Publications.
Von Schomberg, R. (1995) *Contested Technology. Ethics, Risks And Public Debate*, Tilburg: International Centre for Human and Public Affairs.
Watzlawick, P., Beavin, J. H. and Jackson, D. D. (1967) *Pragmatics of Human Communication. A Study of Interactional Patterns, Pathologies and Paradoxes*, New York: Norton & Co.
Wehrmann, C. and De Bakker, E. P. H. M. (2012) 'How to educate and train scientists and science communication students to perform well in science communication activities', in M. Bucchi and B. Trench

(eds) *Quality, Honesty and Beauty in Science and Technology Communication, PCST 2012 Book of Papers*, Vicenza: Observa Science in Society, 92–95.

Weingart, P. (2012) 'The lure of the mass media and its repercussions on science. theoretical considerations on the "medialization of Science"', in S. Rödder, M. Franzen and P. Weingart (eds) *The Sciences' Media Connection: Communication to the Public and its Repercussions. Sociology of the Sciences Yearbook*, Dordrecht: Springer, 17–32.

Weiss, C. H. (1998) 'Have we learned anything new about the use of evaluation?', *American Journal of Evaluation*, 19, 1: 21–33.

Weiss, C. H. (2001) 'Theory-based evaluation: theories of change for poverty reduction programs', in O. Feinstein and R. Picciotto (eds) *Evaluation and Poverty Reduction*, Washington DC: World Bank.

Index

Please note that page numbers relating to Notes will have the letter 'n' following the page number. Any references to Figures or Tables will be in *italics*.

AAAS (American Association for the Advancement of Science) 76, 155n
The Absent Minded Professor (film) 101
accountability, trust portfolio management 63
Action Plan on Science and Society (European Commission) 160
Adamson, J. 86
Addis Ababa 219
'advice' programmes 31
Africa: global spread of science communication 219; survey research 141, *142*
agenda setting 106–7
agriculture, GM 120, 121
AHHAA (science centre) Estonia, 223
Albaek, E. 189, 191
Alfred P. Sloan Science-in-Film Initiative 99
Algarotti, F. 3
Allan, S. 36
Allum, N. C. 149
Althins, T. 41, 42
Alzheimer's disease 75, 103
American Association for the Advancement of Science (AAAS) 76, 155n
American Chemical Society 43
American Film Institute, Sloan Science Advisor programme 99
angels 151
Annuario Scienza e Societa 11n
Ascent of Man (Bronowski) 85
Asia, survey research *142*
Asilomar Conference on Recombinant DNA (1975) 60–1
Association of Science–Technology Centers (ASTC) 44
astrology 147
Atkins, Peter 89
audience research and media effects: agenda setting 106–7; audience reception studies 104–5; awareness raising 106–7; entertainment education 106; framing 106–7; science education 105–6
audiences 33, 74; *see also* audience research and media effects
Australia 29, 65, 90, 141; spread of science communication 217, 218, 219, 220–1
authenticity 99, 107
autobiography of science 21, 23
Avatar (film) 97, 103
awareness raising 106–7

Backer, T. E. 204
backstage concept 10, 12n
Bacon, Francis 148
bacteriology 101
Bad Pharma (Goldacre) 20
balance, science journalism 33
Bauer, Martin W. 30, 153, 154
Bauman, Z. 163
BBC 31, 85, 86, 235
Beck, U. 74, 163, 164
The Beginning or the End (film) 98
Bennett, T. 42
Bentley, P. 191
Bernal, J. D. 59–60
Big Bang 83 86
The Big Bang Theory (TV series) 97, 98
big data 153
Billig, M. 192
BIOPOP project 236
biotechnology 114, 128; food 71, 174, 183
Bishop, D. 92
black holes 1, 83, 91
Blade Runner (film) 103
Blast project, UK 235
A Blind Bargain (film) 98, 102
blogs 34, 76
Bodansky, D. 115
Böhme-Dürr, K. 189

Index

books, popular science 15–26, 87; children's 16, 17, 19; as educational 18; as entertaining 18; fictional 16–17; larger context 19–21; methods of reading 21–3; non-fictional 16–17, 19; paperbacks 19; Pulitzer Prize winners 18; *see also* popular science
Borchelt, R. 67n
Born Free (Adamson) 86
boundary objects 92
boundary work 23
Bowler, P. J. 18
Boykoff, M. T. and J. M. 33
Boyle, Robert 22
Boys From Brazil (film) 103
Brains Trust (UK television programme) 85
brands 10, 35
Braudy, Leo 88
Brazil 140–1; spread of science communication 217, 220
Breaking Bad (film) 98
Breakwell, G. M. 149
Brewer, Ebenezer Cobham 3
A Brief History of Time (film) 90
Britain *see* United Kingdom
British Council 218
British Film Institute 99
British Particle Physics and Astronomy Research Council 233
Broks, P. 28
Bronowski, Jacob 85, 86
Bryson, Bill 23
BSE (bovine spongiform encephalopathy) 161–3, 168
Bucchi, M. 31, 88
Buckingham, D. 20
Budapest 220
Bulgaria 84
Burawoy, Michael 192
Burgess Shale, fossils of 86, 87
Burnham, J. C. 28
Bush, Vannevar 3, 12n

Cambridge Science Festival, UK 235
Cameron, D. 55
Canada: science journalism 30, 31, 32; science museums and centres (SMCs) 43, 44, 48, 52; spread of science communication 218, 220
Canadian Association of Science Centers 43
Captain America (film) 103
carbon dioxide 114
carbon footprint 118
Carson, Rachel 21 103
causation, sequential versus generative 241
CBD (Center for Biological Diversity), US 117, 123n
CBPR (community-based participatory research) 132

CBS News 83
celebrities, analysing scientists as: celebrity as form of representation 88; celebrity status allowing comment on areas outside realm of expertise 92; celebrity versus visibility 93n; commodities, celebrities as 88; framework 89–92; image featuring a blurring of his/her public and private lives 89; scientists' representations featuring tensions/contradictions inherent in fame 91–2; structural relationship of scientist with ideological tensions 91; symbolic nature of celebrity 88; tradable cultural commodity, scientist as 90
celebrity culture 8, 87–9
Center for Advancement of Informal Science Education 136n
Center for Biological Diversity (CBD), US 117, 123n
Center for Nanotechnology in Society, Arizona State University 169
Chambers, Robert 21
Chao, Z. 149
Chernobyl nuclear disaster (1986) 174
Chicago World's Fair (1933–4) 47
children's films 103
Chimba, M. 87
China 3, 20, 65, 176; environmentalism 118, 122; spread of science communication 215, 216, 217
China Civil Science Literacy Survey (2010) 145
Chinese Research Institute for Science Popularisation (CRISP) 60, 147, 220
CHIRR (Consumer Informatics Research Resource) 204, 206
Chomsky, Noam 191
Cité des Sciences et de l'Industrie (Paris) 45, 50
citizen engagement 4
citizen participation 131, 132; *see also* participation
citizen science 6, 132
civic scientific literacy 145
civil service, and politics 72
climate change: civic capacity and public deliberation, investing in 181–2; disagreement about, in US 175–8; diversifying policy options and technological choices 180–1; empowering the public about 178–83; and expertise 71, 73, 75, 78; in films 105; promoting new frames of reference and cultural voices 179–80; sceptics 115; as science communication challenge 114–19; US debate 173–85; *see also* global warming
Climate Reality Project 178
CMST (Canadian Museum of Science and Technology) 43
cognitive deficit 146
cognitive elaboration 150
collaborative knowledge production 134
Columbia 140–1
Columbia Journalism Review 35

247

Index

comedies 101
Coming of Age in Samoa (Mead) 85
Committee on the Public Understanding of Science 221
Commonwealth Scientific and Industrial Research Organisation, Australia 65
communication: evaluation of effects 237–41; styles 164; top-down 3, 162; training 218–19; *see also* science communication; science communication research; science communication research challenges
communication models 3, 17
Communitarian Egalitarians 177, 178, 181, 182
community-based participatory research (CBPR) 132
competence, trust portfolio management 63
conclusion evaluation phase (*ex-post*) 232, 233, 238, 239
Consumer Informatics Research Resource (CHIRR) 204, 206
Contagion (film) 98, 106
contemporary science, SMCs showing 47–8
Convention on Biological Diversity (CBD) 133
Cooper, G. 190
CosmoCaixa (SMC) 42
Cosmos (TV series) 18, 85
counter expertise 74
Cousteau, J. 85, 86
Cox, B. 87
credibility, trust portfolio management 63
Crichton, M. 115, 116
CRISP (China Research Institute for Science Popularisation) 60, 147, 220
critical-constructivist research 151
critical expertise 74
crops, GM 120
Crowther, J. G. 60
CSI (TV series) 97; *CSI* effect 105, 108n
CSTI (*culture scientifique, technique et industrielle*) 8
cultural cognition studies 176
cultural history (of images) 100
culture: celebrity 87–9; film, cultural meanings 100–4; governance cultures 129; popular, scientists in 83–96; science in and science in society 10–11; scientific 8; tradable cultural commodity, scientist as 90
Curie, M. 83
Curtis, R. 22
cyborg bodies 103

Daily Mail 87–8
Danilov, V. 41, 42, 44
Danish Board of Technology 130
Dante's Peak (film) 101
Darwin, Charles 17, 74, 89
Dawkins, Richard 17, 19, 86, 87, 89, 91–2, 191; *The God Delusion* 90, 91, 92

The Day After Tomorrow (film) 103, 105, 106
The Day the Earth Stood Still (film) 103
Dearing, J. 33
de Castro Moreira, I. 104
de Ceglia, F. P. 100
decisionistic model 72
decision-making, and scientific knowledge 71–2
Deep Blue Sea (film) 103
deficit concept/model 4, 16, 160; and BSE crisis 161–3; characteristics 167; first to second order thinking 163–5; old-style thinking 166; and PCST 186, 187; and survey research 151–2
de Kruif, P. 18, 84
De Lalande, Jérôme le François 3
deliberative governance 129
Denmark 5; spread of science communication 218
Department of Labor, Bureau of Labor Statistics (US) 29
Dervin, B. 204
design evaluation phase (*ex-ante*) 232, 238, 239
Design to Win: Philanthropy's Role in the Fight Against Global Warming 181
Destination Moon (film) 103
Deutsches Museum, Munich 47
Dewey, John 125
dialogical turn 4
dialogue 4–5, 6, 11, 216, 222
Dickinson, R. 188
digital divide 61
digital media 9, 58
Dimopoulos, K. 32
disaster films 101
discretionary governance 129
disease groups 133
DNA 86, 103
documentary science programmes 31
dominant technocratic empiricist models 128
Doocy, Steve 176
Dragons of Eden (Sagan) 18, 85
drama: in films 101, 103; in science television programming 31
Dr. Ehrlich's Magic Bullet (film) 98
Dryzek, J. S. 164, 182
dumbing down 16
Dunlap, R. E. 115
Dunning-Davies, J. 93
Dunwoody, S. 78, 189
Dutta, M. J. 207–8
Dutta-Bergman, M. 202, 204 same person as above?

Earth Matters (Friends of the Earth) 116
Eberling, M. 190
E-books 21
ecoAmerica 180
eco-fascists 177
Eddington, A. 17, 18

Edison, T. 84, 101
education: entertainment 106; science 105–6
Eger, M. 23
Ehrlich, P. 85, 98
Eidelman, J. 47
Einsiedel, E. F. 31, 32
Einstein, A. 18, 83, 84
Elsdon-Baker, F. 91–2
Endangered Species legislation 117
The End of Nature (McKibben) 175
Eng, T. 242n
engagement/citizen engagement 4, 5
Enlightenment 28, 55n
Ensia (web magazine) 182
entertainment education 106
Entertainment Industries Council, US 99
Entretiens sur la pluralité des mondes (Fontenelle) 3
Environment Agency, UK 122
environmental citizens 126
environmentalists, as communicators of science 113–24; climate change as science communication challenge 114–19; GMOs, communicating safety and risks in relation to 119–21; participation and public engagement as forums for science communication 121–2
environmental NGOs (ENGOs) 113, 115, 117
epistemological consonance 190
Epstein, S. 133
Equinox series 115, 116
Erin Brockovich (film) 103
errors and inaccuracies, scientific 78
essentialisation, sample surveys 151
establishment science 115
Estonia 223
ethnography 238
Eurobarometer, survey research 140, 141, 150
Euro-Mediterranean and Middle East School of Science Communication, Spain (2013) 219
European Commission 6, 8, 60, 160–1; *Action Plan on Science and Society* 160; and global spread of science communication 217, 219
European Science Foundation (ESF) 168
evaluation of science communication 231–45; conclusion phase (*ex-post*) 232, 233, 238, 239; design phase (*ex-ante*) 232, 238, 239; effects of communication 237–41; experimental design 238, 239; formative evaluation 233; implementation phase (*in itinere*) 232; internal or external evaluators 234; output and outcome, distinguished 233; phases and choices 232–4; public communication of science and technology (PCST) 232, 233, 236–7, 241; and quality 10; science communication as particular kind of social research 234–6; sequential versus generative causation 241; summative evaluation 233; theory-based 232; time factor 240; types of change 237–8

Evans, W. 188, 189, 190
Evans-Cowley, J. 134
evolutionary biology 189
evolutionary psychology 189, 190–1
evolution theory 74, 148, 154
exhibitions, temporary 51–3, 56n
experimental design 238, 239
experimental science, early 59
expertise, scientific 6–7, 28; comments on areas outside realm of expertise 92; counter expertise 74; credibility crisis 74; critical expertise 74; expectations and responsibilities 70–82; expert as advisor to decision-maker 71–2, 75; expert-client relationship 72–3; expert role of scientists in public communication 70, 71–5; interactions of scientific experts and journalists 77; and journalism 74–5, 76, 77, 78, 79; in mass media 75–9; public (re-)creation of scientific expertise 78–9; and public expertise 79; responsibility of public expert 72–3; role of scientists as public experts compared to other possible roles 70; scientific expertise and other forms of knowledge 75; scientific knowledge and decision-making 71–2; social sciences 190; trust and scientific authority 73–5
Exploratorium, San Francisco 48, 49, 216
Eyewitness guide 20

Fahnestock, J. 78
Fahy, D. 21–2, 36
Falk, J. H. 237
fast neutrinos 10
Fenton, N. 189, 190
Ferranti, S. 84
film, science and technology in 97–112; audience research and media effects 104–7; cloning films 103–4; content analysis 99–100; cultural meanings 100–4; genres 100, 101, 102, 103; golden age for science 97; movie scientists 100–4; production 98–9; scientific research fields in popular films 102–4; stereotypes 101, *102*
Finding Nemo (film) 97
first-order thinking *see* deficit model
First Regional Science Promotion conference, Serbia 217–18
focus-group studies 105
Fontenelle, Bernard le Bovier de 3
food biotechnology 71, 174, 183
formative evaluation 233
fossil fuels 180
fossilised science 10
Foucault, Michel 42
Fox News 176
fracking 118
France: anti-globalisation protesters 120; Muscular Dystrophy organisation 133; science

PR 64; and SMCs 45, 47, 50, 53; spread of science communication 219, 221; technology assessment 129
Franco, F. 100
Frankenstein (horror film) 100
Frenette, M. 204
Frewer, L. 127
Friends of the Earth 115, 122; Campaign News section 116
Frogs (film) 103
Fukushima disaster, Japan (2011) 117
Funtowicz, S. O. 72, 173
Fyfe, A. 17, 18

Gamow, G. 16
Gascoigne, T. 29, 232
GATTACA (film) 104
Gauchat, G. 153
'geek chic' 20
geek culture 97
Geek Manifesto (Henderson) 20
gender, and popular culture 87–8, 101
generative causation 241
Genetic Alliance 133
geneticisation 21
genre interpenetration 107
geoengineering 134, 180
German Federal Ministry of Education and Research, MINTiFF initiative 99
German Hygiene Museum 237
Germany 34, 118, 129, 189
Gibbons, M. 75, 131
Giddens, A. 74, 163
Gieryn, T. 12n
Gillieson, K. 20
globalisation 214
global spread of science communication 214–30; emerging centres of science communication 221–6; government programmes, awareness raising 215–17; initiatives to support media attention to science 219; key indicators 215–21; training and supports for scientists 217–19; university research in science communication 220–1; university taught programmes in science communication 219–20
global warming 33, 34, 70, 105, 115, 176, 177; and environmentalists 114, 115, 117, 119; *see also* climate change
GM Nation? initiative, UK 121, 165, 166, 234–5
GMOs (genetically modified organisms), communicating safety and risks relating to 119–21
The God Delusion (Dawkins) 90, 91, 92
Godsend (film) 104
Goffman, E. 12n
Goldacre, B. 20
Goldsmith, O. 17

GONGOs (government organised non-governmental organisations) 122
Goodall, J. 87
Goodell, R. 28, 76, 77, 85, 91, 93n
Google 178
'goose laying golden eggs' metaphor 3
Gore, A. 176, 178
Gould, S. J. 86, 91
governance cultures 129
government programmes, awareness raising 215–17
Grande Galerie de Zoologie/Grande Galerie de l'Evolution du Museum National d'Histoire Naturelle (Paris) 47
Great Exhibition, London (1851) 45
The Great Global Warming Swindle (Channel 4) 116
great men 17, 24n
Greece, science journalism 29
Greenfield, Susan 87, 89, 91, 92
greenhouse gas emissions 177
Greenpeace 113, 117
Gregory, J. 21, 29
Grunig, J. E. 59
Grunwald, A. 75
The Guardian 83, 89, 119
Guide to the Scientific Knowledge of Things Familiar (Brewer) 3

Habermas, Jürgen 72, 164
Haider, M. 204
Haldane, J. B. S. 18, 19
Hansen, A. 188
Happy Feet (film) 103
Haraway, Donna 103
Harrington, N. G. 208
Hart, D. M. 115
Harvard Museum of Natural History 236
Hawking, Jane 89
Hawking, Stephen 83, 87, 88, 91, 92, 93
Hawking Incorporated (Mialet) 90
Hayden Planetarium, New York 87, 93
Haynes, Roslynn 22, 100
health campaign research 198–213; basic, perspective provided by 204–6; basic research 206; channel-based barriers 201; common outcomes 203–4; contemporary 198; dependent variables 204; destination-based barriers 201; diseases/conditions addressed 200; expansion 198–9; impacts of health literacy and health information technology 206–8; independent variables 203, 204; intent of health campaigns and research topic areas 200; intermediate variables 204; message-based barriers 201; public health issues 200; receiver-based barriers 201; results 204–6; social demographic variables 203; sociocultural dimensions (phase two)

202; source-based barriers 201; underlying conceptual frameworks, phases 201–3
health literacy 206–8
Hedgecoe, A. 21
Hein, H. 48
Heinrichs, H. 78, 79
Henderson, Mark 20
heroic phase (public communication of science) 10
hierarchical communication models 3
Hierarchical Individualists 176–7, 178, 180, 181, 182
Hilgartner, S. 16
historical knowledge 42
Hockney, David 83
Hollander, J. 134
Hollywood Black Film Festival 99
Hollywood film industry 101, 108
Holmes, S. 88
Horisont (journal) 223
Horizon (UK TV programme) 31
Horizon 2020 6
hormone-mimicking substances, environmental release 113
Hornig, S. 31
Horrible Science books 20
horror films 100, 101, 102, 103
House (film) 97
House of Lords, UK 4; Select Committee on Science and Technology 43, 50, 148, 150, 160
Howard, S. 153
Hoyle, Fred 21, 86
Hudson, K. 43, 44
The Huffington Post 83
Hulk (film) 103
Hulme, Mike 119
Hungary 220
Hunt, T. 59
Huxley, T. H. 18
hydraulic fracturing (fracking) 118

I, Robot (film) 103
image of scientists: featuring a blurring of public and private lives 89; public, as constructed around discourses of truth, reason and rationality 90–1; *see also* celebrities, analysing scientists as
The Immortal Life of Henrietta Lacks (Skloot) 22
implementation evaluation phase (*in itinere*) 232
The Independent 87
independent scientists 162
in-depth interviews 179
India: Department of Biotechnology 219; spread of science communication 215, 220
Indian Journal of Science Communication 221
informal science education (ISE) 105, 106, 108n
information-seeking 34

Inhofe, James 176
InsideClimate News (non-profit organisation) 36
Institute of Medicine of the National Academies, US 199
institutional neuroticism 149
Institut Royal des Sciences Naturelles de Belgique (Brussels) 53
Integrating Scientific Expertise in Public Media Discourse (INWEDIS), German research project 64
integrity, trust portfolio management 63
interdisciplinary research 75
Intergovernmental Panel on Climate Change (IPCC) 115, 122, 176; Science Report 116
International Association for Public Participation (IAP2) 127
International Centre for Theoretical Physics 217
Internet: geek culture on 97; shift of science journalism to 34–5
INWEDIS (Integrating Scientific Expertise in Public Media Discourse), research project 64
IPCC *see* Intergovernmental Panel on Climate Change (IPCC)
Irwin, A. 97, 151
The Island of Dr Moreau (film) 103
Israel, B. A. 132
Italy 84; spread of science communication 219; Third World Academy of Sciences, Trieste 217

Jacobi, D. 56n
Japan 65, 117, 129; *Science and Technology Basic Plan* (2011–15) 215; spread of science communication 215, 216, 218
Japanese Journal of Science Communication (Kyoto) 221
Japan Science and Technology Agency 215
Jasanoff, S. 151
JCOM – Journal of Science Communication 221
Jeans, J. 17, 18
Journalism (journal) 36
journalism, science 9, 27–39, 186; coverage of science following journalistic norms 32–3; in current period 30–4; and expertise 74–5, 76, 77, 78, 79; global 30; history 27–9; importance 27; Internet, shift to 34–5; medicine and health focus of science news 30–1; and PR 58, 59–60; role of science journalists 35–7; whether scientists losing or gaining control 35; service model of journalism 77; television, scarcity of science news on 31; training as contentious and under-studied 33–4
Journal of Scientific Temper (India) 221
journals and magazines 16, 20, 178, 189; global spread of science communication 219, 221; popular culture, scientists in 83, 84, 86, 87, 90, 91; science journalism 28, 35, 36; *see also* newspaper industry

Jurassic Park (film) 101, 104, 107
Jurdant, B. 21

Kahan, D. 176, 178, 180–1
Kepplinger, H. M. 76
Keselman, A. 208
Kincaid, D. L. 204
King, Sir David 117
Kitzinger, J. 87
Klein, N. 178
knowledge: backstage of knowledge production processes 10, 12n; expert 71; methodological 147; non-knowledge 72; production of and public participation 131–3; scientific *see* scientific knowledge; scientific literacy 145, *146*; and SMCs 41, 42, 45, 47–9; socially robust 75; translation and transfer 11; *see also* expertise, scientific
Knowledge, as Science Communication (journal) 221
KOFAC (Korea Foundation for the Advancement of Science and Creativity) 60
Kohn, M. 89
Korea Foundation for the Advancement of Science and Creativity (KOFAC) 60
Koster, E. 50, 51
Koulaidis, V. 32
Kreps, G. L. 204, 207–8
Kuhn, T. S. 10
Kyoto Protocol 118
Kyvik, S. 191

Lab Coats in Hollywood (Kirby) 98
LaFollette, M. C. 86, 87
L'Aquila earthquake, Italy (2009) 73, 80n
Large Hadron Collider (LHC) Communication Project 233
Laslo, E. 35
L'Astronomie des Dames (de Lalande) 3
Latin America 64; spread of science communication 220, 221–2; survey research 140–1, *143*
Law of the People's Republic of China on Popularisation of Science and Technology 65
lay expertise/knowledge 7
Leane, E. 22
Lebart, L. 153
Lee, L. 100–1
legitimacy, trust portfolio management 63
Lehmkuhl, M. 31
Le Monde 87
Léonard de Vinci: Projets, Dessins, Machines show, Cité des Sciences et de l'Industrie (Paris, 2013) 45
Levi-Montalcini, R. 83
Lewenstein, B. 17, 18, 107
Lightman, B. 17
Lindee, S. M. 103
Lipkin, I. 98

Lippmann, W. 125
literary technology 22
Logan, R. A. 208, 209n
London Science Museum 22, 45, 47, 83
longitudinal research 154
long tail (of online markets) 21
Losh, S. 105
Lovins, Amory 176
Lowenthal, D. 50
Lumière, Louis 102
Lynas, Mark 114

Macdonald, S. 42
Machado, A. 170
Magic School Bus series 17
Maibach, E. 179
Malaysia 216 224
Malle, L. 85
Malone, A. 85
Manhattan Project 98, 101
Mantegazza, P. 3
Marchant, J. 24n
Marey, E. J. 97
Maritime Museum in the Seaplane Harbour (Tallinn, Estonia) 223
market governance 129
MASIS project (assessment of science-in-society practices) 214–15, 217–18
Massarani, L. 104
mass media: health campaign research 199; monitoring 154; natural and social sciences, coverage of 189; and science journalism 28; scientists as experts in 75–9; *see also* media
Mazzolini, R. G. 31
McCall, R. S. 190
McCright, A. M. 115
McGuire, W. J. 201
McKibben, B. 175–6, 178, 182
Mead, M. 83, 85, 87
Meat and Livestock Commission, UK 161
media: artistic 130–1; audience reception studies 104–5; audience research and media effects 105–7; digital 9, 58; initiatives to support attention to science 219; mass media 28, 75–9, 154; and social sciences 186, 187–9; tactical 131; and trust portfolio management 63
media-orientated characteristics 85
media reception studies 74
media skills training 218
media society 76
mediatisation 9, 221
medicine and health, science news 30–1
Medina-Doménech, R. 100
Mejlgaard, N. 153
Mellor, F. 15, 22
Menéndez-Navarro, A. 100
Merton, R. K. 59

metaphors 1–2, 3, 11n; popular science books 21, 22
Metcalfe, J. 29, 232
Meyer, M. 54
Mialet, H. 90
Michael, M. 146
Michelsen, B. 60
Microbe Hunters (de Kruif) 18, 84
Midgley, M. 21
Millennium Project 216
Miller, J. D. 147
Miller, S. 21, 29
Ministry for Science, Technology and Productive Innovation, Argentina 221
Ministry of Agriculture, Fisheries and Food (MAFF), UK 161, 162, 163
Ministry of Science, Technology and Innovation (MOSTI), Malaysia 224
MINTiFF initiative, German Federal Ministry of Education and Research 99
miracles 8, 101
Miraikan (National Museum of Emerging Science and Innovation), Japan 216
Mlodinow, Leonard 92
Mode 1 science 131
Mol, A. P. J. 122
Molella, A. 43
Montreal Science Centre 43
Mooney, C. 33
Moore, J. 43
MOSTI (Ministry of Science, Technology and Innovation), Malaysia 224
Mothers Against Drunk Driving (MADD) 135n
Mouffe, C. 164
Murphy, G. 22, 23
Murphy, P. 21
Musée de la Civilisation (Québec) 52
Museum and Gallery Strategic Commissioning Programme, UK 235–6
Muséum National d'Histoire Naturelle (Paris) 53
Museum of Science and Industry, Chicago 47
museums 10, 42; communications turn 48–9; museum field solution (temporary exhibition) 51–3, 56n; and society 40–1; *see also* SMCs (science museums and centres)
Music to Move the Stars (Jane Hawking) 90
Muybridge, E. 97
Myers, G. 15

Nagoya Protocol on Access and Benefit Sharing 136n
nanotechnology 103, 129, 134, 169, 240
narrative structure 22
NASA 99
National Academy of Sciences, US 97; Science and Entertainment Exchange 99, 107, 108
National Association of Science Writers (NASW), US 60, 61
National Autonomous University, Mexico 220
National Cancer Institute (NCI), US 205–6, 242n
National Commission for the Forecast and Prevention of Major Risks 73
National Conservatory of Arts and Trades (Paris) 45
National Geographic 85
National Institute of Science and Technology Policy (NISTEP) 60
National Institutes of Health, US 104
The National Lottery, UK 216
National Museum of American History, Science in American Life exhibition (1995) 43
National Natuurhistorisch Museum Naturalis (Leiden) 53
National Physics Society, Estonia 223
National Science Foundation (NSF), US 99, 140, 145, 148
National Teachers Association (NSTA) 105
The Nation magazine 178
Natural History (magazine) 86
natural philosophers, seventeenth-century England 59
Natural Resources Defense Council 117
natural sciences, versus social sciences 187, 189, 190
Nature (journal) 32, 76, 87, 155n, 189, 218, 219
NCI (National Cancer Institute), US 205–6, 242n
Nelkin, D. 33, 63, 103, 173, 183
Nelson, D. 98, 203
neo-liberalism 56n
The Netherlands 118, 161; consensus conferences 236–7
Neuhauser, L. 203
neutral transmitter mode, of reporters 33
new atheism 91
Newbery, J. 17
New Scientist 19, 83, 91
newspaper industry 29, 32, 113; broadsheets 188; Malaysia 224; *see also* journals and magazines; media; *specific newspapers*
news pegs 32
Newton (Australian TV programme) 31
Newton, I. 17, 84, 90
Newtonianism for Ladies (Algarotti) 3
The New York Times 90, 177, 182; bestseller lists 18, 84, 85, 86
New Zealand, spread of science communication 218, 220
NGOs (non-governmental organisations) 60, 61, 74, 76; environmental 113, 115, 117; GONGOs (government organised non-governmental organisations) 122
Nigeria 141, 224–5
Nigeria Academy of Science 225

Index

Nine Lessons and Carols for Godless People (Rationalist Association) 23
Nisbet, M. C. 21–2, 33, 36
NISTEP (National Institute of Science and Technology Policy, Japan) 60
Noar, S. M. 208
Nobel laureates in the media 92
non-governmental organisations *see* NGOs (non-governmental organisations)
non-profit organisations 36, 37
normative decision-making 72
Nova documentaries 31
Nowotny, H. 74, 135
nuclear energy 118, 180
The Nutty Professor (film) 101

objectivity, science journalism 33
The Observer 91
Office of Institutional Studies, US 43
online environment 61–2, 134–5
Ontario Science Centre, Toronto 44, 48
open science 6
Open University, UK 235
Oppenheimer, R. 48, 49
Ordway, F. 98
Origins (Tyson) 90
The Origin of Species (Darwin) 17
Oxford Handbook on Climate Change and Society (Dryzek) 175

Pacific Science Center, Seattle (1962) 44
Pakistan Biotechnology Information Centre 219
Palais de la Découverte (Palace of Discovery), Paris 45, 47, 49
PAOs (Public Affairs Officers) 62, 64
paperbacks 19
Paquette, J. 52
Park, D. W. 191
participation: categorising *126*; versus communication 127; deliberative participation practices and policy making 128–30; as forum for science communication 121–2; in front-end problems solving 132; and knowledge production 131–3; ladder of 127; of publics in science and technology 125–39; and science-in-society 5–6; spaces and sponsors 134–5; true 127; understanding 126–8
participatory communication/democracy 5
participatory technology assessment 128
participatory turn 128
partnerships 132
Pasteur, L. 101
Pawson, R. 241
PCSS (public communication of the social sciences) 186, 187, 190; in changing academy 192–4

PCST (public communication of science and technology) 186, 187, 214; evaluation 232, 233, 236–7, 241
PCST conferences 214
Pellechia, M. G. 29
Pepper, J. H. 3, 19
periodic table 85
Perkowitz, S. 104
Pernick, M. 98
Perrin, J. 45–6, 47
PEST (public engagement with science and technology) 5
Peters, H. P. 78, 79, 189
Phillips, L. 162
Pielke, R. (Jr.) 175, 180, 181
PIO (Public Information Officer), science 58, 61, 64
Piotrow, P. T. 204
PISA (Programme for International Student Assessment of the OECD) 216
Playbook of Science (Pepper) 19
popular culture, scientists in 83–96; analysing scientists as celebrities 89–92; and gender 87–8; general culture 84; historical perspectives 84–7
popularisation 3, 7, 10, 11, 12n, 28, 147; popular science books 17, 21
popular science: books 15–26, 87; dominant view 16; evolution 28; meaning of 'popular' 15, 16–17; narrative structure, focus on 22; outline history 17–19; traditional model 16
Popular Science magazine 19
Popular Science Monthly (magazine) 28
Poster, M. 134
post-Second World War period 29
PR *see* public relations (PR)
pragmatistic model 72
Prewitt, K. 147
Primary Science Teaching Trust 20
productivity, trust portfolio management 63
proteins 6, 132, 136n
PSN (National Science Centre), Malaysia 224
Public Affairs Officers (PAOs) 62, 64
public anthropology 192
public communication of science and technology (PCST) *see* PCST (public communication of science and technology)
public communication of the social sciences (PCSS) *see* PCSS (public communication of the social sciences)
public deficit 148, 151
public engagement 5; as forum for science communication 121–2
public engagement with science and technology (PEST) 5
public expertise 79
public experts, scientists as *see* expertise, scientific
Public Health Service, US 98

public information, and PR 60
Public Information Officer (PIO), science 58, 61, 64
public intellectual 187, 191–2
publicity, and PR 60
public opinion surveys 73
public relations (PR): asymmetric 60; contextual nature 64; functional level 66; future research, in science 65–7; historical outlook 59–61; ideal types 60; meaning 58–9; online challenges (promoting and debating science on the web) 61–2; organisational level 66; in practice (trends and contexts) 64–5; professionalisation and institutionalisation of 58; programme level 65–6; science communication 58–69; and science journalism 58, 59–60; societal level 66; symmetric 61; trust portfolio management 59, 62–3; two-way asymmetric model 60; uses of 58
publics: active 60; construction of 125–6; participation in science and technology 125–39; public dialogue as an end 130–1; Publics-in-Particular versus Publics-in-General 126; of science, research paradigms 144–5; terminology 6, 9, 16
public scientists 7
public sphere 18, 84
Public Understanding of Science (journal) 20, 221
public understanding of science (PUS) 5, 140–59, 186, 192; scientific literacy (1960s to mid-1980s) 145–8; after 1985 to mid-1990s 148–50; science in-and-of society (mid-1990s to present) 150–1; cohort analysis 153; complementary data streams, developing 154; constructing dynamics models over time 153; growing longitudinal database, providing sophisticated secondary analysis 153; institutions sponsoring and conducting surveys *144–5*; integrating national and international surveys into more global database 153; mass media monitoring and longitudinal research 154; paradigms of researching the publics of science 144–5; qualitative diversity of 'cultures of science,' working towards 153–4; quasi-panel models and testing 153; survey research 140–1, *142–3*, 151–2
Puritanism 59
PUS (public understanding of science) *see* public understanding of science (PUS)

qualitative research 151, 235
quantitative research 235
Quark (magazine) 221
quasi-experimental design 239

Rakow, L. F. 202
Rationalist Association 23
Ravetz, J. R. 72, 173

Rawls, J. 164
Raza, G. 147
ready-made science 6
realism 22, 149
reception studies, audience 104–5
Reeves, Hubert 87
reflexive sciences 190–1
relativity theory 84
research: audience 104–7; films, scientific research fields 102–4; health campaign 198–213; interdisciplinary 75; longitudinal 154; quantitative versus qualitative 151, 235; risk perception 141; science communication 1–14; scientific expertise 74; scientific literacy 145–6; social sciences and the media 187–9; survey *see* survey research; transdisciplinary 75
Reversing Darwin's Theory (film) 101
Revkin, Andrew 182
Richard Dawkins Foundation for Reason and Science 90, 91
Richards, Graham 190
Riesch, H. 35
risk management and science communication 160–72; first to second order thinking 163–5; second-order thinking 163–6; third-order thinking 167–9
risk perception research 141
risk technocracy 74
Robertson, T. 149
Robocop (film) 103
Rockefeller Foundation 60
Roentgen, William 102
Rolling Stone magazine 178
Rossi, P. H. 234
Rothman, S. 76
Rowan, K. E. 75
Rowe, G. 127
Royal Institution of Great Britain 35
Royal Society of London 20, 148
RTD culture 8

'sacredness' of science 31
Sagan, C. 18, 83, 85, 86, 87
Said, E. 191
Salmon, C. 202
Saltzberg, D. 98
Schell, H. 107
Schibeci, R. 100–1
Schlumberger Excellence in Education Department (SEED) programme, Nigeria 225
Schmierbach, M. 189
science: ambiguity of term 15; autobiography of 21, 23; defined 2; ebb and flow of 22, 23; establishment 115; forms of public participation in 6; fossilised 10; *Lakatosian* view of 22; making accessible (by SMCs) 48–9; 'new epic' of 23; public awareness 4

Index

Science (journal) 219
Science and Engineering Indicators 60
Science and Entertainment Exchange, National Academy of Sciences (US) 99, 107, 108
Science and Human Values (Bronowski) 85
science and society relationship: boundaries 2; challenges for science communication research 9; climate change as challenge 114–19; empirical reality 72; and publics 125; and science in culture 10–11; science in society and in culture 10–11; SMCs relating science and technology to society 49–51; stereotype 2; theatrical dialogue metaphor 1–2; *see also* science-in-society
science and technology studies (STS) 164, 193, 194
Science Bus project, Estonia 223
science cafés 5, 130, 227
Science Center of Pinellas County (1959) 44
science centres 41
science citizenship 97, 100–1
science communication: definitions 186; evaluation 231–45; global spread 214–30; models 3–4; as particular kind of social research 234–6; postgraduate teaching 221; public engagement as forum for 121–2; public relations 58–69; research *see* science communication research; science communication research challenges; and risk management 160–72; university research 220–1; university taught programmes 219–20; *see also* PCST (public communication of science and technology)
Science Communication in the World 214
science communication research 1–14; conceptual review 2–8; as particular kind of social research 234–6; science and society 1–2, 10–11; theatrical dialogue metaphor 1–2, 11n; *see also* science and society
science communication research challenges: collapsing communication contexts 10; global 11; new mediations 9; plural science/public 9; quality and evaluation 10
Science Communicator (India) 221
science culture 153
science education 105–6
science fiction films 101, 102
Science in American Life exhibition (1995) 43, 50
science-in-society: deficit concept 4; MASIS project (assessment of science-in-society practices) 214–15, 217–18; and participation 5–6; popularisation 3; science-in-and-of society 150–1; and science in culture 10–11; *see also* science and society relationship
science-in-the-making 6
Science in the Marketplace (Fyfe and Lightman) 17
science journalism *see* journalism, science
Science Media Centre, UK 35, 189, 218

science movement (1930s) 59
Science Museum (London) Pattern Pod gallery 45
Science Museum Media Monitor, US 188
science museums and centres (SMCs) *see* SMCs (science museums and centres)
science news/science stories 27–38, 29, 30–1, 37, 78
Science Service, US 59, 60
Science: The Endless Frontier report (1945) 12n
scientific citizens 126
Scientific American (magazine) 28
Scientific and Technological Research Council of Turkey (TUBITAK) 225–6
scientific citizenship 242
scientific community 9
scientific culture 8
scientific knowledge: and decision-making 71–2; expertise 75; lacking by public 4; per se 71; and SMCs 42, 44; *see also* knowledge
scientific literacy (1960s to mid-1980s): critique 147–8; recommended steps 146–7; research agenda 145–6
'scientific temper' 221
Scripps, E. W. 59
Seaborg, Glenn 85
Seale, C. 189
second-order thinking 160, 163–5; characteristics 167; putting into practice 165–6; second-order science 72
Second World War 29
Secord, J. 21
SEED (Schlumberger Excellence in Education Department) programme, Nigeria 225
Seeko, D. M. 35
Seely, R. 36
self-determination theory 208
Selfish Gene (Dawkins) 19
Selling Science (Nelkin) 33
semiotics 22
Serbia, First Regional Science Promotion conference 217–18
Shapin, S. 22
Shepherd, R. G. 76
showmen of science 3
Shukla, R. 153–4
Silent Spring (Carson) 21, 103
Silent World (Cousteau) 85
Silkwood (film) 103
Silverstone, R. 31
Singapore, ArtScience Museum 216
Singer, E. 188
Sjoeberg, S. 153
Sjöström, A. 189
Skloot, R. 22
The Sky is Not the Limit (Tyson) 91
Smart, C. 17

SMCs (science museums and centres) 40–57; contexts 42–4; definitions 40–2; developmental phases 44–51, 53; displaying history of technology 45; dissemination activities 44, 49; evolution 40, *46*; facilitating knowledge appropriation 48–9; making science accessible 48–9; mediation strategies 42–3; museum field solution (temporary exhibition) 51–3, 56n; numbers 41, 53; paradigm shift 53–5; relating science and technology to society 49–51; showing contemporary science/enhancing knowledge 45, 47–8; versus society museums 40
Snyder, L. B. 200
social inoculation 202
social media 36, 76, 177
social sciences, communicating 186–97; disciplinary status and public expertise 191–2; funding bodies and professional associations 187–8; and media 186, 187–9; natural sciences versus social sciences 187, 189, 190; public social sciences in changing academy 192–4; reflexive sciences 190–1; research literature on social sciences and media 187–9
society, and science *see* science and society
society museums 40
sociobiology 86
soft capitalism 20
'soft news' 189, 190
Solieraj, S. 177
sound scientific practice 131
South Africa 141
Soylent Green (film) 103
space science 103
Spain 20; spread of science communication 219, 220
SPICE (Stratospheric Particle Injection for Climate Engineering) 119
Spider-Man (film) 97, 103
Spinner, H. F. 79
standardised questionnaire 238
Stanford Heart Health campaign 205
Stannard, R. 16–17
Stares, S. 153
State of Fear (Crichton) 115
Steinke, J. 101, 105
STEM (science, technology, engineering and mathematics) subjects 216
Stevenson, H. 182
STI (science, technology and innovation), Nigeria 225
Stocking, R. S. 190
Stordsdieck, M. 237
story frames 78
The Story of Louis Pasteur (film) 101
storytelling 218
Stratospheric Particle Injection for Climate Engineering (SPICE) 119

STS (science and technology studies) 164, 193, 194
Sturgis, P. J. 149
Šuljok, A. 189
summative evaluation 233
Sundance film festival 99
superstition 145, 147, 154
survey research: and deficit concept 151–2; 'don't know' ('DK') responses 146, 148; Eurobarometer indicators 140, 141, 150; and expertise 73, 77; future steps 152–4; large-scale surveys 149; public understanding of science (PUS) 140–1, *142–3*, 151–2, 154
Swaminathan, M. S. 83
Switzerland 149
synthetic biology 129, 134
systems of science 97–8

tactical media 131
Taiwan 77; National Science and Technology Museum 216
TeaME programme, Estonia 223
technocratic model 72
technologies of humility 151
technology, museums displaying history 45
technosciences 50
The Telegraph 83
television 7, 31; in United States 85, 86
temporary exhibitions 51–3, 56n
The Terminator (film) 103
Terra X (German TV programme) 31
Terry, the Friendly Fracosaurus 20
theatrical dialogue metaphor (science and society) 1–2, 11n
Them! (film) 102
theories espoused/theories-in-action 150
The Thing From Another World (film) 101
third mission of universities 5, 218
third-order thinking 160, 167–9
Third World Academy of Sciences, Trieste 217
Tilley, N. 241
Time magazine 91
Tonight Show with Johnny Carson 86
top-down communication models 3
Topham, J. R. 15
Topol, E. 207
transdisciplinary research 75
translation 11, 78
transsubjectivity 75
transuranium elements 85
triangulation, research 235
true participation 127
trust: blind 8; and scientific authority 73–5
trust portfolio management, and PR 59, 62–3
Tse, T. 208
TUBITAK (Scientific and Technological Research Council of Turkey) 225–6
Tudor, A. 99

Turkey 225–6
Turkish Ministry of National Education 226
Turkish Radio and Television Corporation 226
Turner, Graeme 88
Turner, J. 146
Turner, Stephen 73, 192
Turney, Jon 90, 100
Turow, Joseph 98
Twins (film) 103
2001: A Space Odyssey (film) 98
Tyson, Neil deGrasse 87, 89, 90, 91, 93

uncertainty 72
Uncle Albert books 16
UNESCO 60, 217, 219
unfinished stories 35
UN Framework on Climate Change Conference, Copenhagen 130
United Kingdom: BBC 31, 85, 86, 235; GM Nation? initiative 121, 165, 166, 234–5; natural philosophers, seventeenth-century England 59; popular culture, scientists in 84; public understanding of science, research programme 155n; science journalism 28, 29, 30, 31; science PR 64; scientific experts and journalists, interactions 77; social sciences, communicating 189, 190; spread of science communication 217, 219, 227; survey research 151–2; *see also specific institutions*; *specific newspapers*; *specific organisations*
United States: Chicago World Fair (1933–4) 47; climate change debate 173–85; evolution theory 74, 148; newspaper industry 29, 30–1; popular culture, scientists in 84; and PR 59, 60; Public Broadcasting System 85; science journalism 28, 29, 30, 33, 34; and SMCs 43, 48, 49, 50; social sciences, communicating 188, 189, 190; sociology 192; television 85, 86; *see also specific institutions*; *specific organisations*
universities: postgraduate teaching 221; programmes 219–20; research 220–1; third mission of 5, 218
US Army 192

Vanity Fair (magazine) 89, 90, 91
variant CJD (Creutzfeldt-Jakob disease) 162
Venter, Craig 93
Vestiges of the Natural History of Creation (Chambers) 21

Victor, D. G. 115
Vietnam, Café Khoa Huc 218
Villam, Cédric 87
'violent videogames' debate 189
visible scientists 7, 18, 76, 77; celebrity versus visibility 93n; popular culture, scientists in 85, 86, 88, 92
The Visible Scientists (Goodell) 28
Vogt, C. 154
Voltaire (François-Marie Arouet) 17
Voronoff, S. 102
Vuković, M. B. 189

Wagensberg, J. 42, 49
Wall-E (film) 103
War Games (film) 103
Washington Post 177
Watson, J. 86
web communication model 17
web culture 17
Weber, M. 72
Wei, H. 149
Weingart, P. 100
Weiss, C. H. 188, 232
Wellcome Trust, UK 29, 97, 99, 219
White, H. 22
Why is Snot Green? (Murphy) 22–3
wicked problems 128, 175
Wien, C. 189
Wilson, E. O. 86
Wilson, K. 34
wind power 117–18, 123n
Wine Spectator magazine 90
Wisconsin Center for Investigative Journalism 37
Wissenschaft 187
Withey, S. B. 147
women 83, 87
The Wonderful Electric Belt (film) 102
Wood, J. G. 3
World Conference of Science Journalists (2009) 30
World Federation of Science Journalists (WFSJ) 29, 219
World Meteorological Organisation 115
World's Fairs 47
Wright, D. 20
Wynne, B. 75, 127, 147, 174

Young, A. 89